MECHANISMS OF PLATELET ACTIVATION AND CONTROL

ADVANCES IN EXPERIMENTAL MEDICINE AND BIOLOGY

Recent Volumes in this Series

Volume 336
ANCA-ASSOCIATED VASCULITIDES: Immunological and Clinical Aspects
Edited by Wolfgang L. Gross

Volume 337
NEUROBIOLOGY AND CELL PHYSIOLOGY OF CHEMORECEPTION
Edited by P. G. Data, H. Acker, and S. Lahiri

Volume 338
CHEMISTRY AND BIOLOGY OF PTERIDINES AND FOLATES
Edited by June E. Ayling, M. Gopal Nair, and Charles M. Baugh

Volume 339
NOVEL APPROACHES TO SELECTIVE TREATMENTS OF HUMAN SOLID TUMORS:
Laboratory and Clinical Correlation
Edited by Youcef M. Rustum

Volume 340
THE DESIGN OF SYNTHETIC INHIBITORS OF THROMBIN
Edited by Goran Claeson, Michael F. Scully, Vijay V. Kakkar, and
John Deadman

Volume 341
CIRRHOSIS, HYPERAMMONEMIA, AND HEPATIC ENCEPHALOPATHY
Edited by Santiago Grisolía and Vicente Felipo

Volume 342
CORONAVIRUSES: Molecular Biology and Virus–Host Interactions
Edited by Hubert Laude and Jean-François Vautherot

Volume 343
CURRENT DIRECTIONS IN INSULIN-LIKE GROWTH FACTOR RESEARCH
Edited by Derek LeRoith and Mohan K. Raizada

Volume 344
MECHANISMS OF PLATELET ACTIVATION AND CONTROL
Edited by Kalwant S. Authi, Steve P. Watson, and Vijay V. Kakkar

A Continuation Order Plan is available for this series. A continuation order will bring delivery of each new volume immediately upon publication. Volumes are billed only upon actual shipment. For further information please contact the publisher.

MECHANISMS OF PLATELET ACTIVATION AND CONTROL

Edited by

Kalwant S. Authi

Thrombosis Research Institute
London, United Kingdom

Steve P. Watson

University of Oxford
Oxford, United Kingdom

and

Vijay V. Kakkar

Thrombosis Research Institute
London, United Kingdom

SPRINGER SCIENCE+BUSINESS MEDIA, LLC

Library of Congress Cataloging-in-Publication Data

Mechanisms of platelet activation and control / edited by Kalwant S.
  Authi, Steve P. Watson, Vijay V. Kakkar.
      p.   cm. -- (Advances in experimental medicine and biology ; v.
  344)
     "Proceedings of an International Symposium on Mechanisms of
  Platelet Activation and Control, held April 13-14, 1992, in London,
  United Kingdom"--T.p. verso.
     Includes bibliographical references and index.
     ISBN 978-0-306-44631-3   ISBN 978-1-4615-2994-1 (eBook)
     DOI 10.1007/978-1-4615-2994-1
     1. Blood platelets--Activation--Congresses. 2. Cellular signal
  transduction--Congresses. I. Authi, Kalwant S. II. Watson, Steve
  P. III. Kakkar, V. V. (Vijay Vir) IV. International Symposium on
  Mechanisms of Platelet Activation and Control, (1992 : London,
  England) V. Series.
     [DNLM: 1. Blood Platelets--drug effects--congresses. 2. Blood
  Platelets--physiology--congresses. 3. Platelet Activation--drug
  effects--congresses. 4. Platelet Activating Factor--physiology-
  -congresses.   W1 AD559 v.344 1993 / WH 300 M486 1992]
  QP97.M387 1993
  612.1'17--dc20
  DNLM/DLC
  for Library of Congress   93-32065
                                                                  CIP

Proceedings of an International Symposium on Mechanisms of Platelet Activation and Control, held April
13-14, 1992, in London, United Kingdom

ISBN 978-0-306-44631-3

©1993 Springer Science+Business Media New York
Originally published by Plenum Press, New York in 1993

Thrombosis Research Institute
Manresa Road
Chelsea
London SW3 6LR.

International Platelet Symposium

13-14th April 1992

"Mechanisms of Platelet Activation and Control"

The organisers gratefully acknowledge the support of the following organisations
(in alphabetical order);

Abbott Laboratories

British Society for Haemostasis and Thrombosis

Eisai Europe Ltd.

Glaxo Group Research Ltd.

ICI Pharmaceuticals

Merck Research Laboratories

Sandoz Ag.

Wellcome Research Laboratories

PREFACE

Recent years have seen tremendous advances in our understanding of the molecular mechanism of platelet activation. All aspects of signal transduction in platelets from the identification of surface receptors, G proteins, phospholipases, protein kinases and phosphatases, intracellular receptors for inositol phosphates, the Ca^{2+} regulatory machinery, cytoskeletal constituents to the control mechanism employing cyclic nucleotides has seen an explosion of information regarding their importance and for each constituent in the family of molecules to which they belong. This information has been of interest to researchers across a wide spectrum of disciplines including biochemists, pharmacologists, cell biologists and clinicians. In April 1992 an International Symposium bearing the name of this volume was organised at the Thrombosis Research Institute to bring together scientists from across the world whose common interest was the study of platelet activation and its regulation. We were particularly encouraged by the positive response from our speakers and the participants, their detailed contributions and the very lively discussions that took place throughout the two days of the symposium. Almost every aspect of signal transduction in human platelets was represented. Of the invited speakers twelve were from Europe (including the U.K.), eight from North America and one from Japan. This volume is a compilation of chapters submitted by the speakers and represents a concise but informative picture of the present knowledge of the mechanisms of platelet activation and control.

The organisation of the Platelet Symposium involved committed contributions from most of the staff of the Thrombosis Research Institute, in particular the efforts of Mrs. Sue Vost (in the smooth running of the symposium) and of Mrs. Eileen Bayford (for help with the preparation of this volume) are especially acknowledged. Additionally the financial contributions from the listed organisations were also gratefully received as without their support the symposium could not have taken place.

Kalwant S. Authi
Steve P. Watson
Vijay V. Kakkar

1993

CONTENTS

**The Platelet as a Ca²⁺-driven Cell: Mechanisms which may
Modulate Ca²⁺-driven Responses**
Michael C. Scrutton 1
    Introduction 1
    Ca²⁺-driven responses and the measurement of cytosolic [Ca²⁺]
    in human platelets 2
    Is adrenaline a full excitatory agonist for human platelets?. . . 4
    Does aggregation induce an increase in cytosolic [Ca²⁺]? . . . 5
    Factors modulating Ca²⁺-driven responses 7
        1,2-diacylglycerol 7
        MgATP²⁻ 8
        G proteins 10
        Cyclic-3'5'-AMP 10
        Cyclic-3'5'-GMP 11
    References 12

**Agonist Receptors and G Proteins as Mediators
of Platelet Activation**
*Lawrence F. Brass, James A. Hoxie, Thomas Kieber-Emmons,
David R. Manning, Mortimer Poncz and Marilyn Woolkalis* . . . 17
    Summary 17
    Introduction 18
    Agonist receptors 19
        Thrombin receptor structure 20
        Thrombin receptor activation. 23
        Thrombin receptor desensitization 25
        Unresolved issues 27
    G proteins 28
        G proteins in platelets. 29
        Unresolved issues 31
    References 32

**Regulation of Phosphoinositide-Specific Phospholipase C
Activity in Human Platelets**
Yoshinori Nozawa, Yoshiko Banno and Koh-ichi Nagata . . . 37
    Introduction 37
        Purification and identification of human platelet
        phosphoinositide-specific Phospholipase C isoforms. . . 38
    GTP-binding proteins in human platelets 39
        Heterotrimeric type 39

Low molecular weight types	40
Regulation of Phosphoinositide-Specific Phospholipase C Activity.	42
GTP-binding protein	42
Interaction with cytoskeleton	43
Protein phosphorylation	44
References	45

RAP1B and Platelet Function
Eduardo G. Lapetina and Francis X. Farrell 49

Introduction	49
Rap proteins	49
GTPase activating proteins	51
The role of rap proteins on platelet function.	51
Thrombin induces an association of phospholipase C-γ1 and ras GTPase activating protein with rap1b.	52
Erythroprotein induces ras activation and rasGAP tyrosine phosphorylation in HEL cells.	53
Conclusion	53
References	54

Calcium Signalling and Phosphoinositide Metabolism in Platelets: Subsecond Events Revealed by Quenched-Flow Techniques
Adrian R.L. Gear and Sanghamitra Raha 57

Introduction	57
Methods	59
Platelet isolation, [³H]-inositol loading and electroporation.	59
Quenched-flow aggregometry and inositol phosphate Extraction and analysis.	59
Intracellular platelet calcium analysis.	60
Data evaluation and presentation	60
Results	61
Comparison of ADP and thrombin-induced calcium increase within two seconds of platelet activation	61
Kinitics of 1,4,5-InsP₃ formation during shear, ADP and thrombin activation	61
Early synthesis of 1,3,4,5-InsP₄ during shear, ADP and thrombin activation	62
Discussion	64
References	66

Calcium Influx Mechanisms and Signal Organisation in Human Platelets
Stewart O. Sage, Paul Sargeant, Johan W.M. Heemskerk and Martyn P. Mahaut-Smith 69

Introduction	69
Calcium influx	70
Receptor-mediated calcium entry	70
Clues from stopped-flow fluorimetry.	70
Evidence for an ADP receptor-operated channel	72
Evidence for store-regulated calcium entry	73
The route for store-regulated Ca^{2+} entry	73
The coupling of store-depletion to increased Ca^{2+} entry.	74
Second messenger-operated channels.	76

Lack of thrombin-evoked membrane currents. . . . 77
Temporal organisation of the platelet calcium signal. . . . 78
Single platelet calcium signals 78
Mechanism of oscillation 80
References 81

Ca^{2+} Homeostasis and Intracellular Pools in Human Platelets

Kalwant S. Authi 83
Introduction 83
Ca^{2+} homeostasis. 83
Ca^{2+} elevation mechanisms 83
Maintenance of cytosolic Ca^{2+} concentrations. 85
Ca^{2+} removal across the plasma membranes. . . . 85
Intracellular membrane Ca^{2+} stores and pumps . . . 87
Cyclic AMP regulation of the platelet intracellular Ca^{2+} pumps . 89
Mechanisms of Ca^{2+} elevation in platelets 90
Inositol (1,4,5) trisphosphate. 90
Ca^{2+} release from intracellular stores. 93
Mechanisms of Ca^{2+} entry; Involvement of second messengers . 95
Ca^{2+} influx via a receptor-operated Ca^{2+} channel . . . 97
Concluding statement. 98
References 98

The Use of Inhibitors of Protein Kinases and Protein Phosphatases to Investigate the Role of Protein Phosphorylation in Platelet Activation

Steve P Watson, Robert A Blake, Trevor Lane and Trevor R Walker . . 105
Introduction 105
Role of protein kinase C 106
Studies with protein kinase C inhibitors 107
Studies with a pseudosubstrate inhibitor of protein kinase C . . 110
Tyrosine phosphorylation in platelet activation 112
References 116

Serine/Threonine Kinases in Signal Transduction in Response to Thrombin in Human Platelets:
Use of 17—hydroxywortmannin to discriminate signals

Kenneth J. Clemetson, Markus Kocher and Vinzenz von Tscharner . . 119
Summary 119
Introduction 120
Methods and Materials 121
Isolation of human platelets 121
Polyacrylamide Gel electrophoresis 121
Renaturation knases 121
Two-dimensional Gel Electrohoresis, western
blotting and detection of proteins containing
containing phosphothreonine · 121
Labelling of platelet protines with $^{32}PO_4$ 121
Measurement of cytosolic free calcium
using Fura-2 122
Results and Discussion 122
References 127

Tyrosine Phosphorylation in Platelets: its Regulation and Possible Roles in Platelet Functions
Maurice B. Feinstein, Kevin Pumiglia and Lit-Fui Lau 129
   Introduction 129
   Tyrosine phosphorylation in intact platelets and platelet
   tyrosine kinases and phosphatases 130
   Stimulation of tyrosine phosphorylation in platelets by
   agonists and its regulation by second messengers . . . 132
   Activation of platelets by tyrosine phosphatase inhibitors . . 134
   Effects of tyrosine kinase inhibitors: What responses are tyrosine
   phosphorylation-dependent? 136
   Tyrosine phosphorylation specifically related to aggregation and
   the fibrinogen-GpIIb-IIIa interaction. 140
   Tyrosine phosphorylation and the platelet cytoskeleton. . . 142
   References 145

Evidence that Activation of Phospholipase D can Mediate Secretion from Permeabilized Platelets
Richard J. Haslam and Jens R. Coorssen 149
   Introduction 149
   Factors regulating secretion from permeabilized platelets . . 150
      Ca^{2+}-dependent secretion 152
      Implications for the regulation of exocytosis . . 154
   Correlations between secretion and PLD activity . . . 154
      Effects of GTP[S], PMA and Ca^{2+} on PLD activity . 154
      Inhibitory effects of ethanol 155
      Inhibition by BAPTA 157
   Possible roles for PLD in secretion 158
   Extrapolation to intact platelets 160
   Summary 162
   References 163

Inositol Lipid Metabolism, the Cytoskeleton, Glycoprotein IIb IIIa and platelets
Gérard P. Mauco, Claire Sultan, Bernard Payrastre, Monique Plantavid, Monique Breton and Hugues Chap 165
   Introduction 165
   The "classical" phospholipid effect 165
   Platelets produce inositol lipids phosphorylated on the D-3
   position of inositol 168
   Interactions between the cytoskeletal proteins and phospholipids . 168
   Inositol lipid metabolism is associated to cytoskeleton in platelets. 169
   Gp IIb-IIIa is Involved in platelet production of $PtdIns3,4P_2$. 170
   Conclusion 170
   References 172

Regulation of Platelet Function by the Cytoskeleton
Joan E.B. Fox 175
   Introduction 175
      The platelet membrane skeleton 176
   Association of the GPIIb-IIIa complex with the platelet
   membrane skeleton 178

Role of the GPIIb-IIIa-cytoskeleton interaction in regulating
ligand-induced transmembrane signaling 180
Role of the GPIIb-IIIa-cytoskeleton interaction in regulating
the adhesive properties of the GPIIb-IIIa complex 182
Summary 184
References 184

Cytoskeletal Interactions of Rap1b in Platelets
Gilbert C. White, II, Neville Crawford, and Thomas H. Fischer, . . 187
Introduction 187
Cellular localization of Rap1b. 188
Effect of cAMP on Rap1b association with the cytoskeleton . . 190
Summary 193
References 193

Mechanisms Involved in Platelet Procoagulant Response
Edouard M. Bevers, Paul Comfurius, Robert F.A. Zwaal . . . 195
Introduction 195
Phospholipid-dependent coagulation reactions 195
Platelet procoagulant activity. 196
Mechanisms involved in expression of procoagulant activity. . . 198
Platelet microvesicle formation 198
Scott syndrome 201
Regulation of expression of procoagulant activity: Involvement of
aminophospholipid translocase 201
Concluding remarks 203
References 204

Histamine as an Intracellular Messenger in Human Platelets
Jon M. Gerrard, Satya P. Saxena, and Archibald McNicol . . . 209
Introducion 209
Histamine in platelets 210
Limits to current evidence supporting a role for intracellular
histamine in platelet functions 213
Implications in human disease 214
Histamine as an intracellular messenger in other ccll systems . . 214
Studies in neutrophils 214
Histamine and cell growth 215
Relationship to polyamines 215
Summary 216
References 216

**Platelet Activation via Binding of Monoclonal Antibodies to the Fcγ
Receptor II**
J. Michael Wilkinson, Edward J. Hornby and Kalwant S. Authi . . 221
Introduction 221
Platelet activation by PM6/248 222
Activation caused by other antibodies 225
Activation via FcγRII. 226
References 227

Functional Relationship between Cyclic AMP-Dependent Protein Phosphorylation and Platelet Inhibition

Wolfgang Siess, Bernd Grünberg, Karin Luber 229
    Introduction 229
    Target sites for inhibition of platelet activation by cAMP . . . 229
    Proteins phosphorylated by protein kinase A in platelets . . 231
    Phosphorylation of *RAP*1b is not involved in platelet inhibition by cAMP 232
    *RAP*1B phosphorylation is a sensitive marker for the action of cAMP- and cGMP-elevating vasodilators 233
    References 234

Role of Cyclic Nucleotide-Dependent Protein Kinases and their Common Substrate VASP in the Regulation of Human Platelets

Ulrich Walter, Martin Eigenthaler, Jörg Geiger and Matthias Reinhar . . 237
    Introduction 237
    Diversity of cyclic nucleotide regulation and action . . . 237
    Cyclic nucleotide binding proteins and regulation of cyclic nucleotide levels in human platelets 238
    Cyclic nucleotide-dependent protein kinases and their functional roles in human platelets 240
    Role of VASP and other substrates of cyclic nucleotide-dependent protein kinases in human platelets 243
    Summary 246
    References 247

The Biological and Pharmacological Role of Nitric Oxide in Platelet Function

Marek W. Radomski and Salvador Moncada 351
    Introduction 251
    Endothelium-dependent relaxation and its relation to the formation of nitric oxide 251
    Characterization of nitric oxide synthase 252
    Affinity targets for nitric oxide 252
        The soluble guanylate cyclase 252
        Thiol-containing molecules 253
        Enzymes containing nonhaem iron coordinated to sulphur atoms (Fe-S) 253
    Physiological role of the constitutive NO synthase-soluble guanylate cyclase system in regulation of platelet haemostasis . . . 254
        Role of nitric oxide 254
        Synergistic model of regulation of platelet aggregation . . 255
    Nitric oxide synthesis in thrombotic and haemorrhagic disease . . 255
    Clinical pharmacology of the L-arginine to nitric oxide pathway . . 257
        Pharmacology of nitric oxide formation and action . . 257
        Pharmacology of nitrovasodilators 257
    Concluding remarks 258
    References 260

Contributors 265

Index 267

THE PLATELET AS A Ca^{2+}-DRIVEN CELL: MECHANISMS WHICH MAY

MODULATE Ca^{2+}-DRIVEN RESPONSES

Michael C. Scrutton

Division of Life Sciences
King's College
Campden Hill Road
London W8 7AH, UK

INTRODUCTION

Platelets play a central role in the haemostatic response (Gordon and Milner, 1976). They also contribute significantly to the initiation of responses by other cells which occur concomitantly with, or as a necessary sequel to, the initial events that prevent loss of blood (Larsen et al, 1989). Some of the excitatory agonists involved in initiating these responses, and the nature of the responses which can occur, are shown in Fig.1. The response pattern is characteristic of the agonist used (Crawford and Scrutton, 1987), and in Fig.1 the excitatory agonists are grouped according to the type of response pattern which they induce. These response patterns can be summarised as follows:

1.   "Direct". In this pattern which for human platelets is induced by thrombin and PAF, and possibly in some instances by vasopressin, all responses can result directly from the initial agonist-receptor interaction. The responses induced are aggregation, thromboxane A_2 synthesis and secretion from all three storage granules (amine storage, protein storage and lysosomes) present in the platelet (Crawford and Scrutton, 1987). Interplay between the effect of the released products, particularly ADP and thromboxane A_2, leads to enhanced aggregation especially when a low concentration of the initial agonist is used (Packham and Mustard, 1986).

2.   "Adhesion-dependent". This pattern is characteristic of collagen. It is initiated by interaction of platelets with, and their spreading on, collagen fibrils which causes thromboxane A_2 synthesis, secretion from all three storage granules and uniquely an increase in the phosphatidylserine content of the outer leaflet of the plasma membrane. The enhanced extracellular exposure of phosphatidylserine provides the trigger for the assembly of the factor X and prothrombin activation complexes on the activated platelets and so localises thrombin formation to the vicinity of the site of collagen exposure (Bevers et al,

Mechanisms of Platelet Activation and Control, Edited by
K.S. Authi *et al.*, Plenum Press, New York, 1993

1991). Access of collagen to circulating platelets would normally occur only in an area of vascular damage. Aggregation of platelets does not appear to result as a direct consequence of the interaction of platelets with collagen except possibly at very high concentrations of this agonist. Instead, aggregation induced by addition of collagen is caused by the secondary action of the released products with ADP and thromboxane A_2 again being particularly important. Interplay between the effects of these products is crucial in generating an aggregation response to lower concentrations of collagen, as shown for example since such responses can be equally and markedly reduced either by inhibiting thromboxane A_2 synthesis or by removing ADP (Packham and Mustard 1986). Feed-forward amplification is therefore a very prominent feature of the aggregation response induced by addition of collagen and makes this agonist particularly effective.

3. "Aggregation-dependent" Aggregation is the initial event caused by agonist-receptor interaction. If the extent of aggregation is sufficient weak thromboxane A_2 synthesis may result which can then lead to secretion from the amine and protein storage granules. This pattern of response is induced by agonists such as ADP, adrenaline and 5-hydroxy-tryptamine (5HT) and never leds to lysosomal secretion or enhanced phosphatidylserine exposure (Crawford and Scrutton 1987). It characterises agonists which act largely, if not exclusively, as released products in feed-forward amplification loops.

Platelet reactivity is controlled by interaction with inhibitory agonists which can modulate both the level of response obtained and also in some respects the nature of that response. Two groups of inhibitory agonists exist (Fig.1) which use respectively cyclic-3', 5'-(cAMP) (PGI$_2$, PGD$_2$, adenosine). and cyclic-3',5'-GMP(cGMP) (NO), based signal transduction systems (Aktories and Jakobs 1985; Moncada et al 1987). These two systems interact to give for example a synergistic response to NO when added together with PGI$_2$, or adenosine at low concentration (Radomski et al 1987a).

Ca^{2+}-Driven Reponses and the Measurement of Cytosolic [Ca^{2+}] in Human Platelets

Most platelet responses can be induced either by treatment with a divalent cation ionophore in the presence of physiological extracellular [Ca^{2+}] or by exposure of permeabilised platelet preparations to μM Ca^{2+} (Rink and Sage 1990). All platelet excitatory agonists except adrenaline (see below) have the capacity to induce an increase in cytosolic [Ca^{2+}] (Rink and Sage 1990; Smith et al 1991). The platelet appears therefore to be a classic example of a cell which uses a Ca^{2+}-based excitatory signal transduction

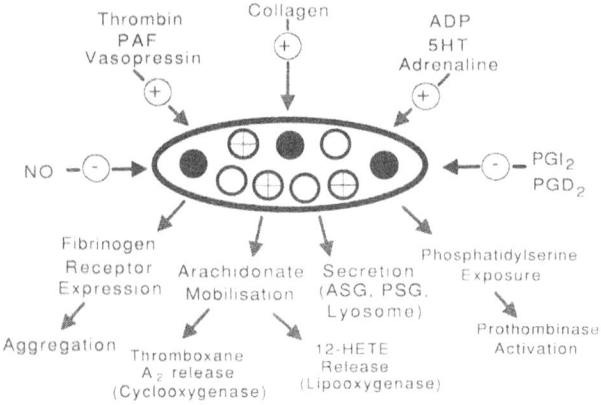

Figure 1   Platelet agonists and responses. The + symbol indicates an excitatory agonist and the - symbol an inhibitory agonist.

pathway. Detailed analysis of this pathway is however complicated by qualitative and quantitative discrepancies between the results obtained when platelet cytosolic $[Ca^{2+}]$ is measured using either fluorescent derivatives of EGTA (Fura 2, Indo 1) or luminescent photoproteins (aequorin). The differences in the results obtained using these two groups of probes are indicated on the left-hand side of Fig.2. The values for both resting and maximally stimulated $[Ca^{2+}]$ are always greater when estimated using aequorin than when using Fura 2 even if the aequorin data are corrected to an appropriate cellular $[Mg^{2+}]$ (Ware et al 1986, 1988). Aequorin with a Kd for Ca^{2+} of 1-2μM is a better probe for measurement of $[Ca^{2+}]$ in the maximally stimulated platelet since for most agonists values are obtained in the low μM range. However in platelets heavily loaded with this probe accurate measurement of cytosolic $[Ca^{2+}]$ is possible for the resting platelet. We have recently confirmed the value of 0.3μM reported by Ware et al (1988) which is clearly greater than that estimated using Fura 2 (0.05-0.1μM) (Figs.2 & 3). For stimulation by adrenaline there is a qualitative discrepancy between the two sets of data since Fura 2 fails to detect any increase in cytosolic $[Ca^{2+}]$ whereas aequorin reports a value which is comparable to that observed on stimulation by ADP (Fig.2). These discrepancies should be resolved or explained since for analysis of data on excitatory agonist signal transduction we need to know which probe is providing relevant information. This requirement has now been made more immediate since measurements using aequorin have been made easily accessible with the commercial availability of a platelet ionised calcium aggregometer.

Fig.2 also contains estimates of the minimal Ca^{2+} concentrations (Ca^{2+} thresholds) for various platelet responses estimated using platelet preparations permeabilised either non-selectively (e.g. by exposure to high voltage electric discharge or agents such as saponin), or selectively to Ca^{2+} using a divalent cation ionophore. Comparison of these thresholds with the values for resting and maximally stimulated $[Ca^{2+}]$ obtained using aequorin or Fura 2 suggests that Fura 2 is measuring the physiologically relevant concentrations of Ca^{2+} since the use of the aequorin data in this comparison predicts responses which are not observed. For example, if Ca^{2+} was present in the appropriate site at the concentration indicated by aequorin, the resting platelet should show significant levels of fibrinogen receptor expression, arachidonate release and P47 phosphorylation (Fig.2). Similarly, ADP

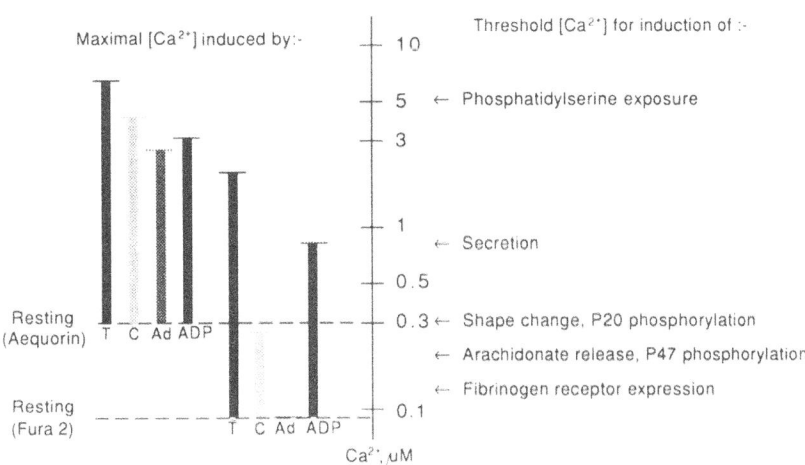

Figure 2   Comparison of Ca^{2+} thresholds for various platelet responses with cytosolic Ca^{2+} concentrations observed in resting and maximally stimulated platelets as measured using Fura 2 and aequorin as probes. T, thrombin; C, collagen; A, adrenaline.

and adrenaline, in addition to collagen and thrombin, should induce secretion directly without a requirement for thromboxane A_2 synthesis. None of these responses are observed. In contrast, such incorrect predictions are not obtained when the Ca^{2+} thresholds are compared with Ca^{2+} concentrations as measured using Fura 2 (Fig.2).

Fig.2 also indicates that for certain responses, e.g. secretion and phosphatidylserine exposure induced by collagen, the maximal increase in $[Ca^{2+}]$ as indicated by Fura 2 is not itself adequate to cause the observed response. For secretion both agonists and synthetic 1,2-diacylglycerols can enhance the sensitivity of the response to Ca^{2+} in permeabilised preparations due probably to an action via protein kinase C (Knight et al 1984; Haslam and Davidson 1984). A similar effect has not been reported for phosphatidylserine exposure and it is unclear at present what factor(s) act together with Ca^{2+} in mediating this response when induced by collagen.

Is Adrenaline a Full Excitatory Agonist for Human Platelets?

The qualitative discrepancy noted above might be resolved if adrenaline does not act as a full agonist but rather has only the ability to enhance the effects of other excitatory agonists either on end-responses and/or on alterations in the concentration of an intermediate of the signal transduction pathway, e.g. an increase in cytosolic $[Ca^{2+}]$ (Kinlough-Rathbone and Mustard 1986). This postulate was developed since responses to adrenaline added as sole agonist are observed somewhat inconsistently in platelet-rich plasma, and are often lost during the manipulations necessary to produce a washed platelet suspension. The aggregation response in such studies has in most instances been estimated from increased light transmittance of the platelet suspension despite the known insensitivity of this method to the formation of small aggregates (Thompson et al 1986).

The apparently inconsistent detection of aggregation induced by adrenaline may be interpreted in two different ways:
(i) The response caused by addition of adrenaline arises from synergistic interaction between added adrenaline and trace amounts of other undefined excitatory agonists present as contaminants in these preparations.
(ii) Adrenaline itself induces aggregation but this response is labile and is readily lost by subjecting the platelets to the various manipulations required to produce a plasma-free preparation or even in some cases by separation from other blood cells.

These possibilities are not easy to distinguish definitively but one approach has used platelets in their natural environment, i.e. whole blood, and has eliminated the contribution of other stimuli by use of appropriate antagonists and/or removal systems. When this approach was applied to fibrinogen receptor expression, fibrinogen binding and aggregation (as measured by a decrease in the platelet count) induced by adrenaline, the response could be reduced, but not eliminated. The major contributions to enhancement of the responses induced by adrenaline were identified as being due to ADP and thromboxane A_2 since marked reductions in response resulted from addition of apyrase (or of PEP+ pyruvate kinase) and of indomethacin. Little effect was seen on addition of antagonists or inhibitors which prevented the action of 5HT, PAF or thrombin. The concentrations of the antagonists/inhibitors used were established as being adequate to eliminate entirely the response induced by the agonist for which they are specific. The extent of reduction in the response was more marked at low concentrations of adrenaline suggesting that, as for other excitatory agonists, feed-forward amplification loops using ADP and thromboxane A_2 can increase the effectiveness of a non-saturating concentration of this agonist (Shattil et al 1989; Petty and Scrutton 1989). For the present discussion it is crucial that the responses were never eliminated when a saturating concentration of adrenaline was used although such elimination could be observed when non-saturating adrenaline concentrations were employed. Although such studies can never be entirely definitive the data indicate that none of the other excitatory agonists which might be present in whole blood can account either alone or in combination for the responses induced by added adrenaline. These data therefore favour the concept that adrenaline induces labile responses which are readily inactivated. This conclusion however raised the dilemna of the signal transduction mechanism by which adrenaline induces fibrinogen receptor expression, and hence aggregation. Adrenaline has no detectable capacity to cause receptor-mediated activation of phosphoinositidase C, and no increases in the concentrations of 1,2-diacylglycerol or inositol-1,4,5-trisphosphate result from stimulation by this agonist (MacIntyre et al 1985). The only well documented effects of adrenaline on platelet signal transducation mechanisms are (i) an increase in cytosolic $[Ca^{2+}]$ detected using aequorin, but not using Fura2 (Ware et al 1986), and (ii) a decrease in [cAMP] observed when the concentration of this second messenger is increased by stimulation with an inhibitory agonist such as PGI_2. The latter effect is due to receptor-induced inibition of adenylate cyclase mediated by one of the isoforms of G_i (Aktories and Jacobs 1985). The former effect is of uncertain significance given the analysis presented above while much evidence indicates that [cAMP] in the resting platelet is at, or very close to, its lowest concentration. No significant reduction in [cAMP] results from addition of excitatory agonists such as adrenaline unless a suitable inhibitory agonist is also present, and there is no evidence that pharmacological inhibition of adenylate cyclase can induce aggregation (Haslam et al 1978). It has been suggested that the α_2-adrenoreceptor may be coupled directly to the fibrinogen receptor (GP IIb/IIIa complex) possibly via G_i (Siess 1989) but little evidence currently supports this postulate.

Does Aggregation Induce an Increase in Cytosolic $[Ca^{2+}]$?

The qualitative and quantitative discrepancies which characterise the measurements of platelet cytosolic $[Ca^{2+}]$ using Fura 2 and aequorin as probes (see above) could be resolved if aggregation itself induced an increase in $[Ca^{2+}]$. This hypothesis is attractive for a number of reasons:

(i) An aggregation-induced increase in cytosolic $[Ca^{2+}]$ would be detected in aequorin-loaded, but not in Fura 2-loaded, platelets due to differences in the properties of these two probes.

(ii) The isolated glycoprotein IIb/IIIa complex can act as a Ca^{2+} channel when incorporated into liposomes (Ryback et al 1988) and monoclonal antibodies to this complex block Ca^{2+} influx into intact platelets (Powling and Hardisty 1985; Yamaguchi et al 1987). Electro-physiological analysis also indicates a role for the glycoprotein IIb/IIIa complex in Ca^{2+} channel activation in platelet plasma membranes (Fujimoto et al 1991).

(iii) Biphasic increases in cytosolic $[Ca^{2+}]$ have been detected in aequorin-loaded platelets but are not observed in these cells when loaded with Fura 2. In some instances the second phase of the increase in cytosolic $[Ca^{2+}]$ as detected by aequorin coincides with induction of aggregation as measured by an increase in light transmittance (LeCompte et al 1990; Potevin et al 1991; Bertolino and Baldiuni 1992).

(iv) An aggregation-induced increase in cytosolic $[Ca^{2+}]$ would provide a partial explanation for the several contact-dependent responses exhibited by the platelet. Such responses include thromboxane A_2 synthesis associated with stimulation by ADP and adrenaline (Charo et al 1977; Smith et al 1973) and increases in [cGMP] associated with stimulation by a wide range of agents all of which induce aggregate formation although by different routes (Edgecombe et al 1992). These responses are believed to require activation of Ca^{2+}-driven enzymes - phospholipase A_2 for thromboxane A_2 synthesis (Loeb and Gross 1986), and NO synthetase for cGMP formation (Radomski et al 1990). However, contact either has other effects relevant to induction of these responses, or causes a localised increase in $[Ca^{2+}]$, since increases in this latter parameter without cellular contact, e.g. in an unstirred system, fail to induce these responses (Edgecombe et al 1992).

We have however been unable to detect contact-induced Ca^{2+} influx or intracellular mobilisation. Stimulation of aequorin-loaded platelets with ADP or thrombin in an unstirred system causes an increase in $[Ca^{2+}]$ but no aggregate formation. On stirring the suspension aggregation is then induced but a second increase in $[Ca^{2+}]$ is not observed, as shown in Fig.3 for ADP. Similarly, in a stirred suspension containing no fibrinogen addition of ADP acuses an increase in $[Ca^{2+}]$ but no aggregation. Subsequent addition of fibrinogen induces aggregation but does not cause an increase in $[Ca^{2+}]$ (Orchard and Scrutton 1993). Further evidence supports the conclusion that contact per se causes neither Ca^{2+} influx not intracellular Ca^{2+} mobilisation. For example:

(i) Stimulation of aequorin-loaded platelets with either 1,2-dioctanoin or ristocetin induces aggregate formation without a detectable increase in $[Ca^{2+}]$ (Orchard and Scrutton 1993). Although ristocetin induces aggregate formation through an interaction involving glycoprotein Ib using Factor VIIIR as cofactor the response especially at lower [ristocetin] also depends on a glycoprotein IIb/IIIa-dependent mechanism (Heinrich et al 1985; Edgecombe et al 1992).

(ii) Platelets which have been pre-treated with chymotryspin can be induced to aggregate by addition of fibrinogen in the absence of an excitatory agonist (Niewarowski et al 1981). In this system using aequorin-loaded platelets aggregate formation but no increase in $[Ca^{2+}]$ results from addition of fibrinogen (Orchard and Scrutton 1993).

Collagen is the other agonist for which aggregate formation does appear to cause an increase in cytosolic $[Ca^{2+}]$ (Bertolino and Balduini 1992). We have similar data showing that addition of collagen to an unstirred platelet suspension induces a small Ca^{2+} transient but that the major increase in cytosolic $[Ca^{2+}]$ occurs in concert with aggregation

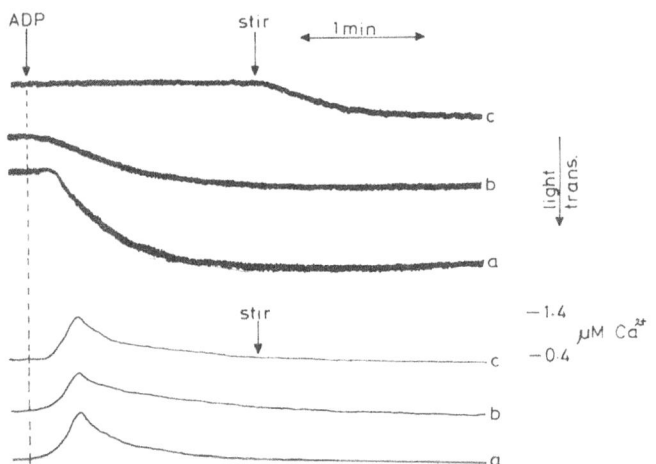

Figure 3   Aggregation responses and increases in cytosolic $[Ca^{2+}]$ in aequorin-loaded platelets. Washed platelets were loaded with aequorin as described by Lees et al (1989) and then suspended in HEPES/Tyrode's albumin pH 7.4 containing 1 mM Ca^{2+} at 2.5 x 10^8 cells/ml. Platelet aggregation and aequorin luminescence were recorded simultaneously using a modified platelet ionised calcium aggregometer (Lees et al, 1989). Signals were processed through a PC-26 analogue to digital processor and analysed using the software package 'Microscope'. Fractional aequorin luminescence was calculated as described by Lees et al (1989) and converted to $[Ca^{2+}]$ by reference to a calibration curve constructed at 37°C and pH 7.2 in the presence of 0.25 mM Mg^{2+}. ADP $(10\mu M)$ was added as indicated by the arrow to a stirred control system (a), to a stirred system containing $5\mu M$ indomethacin (b) or an unstirred system (c). For (c) stirring was started 2 min after addition of ADP as indicated by the arrow.

when stirring is initiated (Orchard and Scrutton 1993). Interpretation of these data as an aggregation-induced increase in cytosolic $[Ca^{2+}]$ requires however that platelets bind effectively to collagen in an unstirred system. This is not the case (Smith 1992). Hence the apparent dependence on aggregate formation reported by Bertolino and Balduini (1992) reflects a requirement for stirring to induce adhesion to the collagen fibres.

Factors Modulating Ca^{2+}-driven Responses

Several factors have the capacity to modulate some or all of the Ca^{2+}-driven responses indicated in Fig.2. The factors presently identified which can fulfil this role are:

(i)   1,2-diacylglycerol
(ii)   MgATP^{2-}
(iii)   G-proteins
(iv)   cAMP
(v)   cGMP

1,2-Diacylglycerol

(a) **Sources** 1,2-diacylglycerol formation in platelets has until recently been considered to result from the breakdown of phsophatidyl-inositol-4,5-bisphohphate as a result of agonist-induced activation of phosphoinositidase C (Siess 1989) (pathway A in Fig.4). This pathway generates primarily 1-stearoyl-2-arachidonylglycerol (Mauco et al 1984). Other routes have now been identified in this cell for generation of 1,2-diacylglycerols which may in some cases have a fatty acid composition differing from that derived by degradation of phosphatidylinositol-4,5-bisphosphate. Some of these alternate routes are shown as pathways B,C and D in Fig.4. At least in electro-permeabilised platelets, thrombin can induce breakdown of phosphatidylinositol-4-phosphate presumably by activation of phosphoinositidase C (pathway C in Fig.4) (Culty et al 1988). Since at least 4 isoforms of phosphoinositidase C are present in platelets (Banno et al 1992) it is not clear whether the same enzyme is responsible for degradation of both phosphoinositides. Furthermore the fatty acid composition of the 1,2-diacylglycerol generated by pathway C needs to be defined, although it will probably resemble that derived by pathway A.

More recently, agonist-induced activation of phospholipase D (pathway D in Fig.4) has been recognised as a mechanism for generation of 1,2-diacylglycerols by breakdown of phosphatidyl-choline (Exton 1990). In certain cells including possibly the platelet (Nakashima et al 1991) this pathway is responsible for longer-term elevation of [1,2-diacylglycerol] and hence for sustained activation of protein kinase C (see below). Although the immediate product of the activation of phospholipase D is phosphatidate the platelet possesses phosphatidate phosphohydrolase (Nozawa et al 1991), and hence can generate 1,2-diacylglycerol by pathway D. In addition, 1,2-diacylglycerol can be generated directly from phosphatidylcholine by activation of phosphocholinidase C (phosphatidyl-choline-specific phospholipase C) (pathway B in Fig.4). Activation of phospholipase D can be detected by production of a phosphatidylalcohol if studies are performed in the presence of an appropriate alcohol acceptor since the enzyme can catalyze trans-phosphatidylation (Dawson 1967). The disadvantage of this assay is its requirement for the addition of relatively high concentrations of the acceptor alcohol (Randall et al 1990) with the consequent risk that cellular metabolism will be perturbed. Alternatively activation of both phospholipase D and phosphocholinidase C can be detected in cells containing phosphatidyl [^{3}H] choline by measurement of the production of [^{3}H] choline and [^{3}H] choline phosphate respectively (Cook and Wakelam 1989). When the latter studies were performed using [^{3}H]choline-labelled platelets activation of both pathways B and D from Fig.4 was detected

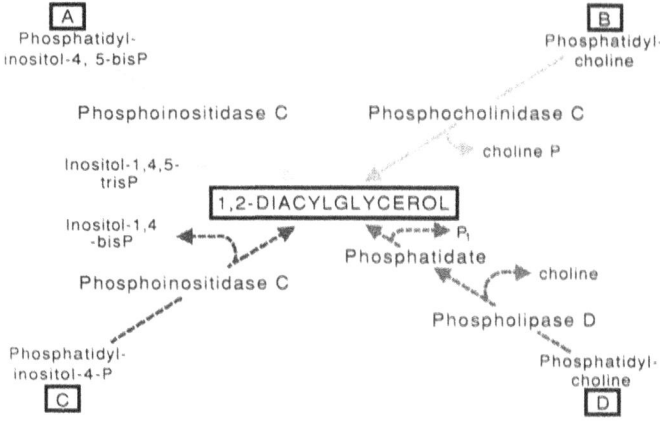

Figure 4   Pathways for formation of 1,2-diacylglycerol.

as a result of stimulation by collagen or thrombin. No such activation was observed when ADP, adrenaline or PAF were used as agonists while a divalent cation ionophore only caused [^{3}H]choline metabolite release when added at relatively high concentration. Comparison of time courses and dose/response curves for [^{3}H]choline and [^{3}H]choline phosphate release induced by collagen and thrombin indicated that choline release was not a result of the dephosphorylation of initially released choline phosphate. Activation of both pathways is also indicated by Fig.5 which shows that in an unstirred system, or in the presence of 5mM EDTA, reduced, but significant, release of [^{3}H]choline was observed, but release of [^{3}H]choline phosphate was abolished (Petty and Scrutton 1992). These data indicate that collagen and thrombin can activate phospholipase D as an immediate consequence of receptor occupancy and, for collagen, explain prior results (Pollock et al 1986) showing that this agonist causes 1,2-diacylglycerol formation as a result of the breakdown of phosphatidylcholine.

Activation of phosphocholinidase C by these agonists appears, in contrast, to be a secondary consequence of aggregate formation (Fig.5) which may explain the failure to observe such activation on stimulation by thrombin (Rittenhouse 1979). It may provide a further example of a contact-induced response (see above). The results obtained using measurement of [^{3}H]choline release are in accord with earlier studies in which stimulation by thrombin was shown to increase phosphatidylethanol formation (Rubin 1988).

(b) **Mechanism of action** Evidence which has been interpreted to support the postulate that 1,2-diacylglycerol modulates Ca^{2+}-driven platelet responses has been obtained for secretion (Knight et al 1984), arachidonate release (Halenda and Rehm 1987) and fibrinogen receptor expression (Shattil and Brass 1987). Such an interpretation is however not based on direct manipulation of the concentration of the physiological 1,2-diacylglycerol resulting for example from alteration of the activity of either 1,2-diacylglycerol kinase and/or 1,2-diacylglycerol lipase in the intact cell. Instead it depends on the use of synthetic, non-physiological 1,2-diacylglycerols, e.g. 1,2-dioctanoin (diC$_8$), at relatively high concentrations, or of other activators and/or inhibitors of protein kinase C. This alternative approach mimics neither the localisation nor, for phorbol esters and related compounds, the transience of the natural 1,2-diacylglycerol signal (Nakashima et al 1991). The interpretation of the data obtained using phorbol esters and related compounds as indicating a role for 1,2-diacylglycerol depends also on the assumption that activation of protein kinase C is involved in generation of the response. Such an assumption has been

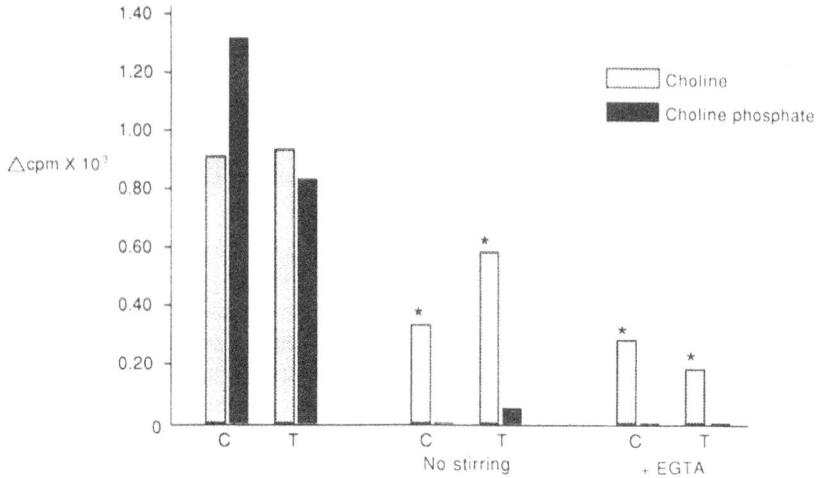

Figure 5   Properties of release of [³H]choline and [³H]choline phosphate from [³H]choline-labelled platelets stimulated with collagen (C) or thrombin (T).

supported for secretion since inhibitors of protein kinase C block the agonist-induced response over the same concentration range as that required to prevent phosphorylation of P47 (pleckstrin), the major substrate for this enzyme in the platelet. A similar correlation has not been observed for aggregation (Watson et al 1988). However P47 phosphorylation is not causal for the secretory response since near-maximal phosphorylation of this polypeptide can occur in the absence of significant 5HT release (Haslam and Davidson 1984). Since the biological role of P47 has not been defined (Tyers et al 1988) it is unclear whether phosphorylation of this polypeptide is an early event in the pathway leading to secretory granule exocytosis or whether it forms part of a parallel pathway having no role in the secretory response.

It has also been proposed that activation of protein kinase C inhibits the formation of 1,2-diacylglycerol from phosphoinositides by promoting the uncoupling of receptors from phosphoinositidase C, possibly as a consequence of phosphorylation of a G-protein. This postulate is based on results obtained by non-physiological activation of protein kinase C as a result of prolonged incubation with phorbol esters (Watson and Lapetina 1985). There is little evidence that such effects are important in normal cellular function since phosphatidate formation by thrombin is not enhanced when protein kinase C is completely inhibited (Watson et al 1988). In addition the only platelet G-protein (G_z) which is a substrate for phosphorylation by protein kinase C (Carlson et al 1991; Lounsbury et al 1991) has no known role in receptor to phosphoinositidase C coupling.

MgATP²⁻

Recent studies have indicated that MgATP²⁻ is a required cofactor for the expression of Ca^{2+}-driven arachidonate mobilisation in electropermeabilised human platelets (Patel and Scrutton 1991). This observation resolves discrepancies between previous observations indicating that this Ca^{2+}-driven response could be observed in platelets treated with a divalent cation ionophore in the presence of mM extracellular Ca^{2+} (Halenda and Rehm 1987), but not on incubation of isolated human platelet membranes with Ca^{2+} (Nakashima et al 1989). Enhancement of arachidonate mobilisation by MgATP²⁻ does not appear to involve phosphorylation since metabolically stable analogues of ATP are also effective (Patel and Scrutton 1991). It may be explained if a nucleotide-binding protein activates the

Ca^{2+}-dependent human platelet phospholipase A$_2$ as has been proposed for the Ca^{2+}-independent human myocardial enzyme (Hazen and Gross 1991).

G Proteins

In addition to their role in transmembrane signalling to phosphoinositidase C (Manning and Brass 1991) G-proteins have been implicated in the modulation of Ca^{2+}-driven responses at other levels. This role is most clearly defined for arachidonate release where enhancement of the Ca^{2+}-driven response by GTPγS can be clearly separated from activation of 1,2-diacylglycerol production by this guanine nucleotide (Nakashima et al 1987). The pertussis toxin sensitivity of the G-protein apparently involved in receptor-mediated activation of phospholipase A$_2$ suggests its identification as one of the three isoforms of G$_i$ which are present in the platelet (Manning and Brass 1991).

For other responses the situation is less clear. In other cells a G-protein appears to act at a site distal to Ca^{2+} in the secretory pathway since for example inhibitors of G-protein function, e.g. GDPßS, block the Ca^{2+}-driven response (Gomperts et al 1990). Such an effect is not observed in the platelet although metabolically stable GTP analogues can induce substantial amine storage granule and lysosomal secretion from permeabilised preparations at very low (<nM) Ca^{2+} (Athayde and Scrutton 1990; Scrutton and Athayde 1991). Ca^{2+}-driven fibrinogen receptor exposure in saponin-permeabilised preparations is only observed in the presence of a metabolically stable GTP analogue (Shattil and Brass 1987) suggesting that a G-protein may modulate coupling of receptors to the glycoprotein IIb/IIIa complex. However, neither the nature of this putative G-protein nor details of its role in this process have been defined.

Cyclic-3',5'-AMP

Actions of cAMP as a second messenger mediating the effects of inhibitory agonists such as prostaglandin I$_2$ and adenosine may involve impairment of (i) the efficacy of the Ca^{2+} signal in driving a given response; or of (ii) the mechanisms responsible for elevation of cytosolic [Ca^{2+} in response to an excitatory agonist. An increase in [cAMP] may also decrease the ability to maintain an elevated cytosolic [Ca^{2+}] by enhancing removal of Ca^{2+} from the cytosol (Adunyah and Dean 1987). However this effect is not consistently observed (O'Rourke et al 1989).

Since inhibitory agonists such as PGI$_2$ can prevent aggregation induced by a divalent cation ionophore, a site (or sites) for the action of cAMP exist which are distal to the site of action of Ca^{2+} in the signal transduction pathway (Siess and Lapetina 1989). This action may be response-specific since addition of cAMP causes no inhibition of Ca^{2+}-driven 5HT secretion in electropermeabilised platelets (Fig.6). In this latter system addition of cAMP prevents enhancement of Ca^{2+}-driven 5HT secretion by thrombin or, at higher [cAMP], by a 1,2-diacylglycerol (Knight and Scrutton 1984). Inhibition is also observed in the presence of GTP but not significantly when the Ca^{2+}-driven response is enhanced by a metabolically stable GTP analogue, e.g. GTPγS (Fig.6) (Knight and Scrutton 1987). These findings suggest that inhibition by cAMP of agonist-induced enhancement of secretion results froman increase in the GTPase activity of the G-protein responsible for receptor-phosphoinositidase C coupling (Fig.7) which reduces the lifetime of the GTP-G-protein (activated) complex. This action of cAMP may however not be direct since none of the G-proteins known to be involved in transmembrane signalling to phosphoinositidase C are phosphorylated by protein kinase A (Manning and Brass 1991). It could therefore involve phosphorylation and activation by cAMP of a GTPase-activating (GAP) protein. Although cAMP-dependent phosphorylation of rap1B has been reported, the characteristics of this effect are not compatible with a proposed role in regulation of phosphoinositidase C (Siess and Lapetina 1990).

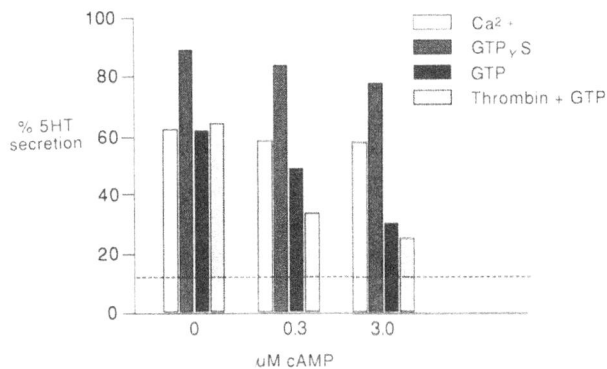

Figure 6   Effect of cAMP on [³H]5HT secretion from electropermeabilised human platelets induced by Ca²⁺ in the presence of various additions. Electropermeabilised platelets were prepared and the studies were performed as described by Knight et al (1984). 5HT secretion was induced by addition of 5 μM Ca²⁺, 100 μM GTP + 5 nM thrombin in the presence of 2 μM Ca²⁺ or 20 μM GTPγS in the presence of 1 μM Ca²⁺. The extent of 5HT release observed in the presence of 1 nM Ca²⁺ is indicated by the dashed line.

Attempts to gain further insight into this system have thus far not been successful. Agonist-induced GTPase activity cannot be detected in electropermeabilised platelet preparations above a rather high level of agonist-independent activity, while in membrane preparations where agonist-activated GTPase can be observed, this activity is insensitive to addition either of cAMP or of the catalytic subunit of protein kinase A (D.E.Knight and M.C. Scrutton, unpublished data).

The mechanism of action proposed in Fig.7 would explain inhibition by an increase in [cAMP] of increases in cytosolic [Ca²⁺] (Rink and Sanchez 1984), and of 1,2-diacylglycerol (Rittenhouse 1979) induced by thrombin since all these effects are a consequence of decreased activation of phosphoinositidase C. It is therefore of interest that increases in cytosolic [Ca²⁺] and in [1,2-diacylglycerol] induced by collagen are not inhibited by an increase in platelet [cAMP] (Smith et al 1992), suggesting that the signal transduction pathway used by this agonist to increase cytosolic [Ca²⁺] differs significantly from that used by thrombin.

Cyclic-3',5'GMP

Although the pattern of inhibition of platelet responsiveness by cGMP differs significantly from that caused by an increase in cAMP (Radomski et al 1987b) the effects of cGMP and cAMP on the excitatory agonist signal transduction pathway described thus far are mostly similar. An increase in [cGMP] inhibits phosphoinositide hydrolysis and the increase in [Ca²⁺] induced by several excitatory agonists although for ADP Ca²⁺ influx is reduced more effectively than extracellular Ca²⁺ mobilisation (Nakashima et al 1986; Morgan and Newby 1989). Ca²⁺ extrusion from the platelet is enhanced but cGMP does not appear to increase Ca²⁺ uptake into the endoplasmic reticulum (Johansson and Haynes 1992). Inhibition of aggregation induced by a divalent cation ionophore or exogenously added 1,2-diacylglycerol is observed (Doni et al 1991) but in permeabilised preparations cGMP does not reduce the extent of Ca²⁺-driven secretion (Knight and Scrutton 1984). Some, or all, of the effects observed in the intact cell may result indirectly from enhancement of the effect of cAMP as a consequence of inhibition by cGMP of cAMP phospho-diesterase (Maurice and Haslam 1991). Actions of cGMP observed thus far in

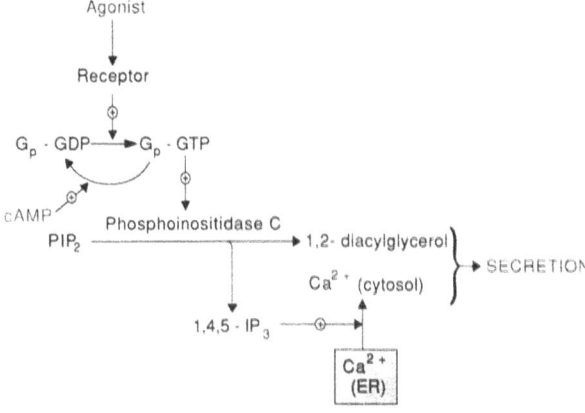

Figure 7   A possible model for inhibition by cAMP of agonist-induced phosphoinositidase C activity. In the figure G_p is used as indicating the class of G-proteins involved in receptor-phosphoinositidase C coupling.

permeabilised preparations occur at high concentrations of this cyclic nucleotide (Haslam et al 1985) and appear unlikely to be of physiological significance.

REFERENCES

Adunayah, S.E., and Dean, W.L., 1987, Regulation of human platelet membrane Ca^{2+} transport by cAMP and calmodulin-dependent phosphorylation, *Biochim. Biophys. Acta.* 930:401.

Athayde, C.M., and Scrutton, M.C., 1990, Guanine nucleotides and Ca^{2+}-dependent lysosomal secretion. *Europ. J. Biochem.* 189:647.

Aktories, K., and Jakobs, K.H., 1985, Regulation of platelet cyclic AMP formation, in: "The platelets: physiology and pharmacology", G.L. Longenecker, ed., *Academic*, New York, p.243.

Banno, Y., Suzuki, T., and Nozawa, Y., 1992, Isolation of a polyphosphoinositide phospholipase C (type ß) from cytosolic and membrane fractions of human platelets, *Platelets*, 3:69.

Bertolino, G., and Balduini, C.L., 1992, Aggregation of human platelets stimulates calcium ion movement and release reaction, *Platelets* 3:79.

Bevers, E.M., Comfurius, P., and Zwaal, R.F.A., 1991, Platelet procoagulant activity: physiological significance and mechanisms of exposure, *Blood Revs.* 5:146.

Carlson, K.E., Brass, L.F., and Manning, D.R., 1991, Thrombin and phorbol esters cause the selective phosphorylation of a G-protein other than G_i in human platelets, *J. Biol. Chem.* 264:13298.

Charo, I.F., Feinman, R.D., and Detwiler, T.C., 1977, Interrelations of platelet aggregation and secretion, *J. Clin. Invest.* 60:866.

Cook, S.J., and Wakelam, M.J.O., 1989, Analysis of the water-soluble products of phosphatidylcholine breakdown by ion exchange chromatography, *Biochem. J.* 263:581.

Crawford, N., and Scrutton, M.C., 1987, Biochemistry of the blood platelet, in: *"Haemostasis and Thrombosis"*, A.L. Bloom and D.P. Thomas, eds., 2nd edition, Churchill Livingstone, Edinburgh, p.47.

Culty, M., Davidson, M.M.L., and Haslam, R.J., 1988, Guanosine-5'-[τ-thiol]-triphosphate and thrombin stimulate the hydrolysis of polyphosphoinositides by phospholipase C in electropermeabilised platelets, *Europ. J. Biochem.* 171:523.

Dawson, R.M.C., 1967, The formation of phosphatidylglycerol and other phospholipids by the transferase activity of phospholipase D, *Biochem. J.* 102:205.

Doni, M.G., Deana, R., Padoin, E., Ruzzene, M., and Alexandre, A., 1991, Platelet activation by diacylglycerol is inhibited by nitroprusside, *Biochim. Biophys. Acta* 1094:323.

Edgecombe, M., Scrutton, M.C., and Kerry, R., 1993, Platelet-platelet contact is required to observed guanylate cyclase activation in stimulated platelets, *Platelets*. In press.

Exton, J.H., 1990. Signalling through phosphatidylcholine. *J. Biol. Chem.* 265, 1.

Fujimoto, T., Fujimura, K., and Kuramoto, K., 1991, Electrophysiological evidence that glycoprotein IIb-IIIa complex is involved in calcium channel activation in human platelet plasma membrane, *J. Biol. Chem.* 266:16370.

Gomperts, B.D., Churcher, Y., Koffer, A., Kramer, I.M., Lillie, T., and Tatham, P.E., 1990, The role and mechanism of the GTP-binding protein G_E in the control of regulated exocytosis, *Biochem. Soc. Symp.* 56:85.

Gordon, J.L., and Milner, A.J., 1976, Blood platelets as multifunctional cells, in *"Platelets in Biology and Pathology"*, J.L. Gordon, ed. Elsevier/North Holland, Amsterdam, p.3.

Halenda, S., and Rehm, A.G., 1987, Thrombin and C-kinase activators potentiate calcium-stimulated arachidonic acid release in human platelets, *Biochem. J.* 248:471.

Hallam, T.J., Daniel, J.L., Kendrick-Jones, J., and Rink, T.J., 1985, Relationship between cytoplasmic free calcium and myosin light chain phosphorylation in intact platelets, *Biochem. J.* 232:373.

Haslam, R.J., and Davidson, M.M.L., 1984, Potentiation by thrombin of secretion from permeabilised platelets equilibrated with Ca^{2+} buffers: relationship to protein phosphorylation and diacylglycerol formation, *Biochem. J.* 222:351.

Haslam, R.J., Davidson, M.M.L., and Desjardins, J.V., 1978, Inhibition of adenylate cyclase by adenosine analogues in broken and intact human platelets: evidence for unidirectional control of platelet function by cyclic-3',5'-adenosine monophosphate, *Biochem. J.* 176:83.

Haslam, R.J., Davidson, M.M.L., Knight, D.E., and Scrutton, M.C., 1985, GTP not cyclic GMP enhances secretion from permeabilised platelets, *Nature (Lond).* 313:821.

Hazen, S.L., and Gross, R.W., 1991, Human myocardial Ca^{2+}-independent phospholipase A_2 is modulated by ATP: concordant ATP-induced alterations in enzyme kinetics and mechanism-based inhibition, *Biochem. J.* 280:581.

Heinrich, D., Scharf, T., Santoro, S., Clementson, K.J., and Mueller-Eckhardt, C., 1985, Monoclonal antibodies against human glycoproteins IIb-IIIa. II. Different effects on platelet function, *Thromb. Res.* 38:547.

Johansson, J.S., and Haynes, D.H., 1992, Cyclic GMP increases the rate of the calcium extrusion pump in intact human platelets but has no direct effect on the dense tubular calcium accumulation system, *Biochim. Biophy. Acta.* 1105:40.

Kinlough-Rathbone, R.L., and Mustard, J.F., 1986, Synergism of agonists, in: *"Platelet responses and metabolism"*, H. Holmsen, ed., CRC Press Inc., Boca Raton, Vol.1, p.193.

Knight, D.E., and Scrutton, M.C., 1984, Cyclic nucleotides modulate a system which controls the Ca^{2+} sensitivity in human platelets, *Nature (Lond).* 309:66.

Knight, D.E., and Scrutton, M.C., 1987, Secretion of 5-hydroxytryptamine from electropermeabilised human platelets: effects of GTP and cyclic-3',5'-AMP. *FEBS Letts.* 233:47.

Knight, D.E., Niggli, V., and Scrutton, M.C., 1984, Thrombin and activators of protein kinase C modulate secretory responses of permeabilised human platelets induced by Ca^{2+}, *Europ. J. Biochem.* 143:437.

Lanza, F., Beretz, A., Kubina, M., and Cazenave, J.P., 1987, Increased aggregation and secretion responses of human platelets when loaded with the calcium fluorescent probes, quin2 and Fura2, *Thromb. Haemostas.* 58:737.

Larsen, E., Celi, A., Gilbert, G.E., Furie, B.C., Erban, J.K., Bonfanti, R., Wagner, D.D., and Furie, B., 1989, PADGEM protein: a receptor which mediates the interaction of activated platelets with neutrophils and monocytes, *Cell* 59:305.

Lecompte, T., Potevin, F., Champeix, P., Morel, M.C.X., Favier, R., Hurtaud, M.F., Schiegel, N., Samama, M., and Kaplan, C., 1990, Aequorin-detected calcium changes in stimulated thrombasthenic platelets: aggregation-dependent calcium movement in response to ADP, *Thromb. Res.* 58:561.

Lees, A.D., Wilson, J., Orchard, C.H., and Orchard, M.A., 1989, Ouabain enhances basal and stimulus-induced cytoplasmic calcium concentrations in platelets, *Thromb. Haemostas.* 62:1000.

Loeb, L.A., and Gross, R.W., 1986, Identification and purification of sheep platelet phospholipase A_2: activation by physiological concentrations of calcium ion., *J. Biol. Chem.* 261:10467.

Lounsbury, K.M., Casey, P.J., Brass, L.F., and Manning, D.R., 1991, Phosphorylation of G_Z in human

MacIntyre, D.E., Pollock, W.K., Shaw, A.M., Bushfield, M., MacMillan, L.J., and McNicol, A., 1985, Agonist-induced inositol phospholipid metabolism and Ca^{2+} flux in human platelet activation, *Adv. Exp. Med. Biol.* 192:127.

Manning, D.R., and Brass, L.F., 1991, The role of GTP-binding proteins in platelet activation, *Thromb. Haemostas.* 66:393.

Mauco, G., Dangelmaier, C.A., and Smith, J.B., 1984, Inositol lipids, phosphatidate and diacylglycerol share stearoylarachidonylglycerol as a common backbone in thrombin-stimulated platelets, *Biochem. J.* 224:933.

13

Maurice, D.H., and Haslam, R.J., 1991, Molecular basis of the synergistic inhibition of platelet function by nitrovasodilators and activators of adenylate cyclase: inhibition of cyclic AMP breakdown by cyclic GMP, *Mol. Pharmacol.* 37:671.

Moncada, S., Palmer, R.M.J. and Higgs, E.A., 1987, Prostacyclin and endothelium-derived relaxing factor: biological interactions and significance, in: *"Haemostasis and Thrombosis 1987"*, M. Verstraete, J. Vermylen, R. Lijnen, J. Arnout, eds., Leuven University Press, Leuven, p.597.

Morgan, R.O., and Newby, A.C., 1989, Nitroprusside differentially inhibits ADP-stimulated calcium influx and mobilisation in human platelets, *Biochem. J.* 258:447.

Naka, S., Tohmatsu, T., Hattori, H., Okano, Y., and Nozawa, Y., 1986, Inhibitory action of cyclic GMP on secretion, polyphosphoinositide hydrolysis and Ca^{2+} mobilisation in thrombin-stimulated human platelets, *Biochem. Biophys. Res. Comm.* 135:1099.

Nakashima, S., Hattori, H., Shirato, L., Takenaka, A., and Nozawa, Y., 1987, Differential sensitivity of arachidonic acid releease and 1,2-diacylglycerol formation to pertussis toxin, GDPßS and NaF in saponin-permeabilised platelets: possible evidence for distinct GTP-binding proteins involving phospholipase C and A_2 activation, *Biochem. Biophys. Res. Comm.* 148:971.

Nakashima, S., Suganuma, A., Matsui, A., Hattori, H., Sato, M., Takenaka, A., and Nozawa, Y., 1989, Primary role of calcium ions in arachidonic acid release from rat platelet membranes: comparison with human platelet membranes, *Biochem. J.* 259:139.

Nakashima, S., Suganuma, A., Matsui, A., and Nozawa, Y., 1991, Thrombin induces a biphasic 1,2-diacylglycerol production in human platelets, *Biochem. J.* 275:355.

Niewiarowski, S., Budzynski, A., Morinelli, T.A., Brudzynski, T.M.M., and Stewart, G.J., 1981, Exposure of fibrinogen receptors on human platelets by proteolytic enzymes, *J. Biol. Chem.* 256:917.

Nozawa, Y., Shigeru, S., and Nagata, K., 1991, Phospholipid-mediated signalling in receptor activation of human platelets, *Biochim. Biophys. Acta.* 1082:219.

O'Rourke, F., Zavoico, G.B., and Feinstein, M.B., 1989, Release of Ca^{2+} by inositol-1,4,5-trisphosphate in platelet membrane vesicles is not dependent on cyclic AMP-dependent protein kinase, *Biochem. J.* 257:715.

Orchard, M.A., and Scrutton, M.C., 1993, Aggregation fails to increase cytosolic $[Ca^{2+}]$ in aequorin-loaded human platelets. *Platelets.* In press.

Packham, M.A., and Mustard, J.F., 1986, Interactions of platelet activating pathways: studies with inhibitors specific for individual pathways, in: *"Platelet responses and metabolism"*, H. Holmsen, ed., CRC Press Inc., Boca Raton, p.268.

Patel, S., and Scrutton, M.C., 1991, Ca^{2+}-driven [^{3}H]arachidonate release in electropermeabilised human platelets shows an absolute requirement for $MgATP^{2-}$, *Biochem. J.* 273:561.

Peltola, K., and Scrutton, M.C., 1990, Guanine nucleotides enhance calcium-driven protein storage granule secretion from electropermeabilised human platelets, *Biochem. Soc. Trans.* 18:466.

Petty, A., and Scrutton, M.C., 1989, Platelet aggregation in whole blood: is the response to adrenaline, 5-hydroxytryptamine and PAF a direct consequence of stimulation by these agonists? *Thromb. Res.* 54:151.

Petty, A., and Scrutton, M.C., 1992, Release of choline metabolites from human platelets: evidence for activation of phospholipase D and of phosphatidylcholine-specific phospholipase C, *Platelets.* 4, 23.

Pollock, W.K., and Rink, T.J., 1986, Thrombin and ionomycin can raise platelet cytosolic Ca^{2+} to micromolar levels by discharge of internal Ca^{2+} stores: studies using Fura2, *Biochem. Biophys. Res. Comm.* 139:308.

Pollock, W.K., Rink, T.J., and Irvine, R.F., 1986, Liberation of [^{3}H]arachidonate and changes in cytosolic free calcium in Fura2-loaded human platelets stimulated by ionomycin and collagen, *Biochem. J.* 235:869.

Potevin, F., Lecompte, T., Favier, R., and Samama, M., 1991, Rapid aequorin loading into platelets in the presence of DMSO - characterisation of the responses (changes in light transmission and in calcium) to various agonists, *Thromb. Haemostas.* 66:334.

Powling, M.J., and Hardisty, R.M., 1985, Glycoprotein IIb-IIIa complex and Ca^{2+} influx into stimulated platelets, *Blood* 66:731.

Radomski, M.W., Palmer, R.M.J., and Moncada, S., 1987a, The anti-aggregating proeprties of vascular endothelium: interactions between prostacyclin and nitric oxide, *Brit. J. Pharmacol.* 92:639.

Radomski, M.W., Palmer, R.M.J. and Moncada, S., 1987b, The role of nitric oxide and cGMP in platelet adhesion to vascular endothelium, *Biochem. Biophys. Res. Comm.* 148:1482.

Radomski, M.W., Palmer, R.M.J., and Moncada, S., 1990, An L-arginine/nitric oxide pathway present in human platelets regulates aggregation, *Proc. Natl. Acad. Sci. USA* 87:5193.

Randall, R.W., Bonser, R.W., Thompson,N.T., and Garland, L.G., 1990, A novel and sensitive assay for phospholipase D in intact cells, *FEBS Letts.* 264:87.

Rink, T.J., and Sanchez, A., 1984, Effects of prostaglandin I_2 and forskolin on secretion from platelets evoked at basal concentrations of cytoplasmic free calcium by thrombin, collagen, phorbol ester and exogenous diacylglycerol, *Biochem. J.* 222:833.

Rink, T.J., and Sage, S.O., 1990, Calcium signalling in human platelets, *Ann. Rev. Physiol.* 52:431.

Rittenhouse, S.E., 1979, Production of diglyceride in activated platelets, *J. Clin. Invest.* 63:580.

Rubin, R., 1988. Phosphatidylethanol formation in human platelets: Evidence for thrombin induced activation of phospholipase D. *Biochem. Biophys. Res. Commun.* 156, 1090.

Rybak, M.E., Renzulli, L.A., Bruns, M., and Cahaly, D.P., 1988, Platelet glycoprotein IIb and IIIa as a calcium channel in liposomes, *Blood* 72:714.

Scrutton, M.C., and Athayde, C.M., 1991, The biochemical basis for the regulation of platelet responsiveness, in: *"The Platelet in Health and Disease"*, C.P. Page, ed., Blackwell, Oxford, p.6199.

Shattil, S.J., and Brass, L.F., 1987, Induction of the fibrinogen receptor on human platelets by intracellular mediators, *J. Biol. Chem.* 262:992.

Shattil, S.J., Budzynski, A., and Scrutton, M.C., 1989, Epinephrine induces platelet fibrinogen receptor expression, fibrinogen binding and aggregation in whole blood in the absence of other excitatory agonists, *Blood* 73:150.

Siess, W., 1989, Molecular mechanisms of platelet activation, *Physiol. Rev.* 69:58.

Siess, W., and Lapetina, E.G., 1989, Prostacyclin inhibits platelet aggregation by phorbol ester or Ca^{2+} ionophore at steps distal to activation of protein kinase C and Ca^{2+}-dependent protein kinases, *Biochem. J.* 258:57.

Siess, W., and Lapetina, E.G., 1990, Functional relationship between cyclic AMP-dependent protein phosphorylation and platelet inhibition, *Biochem. J.* 271:815.

Smith, J.B., 1992, Personal communication.

Smith, J.B., Ingerman, C., Kocsis, J.J., Silver, M.J., 1973, Formation of prostaglandins during the aggregation of human platelets, *J. Clin. Invest.* 52:965.

Smith, J.B., Dangelmaier, C.A., Selak, M.A., and Daniel, J.L., 1991, Facile platelet adhesion to collagen requires metabolic energy and actin polymerisation, and evokes intracellular free calcium mobilisation, *J. Cell. Biochem.* 47:54.

Smith, J.B., Dangelmaier, C.A., Selak, M.A., Ashby, B., and Daniel, J.L., 1992, Cyclic AMP does not inhibit collagen-induced platelet signal transduction, *Biochem. J.* 283:889.

Thompson, N.T., Scrutton, M.C., and Wallis, R.B., 1986, Particle volume changes associated with light transmittance changes in the platelet aggregometer: dependence upon aggregating agent and effectiveness of stimulus, *Thromb. Res.* 41:615.

Tyers, M., Rachubinski, R.A., Stewart, M.I., Varrichio, A.M., Shorr, R.G.I., Haslam, R.J., and Harley, C.B., 1988, Molecular cloning and expression of the major protein kinase C substrate of platelets, *Nature (Lond.)*, 333:470.

Verhallen, P.J.F., Bevers, E.M., Comfurius, P., and Zwaal, R.F.A., 1987, Correlation between calpain-mediated cytoskeletal degradation and platelet procoagulant activity, *Biochim. Biophys. Acta.* 903:206.

Ware, J.A., Johnson, P.C., Smith, M., and Salzman, E.W., 1986, Effect of common agonists on cytoplasmic ionised calcium concentrations in platelets, *J. Clin. Invest.* 77:878.

Ware, J.A., Smith, M., Fossel, E.T., and Salzman, E.W., 1988, Cytoplasmic Mg^{2+} concentration in platelets: implications for determination of Ca^{2+} with aequorin, *Am. J. Physiol.* 255:H855.

Watson, S.P., and Lapetina, E.G., 1985, 1,2-Diacylglycerol and phorbol ester inhibit agonist-induced formation of inositol phosphates in human platelets: possible implications for negative feedback regulation of inositol phospholipid hydrolysis, *Proc. Natl. Acad. Sci. USA* 82:2623.

Watson, S.J., McNally, J., Shipman, L.J., and Godfrey, P.P., 1988, The action of the protein kinase C inhibitor, staurosporine, on human platelets: evidence against a regulatory role for protein kinase C in the formation of inositol trisphosphate by thrombin, *Biochem. J.* 249:345.

Yamaguchi, A., Yamamoto, N., Kitigawa, H., Tanaue, K., and Yamazaki, H., 1987, Ca^{2+} influx mediated through the GP IIb/IIIa complex during platelet activation, *FEBS Lett.* 225:228.

AGONIST RECEPTORS AND G PROTEINS AS MEDIATORS

OF PLATELET ACTIVATION

Lawrence F. Brass, James A. Hoxie, Thomas Kieber-Emmons, David R. Manning, Mortimer Poncz and Marilyn Woolkalis

Departments of Medicine, Pathology, Pharmacology and Pediatrics of the University of Pennsylvania and the Wistar Institute. Address correspondence to Dr. Brass at the Hematology-Oncology Section, Silverstein 7, University of Pennsylvania, 3400 Spruce St., Philadelphia, PA 19104. Telephone 215-662-3910. FAX 215-662-7617

SUMMARY

Recent studies have helped to define the earliest events of signal transduction in platelets, particularly those involved in the generation of second messengers. The best-understood of these events are those which involve guanine nucleotide binding regulatory proteins. G proteins are heterotrimers comprised of α, β and γ subunits, each of which can exist in multiple forms. Some, but not all, of the known variants of $G\alpha$ are substrates for ADP-ribosylation by pertussis toxin, a modification which disrupts the flow of information from receptor to effector. The G proteins that have been identified in platelets to date are Gs, Gi_1, Gi_2, Gi_3, Gz and Gq. Gs and one or more of the Gi family members regulate cAMP formation by adenylylcyclase. Gi may also be responsible for the pertussis toxin-sensitive activation of phospholipase C which occurs when platelets are activated by thrombin. Gq is thought to be responsible for the pertussis toxin-resistant activation of phospholipase C by TxA_2. Gz does not have an established role, but has the unique property of being phosphorylated by protein kinase C during platelet activation. Recent efforts to clone the receptors that interact with G proteins in platelets have been successful for epinephrine, thrombin, TxA_2 and platelet activating factor. Each of these resembles other G protein-coupled receptors, being comprised of a single polypeptide with 7 transmembrane domains. In the case of thrombin, receptor activation is thought to involve a unique mechanism in which thrombin cleaves its receptor, creating a new N-terminus that can serve as a tethered ligand. Peptides corresponding to the tethered ligand can mimic the effects of thrombin, while antibodies to the same domain inhibit platelet activation. Shortly after activation, thrombin receptors become resistant to re-activation by thrombin. This desensitization, which appears to be due to a combination of proteolysis, phosphorylation and internalization, provides a potential mechanism for limiting the duration of thrombin-initiated signals in platelets.

Mechanisms of Platelet Activation and Control, Edited by
K.S. Authi *et al.*, Plenum Press, New York, 1993

INTRODUCTION

A variety of agents have been shown to cause platelet activation, including collagen, ADP, thromboxane A₂, epinephrine and thrombin. According to current models, the receptors for many of these agonists are formed by proteins which cross the plasma membrane at least once. The extracellular and transmembrane domains of the receptor form the agonist binding site, while the cytosolic domains interact with second-messenger-producing enzymes whose activity is modulated by receptor occupation. Often at least one additional protein is required for this receptor:effector interaction. Typically this protein is a member of a family of GTP-binding regulatory proteins called G proteins. These proteins mediate the interaction between receptors and effectors and, depending upon the particular G protein, may either stimulate or inhibit the effector (Fig. 1).

The ability of platelets to respond quickly to vascular injury is dependent upon a number of internal amplification mechanisms. These help to insure that, once initiated, the process of platelet activation will continue and become self-reinforcing. Two intracellular pathways play a central role in platelet activation by most agonists. Each begins with the enzymatic hydrolysis of specific membrane phospholipids. The phosphoinositide pathway starts when phosphatidylinositol 4,5-bisphosphate (PIP₂) is cleaved by phospholipase C to form inositol 1,4,5-trisphosphate (1,4,5-IP₃) and diacylglycerol, both of which serve as second messengers in platelets (Fig. 1). IP₃ releases Ca^{++} from the platelet dense tubular system and contributes to the rise in the cytosolic free Ca^{++} concentration that usually accompanies platelet activation. In turn, this promotes the activity of enzymes which are not fully functional at the low Ca^{++} concentration present in resting platelets. Diacylglycerol activates protein kinase C, leading to protein phosphorylation, granule secretion and fibrinogen receptor expression. A second pathway begins when arachidonate is released from membrane phospholipids either by the direct action of phospholipase A₂ or by the sequential action of phospholipase C and diacylglycerol lipase. The newly-liberated arachidonate is then metabolized to thromboxane A₂ (TxB₂) which is a potent stimulus for platelet activation. Since TxA₂ can diffuse across the platelet plasma membrane, it can serve as a messenger between platelets as well as within platelets.

Another second messenger whose synthesis is affected by platelet agonists is adenosine 3',5'-cyclic phosphate (cAMP), which is synthesized by adenylylcyclase. Agents which raise intracellular cAMP levels, such as the prostaglandin I₂ generated by endothelial cells, dampen platelet responsiveness. Most platelet agonists suppress cAMP formation, although by itself this effect is insufficient to trigger platelet activation. Presumably, the inhibitory effects of cAMP are due to the phosphorylation of key proteins by cAMP-dependent protein kinases (protein kinase A). Platelet substrates for this enzyme include glycoprotein Ibß (Fox et al, 1987; Wardell et al, 1989), actin binding protein (Wallach et al, 1978), myosin light chain (Hathway & Adelstein, 1979; Hallam et al, 1985), rap1b (Matsui et al, 1990; White et al, 1990; Fischer et al, 1990), and several unidentified proteins (Haslam et al, 1979; Kaser-Glanzmann et al, 1979). However, it is not yet clear how the phosphorylation of any of these proteins is responsible for inhibition of platelet activation.

Once an agonist binds to its receptor, platelet shape change, secretion and aggregation usually follow rapidly. Since several of the secreted granule constituents are capable of supporting platelet activation, the release of stored materials, like the formation of TxA₂, helps to recruit additional platelets into expanding platelet aggregates. In general, a maximal platelet response to injury requires granule secretion. Therefore, a defect in either signal transduction or the storage granules themselves adversely effects platelet function.

The hallmark of platelet activation is aggregation. Platelets form aggregates because fibrinogen binds to the platelet surface and acts as a bridge between adjacent platelets. Platelet fibrinogen receptors are located on the plasma membrane glycoprotein IIb-IIIa complex and belong to the integrin family of adhesive protein receptors. The glycoprotein IIb-IIIa complex is present on the surface of resting as well as activated platelets, but appears to undergo a transformation during platelet activation which enables it to serve as a binding site for fibrinogen and other adhesive proteins. The precise nature of this transformation is still under investigation, as are the intracellular events that trigger it.

AGONIST RECEPTORS

Platelets can be activated by a variety of agents, but not to equal effect. Platelet agonists are commonly classified as strong and weak, but the distinctions between these

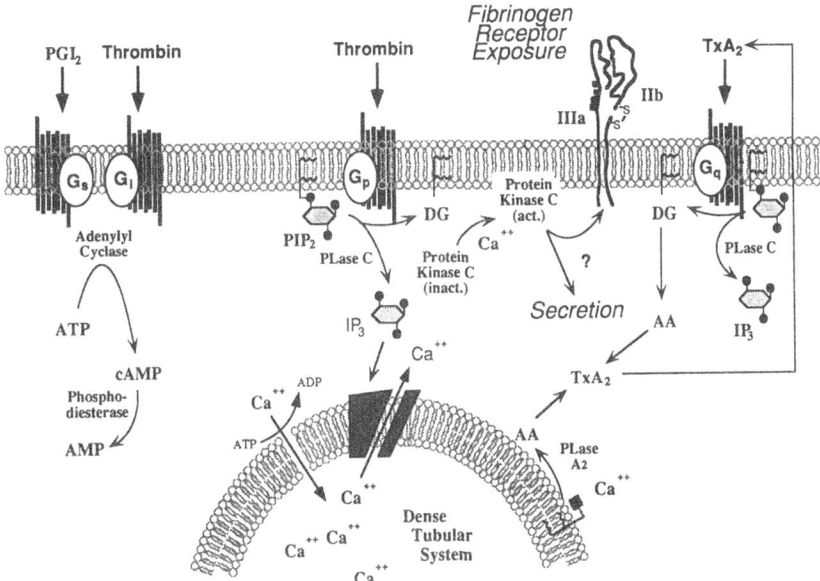

Figure 1 <u>Signal transduction during platelet activation.</u> The binding of agonists to receptors on the platelet surface initiates cascades of intracellular second messengers, including inositol 1,4,5-trisphosphate (IP$_3$) and diacylglycerol (DG). IP$_3$ releases Ca^{++} from the platelet dense tubular system. Diacylglycerol activates protein kinase C, triggering granule secretion and fibrinogen receptor exposure. At the same time, the rising cytosolic Ca^{++} concentration facilitates arachidonate (AA) release by phospholipase A$_2$, a process that may occur at both the plasma membrane and the dense tubular system membrane. Arachidonate is metabolized to thromboxane A$_2$ (TxA$_2$), which diffuses out of the cell, interacts with receptors on the platelet surface and causes further platelet activation. At some point during this process, tyrosine kinases are activated in platelets. It is not yet clear whether this occurs early in platelet activation or predominantly as a consequence of fibrinogen receptor expression and platelet aggregation. In many cases, the interaction between agonists and the enzymes responsible for second messenger generation is mediated by a G protein. In platelets, G proteins have been shown to regulate phosphoinositide hydrolysis and cAMP formation, and are probably involved in the activation of phospholipase A$_2$. Phospholipase C is activated in a pertussis toxin-sensitive manner by the unidentified G protein, and in a pertussis toxin-resistant manner by Gq. Adenylylcyclase is stimulated by the G protein, Gs, and inhibited by the G protein, Gi.

categories are often blurred. By one set of definitions, strong agonists are those which can trigger granule secretion even when aggregation is prevented by measures such as removing the extracellular Ca^{++} needed for fibrinogen binding. Thrombin and collagen are examples of strong agonists. In contrast, weak agonists, such as ADP and epinephrine, require aggregation for secretion to occur. Presumably, the stronger platelet agonists play the primary role in platelet activation *in vivo*, while the weaker agonists play a supportive role - particularly ADP, which is released from platelet dense granules during the early phases of platelet aggregate formation. Another way to classify agonists is to consider the sets of intracellular effectors that are coupled to their receptors. By this classification, strong agonists are those which stimulate phosphoinositide hydrolysis and eicosanoid formation, raise the cytosolic free Ca^{++} concentration and cause aggregation and secretion that are relatively unimpaired by inhibitors of TxA_2 formation. This suggests that their receptors are coupled to phospholipase C as well as phospholipase A_2. It also suggests that the phosphoinositide hydrolysis pathway is the dominant mediator of platelet responses to strong agonists. Weak agonists, on the other hand, have little or no ability to cause phosphoinositide hydrolysis and are more dependent on TxA_2 formation for their effects. This suggests that their receptors are linked to phospholipase A_2, but not to phospholipase C.

Until recently, far more was known about platelet responses to agonists than about the structure of the agonist receptors. Recently, however, there has been a great deal of progress in this area, highlighted by the successful cloning of the receptors for TxA_2 (Hirata *et al*, 1991), platelet activating factor (Honda *et al*, 1991), and thrombin (Vu *et al*, 1991a; Rasmussen *et al*, 1991). Each of these receptors has proven to be a member of the G protein-coupled family of receptors, a family already known to include platelet α_2-adrenergic receptors (Kobilka *et al*, 1987). Each is formed by a single polypeptide chain with multiple transmembrane domains. In several respects, the thrombin receptor has proven to be the most novel of these receptors, and it is upon the thrombin receptor that the remainder of this discussion will concentrate.

Thrombin receptor structure: Thrombin is generated *in vivo* by the sequential activation of the enzymes of the coagulation cascade. When added to platelets *in vitro*, thrombin causes phosphoinositide hydrolysis, eicosanoid formation, protein phosphorylation and an increase in cytosolic Ca^{++}. Thrombin also suppresses cAMP synthesis. These responses can be detected at thrombin concentrations as low as 0.1 nM, which is equivalent to approximately 0.01 NIH units/ml of purified thrombin. Although arachidonate metabolites contribute to platelet responses to thrombin, increasing the thrombin concentration overcomes the blockade of platelet activation caused by aspirin without restoring thromboxane formation.

Until recently, relatively little was known about the structure of the thrombin receptor. Studies by Harmon and Jamieson (Harmon & Jamieson, 1986) suggested that there are approximately 50 high affinity thrombin binding sites per platelet with a K_d of 0.3 nM and 1700 moderate affinity sites/platelet with a K_d of 11 nM. Essentially identical results were obtained by De Marco, *et al.* (De Marco *et al*, 1991) who found 100-300 sites/platelet with a mean K_d of 0.5 nM and 3400-8200 sites/platelet with a mean K_d of 50 nM. Indirect evidence suggested that the high affinity binding sites might be associated with membrane glycoprotein Ib since they were absent from Bernard-Soulier platelets (Jamieson & Okumura, 1978; De Marco *et al*, 1991), could be removed by *Serratia marcesans* protease (Harmon & Jamieson, 1988) and could be blocked with anti-glycoprotein Ib antibodies, as could some platelet responses to thrombin (Dennis *et al* 1991; De Marco *et al*, 1991). However, the large disparity between the number of high affinity thrombin binding sites and the number of copies of GP Ib (25,000/platelet) was unexplained. There was also no satisfactory explanation for the mandatory role for thrombin's proteolytic activity in causing platelet activation and no evidence that either of the binding sites defined with radioiodinated thrombin was a receptor in the functional sense.

Since data from a number of laboratories had established that the effects of thrombin on phospholipase C and adenylylcyclase were mediated by G proteins, the presumption was that the thrombin receptor would prove to be related structurally to other G protein coupled receptors. With certain exceptions, these consist of a single peptide chain with an extracellular N-terminus, an intracellular C-terminus and 7 hydrophobic domains thought to span the plasma membrane. Analysis of several of these receptors suggests that receptor:G protein interactions are specified predominantly by portions of the third cytoplasmic loop of the receptors. Receptor:agonist interactions, on the other hand, are thought to be specified by the transmembrane domains and the extracellular loops of the receptor. Adrenergic receptors have been studied in great detail. In this case, the agonist appears to enter a pocket formed by the transmembrane domains of the receptor, well beneath the surface of the plasma membrane (Salter et al, 1985; Cichowski et al, 1992). However, prior to the cloning of the thrombin receptor, there was no precedent for a G protein coupled receptor being activated by a protease or proteolysis.

Attempts to identify the thrombin receptor by methods based upon its interaction with thrombin failed to give definitive results. Usually the proteins isolated by such methods proved to be non-receptor thrombin-binding proteins, such as protease nexin. The approach which ultimately proved successful involved expression cloning, a technique that had been used to identify other G protein coupled receptors. Three laboratories reported that the injection of RNA from megakaryoblasts, fibroblasts or endothelial cells into Xenopus laevis oocytes would render the oocytes responsive to thrombin (Coughlin & Lingappa, 1989; Van Obberghen-Schilling et al, 1990; Pipili-Synetos et al, 1990). Subsequently, clones capable of encoding functional thrombin receptors were isolated from the human megakaryoblastic Dami cell line (Vu et al, 1991a) and from hamster fibroblasts (Rasmussen et al, 1991). The differences between them are probably attributable to species rather than tissue differences. To date, there is no hard evidence available for the existence of more than one class of human thrombin receptors.

The structure of the Dami cell thrombin receptor, which is presumed to be identical to the platelet thrombin receptor, is illustrated in Figure 2 (Vu et al, 1991a). As predicted, it is comprised of a single polypeptide chain with 7 hydrophobic domains. An additional hydrophobic domain near the site of translation initiation is thought to be a signal peptide which is cleaved from the mature protein. This has not, however, been proven. Other notable features include an extracellular N-terminus and several potential sites for N-linked glycosylation. A potential site for cleavage by thrombin exists between Arg^{41} and Ser^{42}. Mutations at this site prevent activation of the expressed receptor by thrombin (Vu et al, 1991a; Vu et al, 1991b). A series of negatively-charged amino acids in the N-terminus distal to the site of cleavage by thrombin is thought to interact with the anion-binding exosite on thrombin, in a manner analogous to hirudin (Vu et al, 1991a; Liu et al, 1991).

Thus, a model exists for the structure of the thrombin receptor which accounts for at least some of the known features of the interaction of thrombin with platelets. The receptor protein itself has not, however, been isolated and there is at present no published confirmation that the amino acid at the N-terminus of the receptor changes after activation by thrombin. To begin to obtain information about the receptor at the protein level, we have developed a series of peptide-directed monoclonal antibodies against epitopes within the SFLLRNPNDKYEPF domain (residues 42-55) of the receptor (Vassallo et al, 1992a).

These antibodies bind to the surface of thrombin-responsive cells, inhibit platelet activation by thrombin, and recognize a 65 kDa protein on Western blots of platelet membranes (Vassallo et al, 1992a; Brass et al, 1992). This difference in molecular weight between the apparent size of the receptor on SDS gels and the 47 kDa predicted from the predicted amino acid content of the receptor is presumably due to glycosylation, as it is with other G protein-coupled receptors. The antibodies also inhibit platelet activation by γ-thrombin and trypsin, but have no effect on platelet responses to ADP, epinephrine or U46619 (Vassallo *et al,* 1992a; Brass *et al,* 1992).

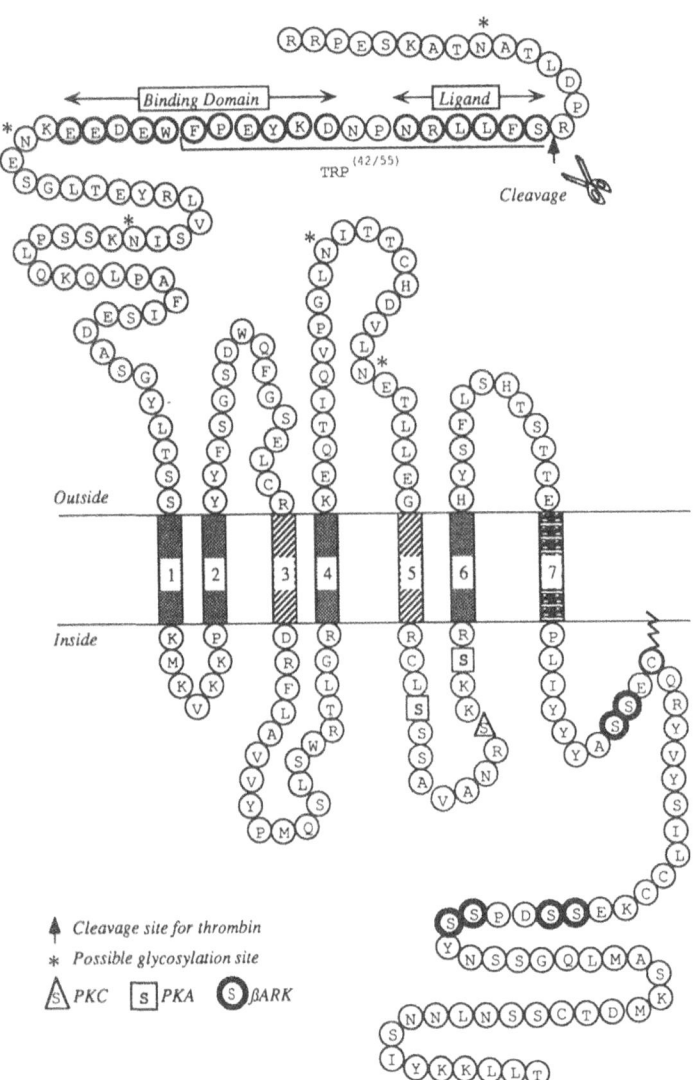

Figure 2. Thrombin receptor structure. Some features of the human platelet thrombin receptor (Vu *et al,* 1991 a) are shown, including potential sites for cleavage by thrombin, N-linked glycosylation, and phosphorylation by either protein kinase C (PKC), cAMP-dependent protein kinase (PKA) or a kinase analogous to ß-adrenergic receptor kinase (ßARK). The acidic domain in the N-terminus thought to bind to thrombin's exosite is indicated, as is the SFLLRN... sequence thought to serve as the receptor's tethered ligand. For simplification, the putative transmembrane domains have been omitted from the drawing. The peptide TRP[42/55] corresponding to residues Ser[42] through Phe[55] was originally shown by Vu, *et al.* to cause platelet aggregation. Subsequent studies have shown that this and other platelet responses to thrombin can be mimicked with peptides as short as the hexapeptide SFLLRN.

Thrombin receptor activation: Based upon the existence of a thrombin cleavage site within the N-terminus of the receptor, Vu, et al. (Vu et al, 1991a) proposed that the N-terminus distal to the point of cleavage might form a tethered ligand capable of activating the receptor. In support of this hypothesis, they were able to show that a 14 residue peptide corresponding to receptor residues Ser[42] thru Phe[55] (SFLLRNPNDKYEPF or TRP[42/55] in Figure 2) could cause platelet aggregation (Vu et al, 1991a). Based upon these observations, they proposed a model for receptor activation in which thrombin cleaves its receptor after Arg[41], creating a new N-terminus that serves as the actual ligand for the receptor (Figure 3). In the year since its publication, key features of this model have been confirmed by several laboratories. In particular, it has been shown that receptor-derived peptides can evoke many, if not all, of the thrombin responses observed in platelets, fibroblasts and endothelial cells (Vu et al, 1991a; Rasmussen et al, 1991; Huang et al, 1991; Ngaiza & Jaffe, 1991; Brass & Woolkalis, 1992; Vouret-Craviari et al, 1992; Hung et al, 1992; Brass, 1992; Vassallo et al, 1992b; Seiler et al, 1992) (Table 1).

Table 1. Platelet and HEL Cell Responses to Peptides Modeled after the Thrombin Receptor Tethered Ligand

1.	Activation of phospholipase C	5. Inhibition of adenylylcyclase
2.	Increased cytosolic Ca^{++}	6. Fibrinogen receptor expression
3.	Activation of PI-3-kinase	7. Aggregation
4.	Activation of tyrosine kinases	8. Alpha and dense granule secretion

Vu et al, 1991a; Huang et al, 1991; Brass & Woolkalis, 1992; Vouret-Craviari et al 1992; Brass, 1992; Vassallo et al, 1992b. The data on peptide-induced secretion of α- and dense granule constituents are unpublished observations by S. Shattil, M. Cunningham and L. Brass.

To further define the structure of the tethered ligand domain and the mechanism by which it activates the thrombin receptor, we prepared two additional series of peptides (Vassallo et al, 1992b). The first included fragments of SFLLRNPNDKYEPF and was designed to identify the portions of this domain that are needed for receptor activation (Fig. 4A). The second series included peptides in which substitutions were made from the sequence predicted to be present in the cloned receptor (Fig. 4B). The ability of these peptides to cause platelet aggregation was compared. The results show that the activity of the peptide resides in its first 6 residues, SFLLRN. A peptide comprised of these residues was several-fold more potent as a stimulus for platelet aggregation than was the parent peptide. Reversal of the Ser and Phe residues (as originally reported (Vu et al, 1991a)) abolished activity, as did 1) omission of Ser, 2) substitution of Ala for Phe, 3) omission of Arg and Asn or 4) substitution of Ala for Arg. On the other hand, substitution of Thr or Ala for Ser, or Ala for the initial Leu residue had a comparatively modest effect (Vassallo et al, 1992b).

Figure 5 summarizes these observations and compares the human thrombin receptor sequence with the corresponding domain of the hamster and murine sequences. Within the critical first 6 or 7 residues distal to the site of cleavage by thrombin, the three sequences are identical except for the substitution of Phe for Leu, a residue which the peptide studies suggest is non-critical. Outside this region, the sequence show greater divergence. In an attempt to predict the structure of the receptor N-terminus, a search was made in the Brookhaven crystallographic database for proteins containing SFLLRNPN (Vassallo et al, 1992b). The results of this analysis suggested that when part of a larger protein, the

SFLLRNPN sequence might form a β-strand followed by a turn. On the other hand, the free hexapeptide SFLLRN or the thrombin-cleaved N-terminus might form either a β structure or a helix. Both of these possibilities were tested by computer-based modeling. This approach led to an extended helical structure for the peptide in which the side chains of the Arg and Phe residues are on the same side looking down the axis of a helix. This configuration was preferred over the lowest energy structure derived from the β-strand

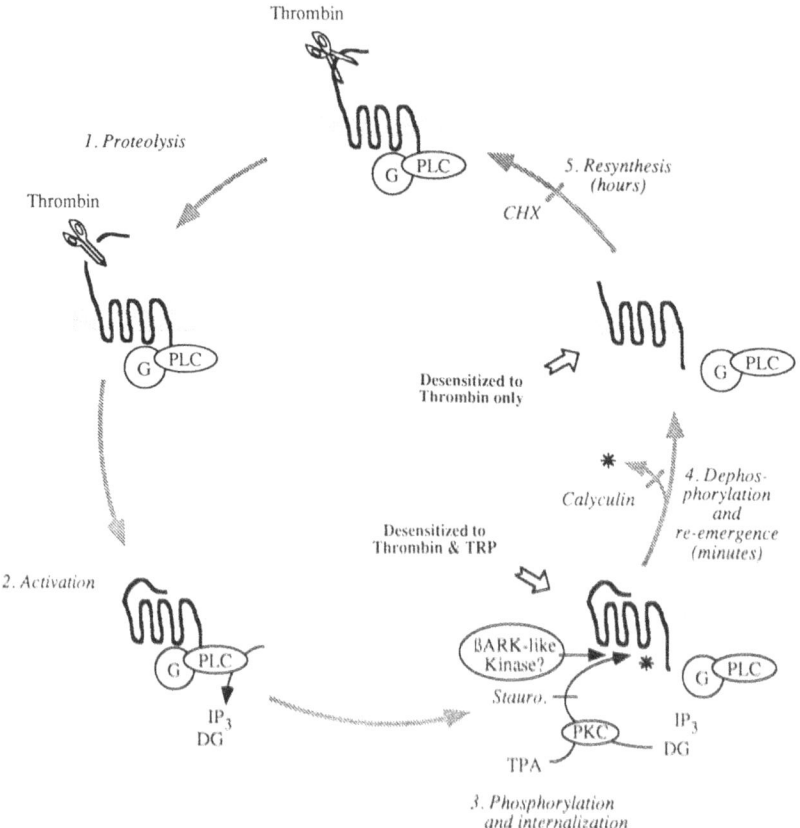

Figure 3   A model for thrombin receptor activation, desensitization and recovery. Interaction of the thrombin receptor with either thrombin or the receptor-derived $TRP^{42/55}$ peptide leads to the G-protein-mediated, pertussis toxin-sensitive activation of phospholipase C. With thrombin, desensitization of the receptor is due to a combination of proteolysis, internalization and phosphorylation, the latter potentially involving a receptor-specific kinase analogous to bARK. Protein kinase C may also play a role in receptor phosphorylation, but the data suggest that this is more likely to be the case with a direct activator of protein kinase C such as TPA. Recovery of the thrombin response in thrombin-desensitized cells requires protein synthesis and dephosphorylation, and occurs over many hours. Recovery of the $TRP^{42/55}$ response in thrombin-desensitized cells only requires dephosphorylation and can occur far more rapidly (Brass, 1992).

starting geometry. In other words, these results suggest that in the intact receptor SFLLRN may be part of an extended β-structure, but that proteolysis by thrombin causes it to assume a modified a-helical configuration in which the sidechains of the critical Phe and Arg residues point in the same direction - potentially facing into a pocket formed by the remainder of the receptor. Although this model can only be thought of as a working hypothesis, it does fit with the observed consequences of amino acid substitutions within the SFLLRN sequence and provides a basis for future experimental design (Vassallo *et al*, 1992b).

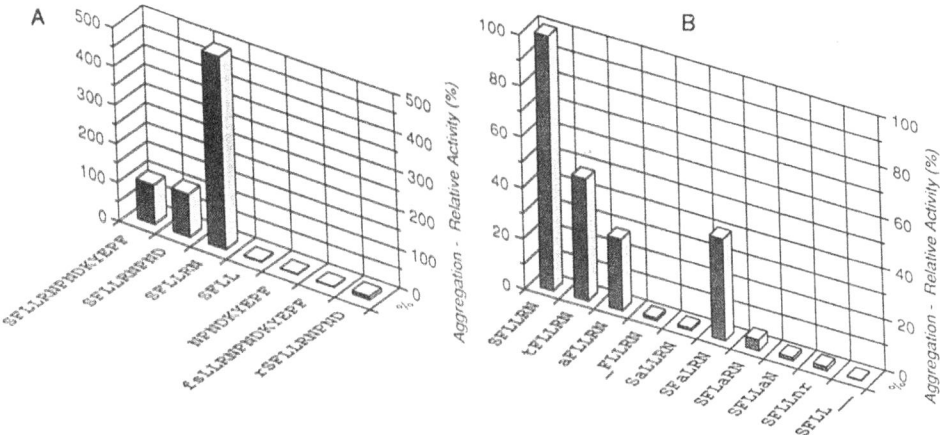

Figure 4   Platelet activation by thrombin receptor "tethered ligand" domain peptides. The peptides shown in the figure are fragments or variants of the 14 residue tethered ligand domain (TRP[42/55] or SFLLRNPNDKYEPF) shown (Vu et al, 1991a) to activate platelets. For each peptide the concentration needed to cause half-maximal platelet aggregation was determined. Part (A) of the figure compares fragments of TRP[42/55] and demonstrates that peptides comprised of the first 6 amino acid residues (SFLLRN) are several-fold more potent than TRP[42/55]. The results are expressed as % of the activity of TRP[42/55]. Part (B) of the figure compares peptides in which amino acid substitutions have been made in the SFLLRN sequence. The figure is derived from data in reference (Vassallo et al, 1992b).

<u>Thrombin receptor desensitization:</u>   It has been shown in a variety of cells that an initial encounter with thrombin produces a state of homologous desensitization in which a subsequent encounter evokes little or no response (Huang & Detwiler, 1987; Paris et al, 1988; Jaffe et al, 1987; Jones et al, 1989; Brass et al, 1991). In theory thrombin desensitization could involve any of the mechanisms known to affect other G protein-coupled receptors and, in addition, might arise from proteolysis of the receptor itself. Use of the tethered ligand peptides made it possible to distinguish proteolytic from non-proteolytic mechanisms for desensitization. The studies used to define the mechanisms were performed with the megakaryoblastic HEL and CHRF-288 cell lines. Although thrombin and the receptor-derived peptide TRP[42/55] desensitized HEL cell thrombin receptors against the effects of both, marked differences were found in the rate at which the cells recovered thrombin receptor function (Brass et al, 1991; Brass, 1992). When the cells were desensitized by exposure to thrombin, recovery of thrombin responsiveness took 20 hrs and could be inhibited with the protein synthesis inhibitor, cycloheximide. When the cells were desensitized with TRP[42/55], on the other hand, recovery of the thrombin response was detectable within 15 min and was complete within approximately 3 hrs. A rapid rate of recovery was also observed when the cells were desensitized with thrombin or TRP[42/55] and then re-challenged with the peptide. In each of these cases recovery was unaffected by cycloheximide, but could be delayed by incubating the cells with inhibitors of serine/threonine phosphatases such as calyculin or okadaic acid (Brass, 1992).

These results suggest that at least two mechanisms are responsible for thrombin receptor desensitization (Fig. 3). The first involves proteolysis and presumably includes, but is not necessarily limited to, cleavage of the receptor by thrombin between Arg[41] and Ser[42]. This activates the receptor, but also may leave it in a state in which it is unable to

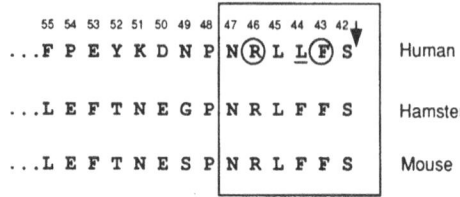

	55	54	53	52	51	50	49	48	47	46	45	44	43	42	
...F	P	E	Y	K	D	N	P	N	Ⓡ	L	L	Ⓕ	S	Human
...L	E	F	T	N	E	G	P	N	R	L	F	F	S	Hamster
...L	E	F	T	N	E	S	P	N	R	L	F	F	S	Mouse

Figure 5   N-terminal sequence comparison. The figure shows the predicted N-terminal amino acid sequence of the thrombin receptor following cleavage by thrombin (*arrow*). This domain of the human sequence is identical for at least two megakaryoblastic cell lines (Dami cells (Vu *et al*, 1991a) and HEL cells) and umbilical vein endothelial cells. The numbering refers to the published Dami cell sequence. Studies with peptides derived from the human sequence (Fig. 4) suggest that the SFLLRN is sufficient for receptor activation. Within this domain, the Phe and Arg residues appear to be relatively more important for receptor activation, while the initial Leu residue and the serine residue are relatively less important. A comparison of the human, hamster (Rasmussen *et al*, 1991) and murine (Vu *et al*, 1991b) sequences show that the SFLLRN is highly conserved, except for the apparently less critical initial Leu residue. Peptides containing SFFLR... have been shown to activate human platelet thrombin receptors (Vouret-Craviari *et al*, 1992).

interact with thrombin a second time. The second mechanism appears to be phosphorylation, possibly of the receptor itself. Since the thrombin receptor has been cloned, but not yet isolated, the evidence that it can be phosphorylated is necessarily indirect. Nevertheless, the structure of the thrombin receptor is similar to other G protein-coupled receptors, including several which are known to be substrates for phosphorylation (Fig. 2).

If phosphorylation plays a role in thrombin receptor desensitization, what can be said about the protein kinase(s) that are involved and about potential sites for phosphorylation? Present evidence suggests that at least two different kinases may play a role. The first of these is protein kinase C, which has been shown to phosphorylate and desensitize several G protein-coupled receptors, including those for serotonin and adrenergic agonists (Raymond, 1991; Kelleher *et al*, 1984; Leeb-Lundberg *et al*, 1985). Protein kinase C has also been shown to phosphorylate adenylylcyclase in platelets (Simmoteit *et al*, 1991), at least one form of phospholipase C (Ryu *et al*, 1990) and Gzα (Carlson *et al*, 1989; Lounsbury *et al*, 1991). Thrombin receptors contain at least one consensus sequence for phosphorylation by protein kinase C, a serine residue within the third cytoplasmic loop (Fig. 2) (Kemp & Pearson, 1990; Vu *et al*, 1991a). Could activation of protein kinase C account for thrombin receptor desensitization? Both thrombin and TRP[42/55] stimulate phospholipase C and raise the cytosolic free Ca^{++} concentration in HEL cells, conditions that promote the activity of protein kinase C. Direct activation of protein kinase C with TPA also desensitizes HEL cells. However, while the desensitization caused by thrombin is homologous, that caused by TPA is heterologous, affecting HEL cell responses to agonists other than those that work *via* the thrombin receptor (Brass, 1992). Furthermore, staurosporine, while completely preventing the desensitization caused by TPA, had only a minimal effect on the desensitization caused by thrombin and TRP[42/55]. Therefore, it is unlikely that protein kinase C plays a major role in the desensitization caused by thrombin and TRP[42/55].

One alternative to protein kinase C is a receptor-specific protein kinase such as the ß-adrenergic receptor kinase (ßARK) or rhodopsin kinase.

At present, there is no direct evidence that a kinase with specificity for the thrombin receptor exists. However, ßARK has been shown to phosphorylate α_2-adrenergic receptors, as well as ß-adrenergic receptors and both HEL cells and platelets contain an ≈ 80 kDa protein recognized by rabbit polyclonal anti-βARK antiserum (unpublished observations). It is conceivable that it or a related kinase could phosphorylate thrombin receptors (Benovic *et al*, 1987). In this case, the homologous nature of the desensitization caused by thrombin would be attributable to the preference of a ßARK-like kinase for the occupied form of the receptor. The sites for phosphorylation by ßARK in ß- and α_2-adrenergic receptors are located in the third cytoplasmic loop and in the C-terminal tail. Studies with synthetic substrates suggest that ßARK prefers serine residues to threonine residues and that an acidic residue on the amino side of the serine residue enhances phosphorylation (Onorato *et al*, 1991). There are at least 4 such serine residues in the C-terminal tail of the thrombin receptor (Fig. 2).

If desensitization of the thrombin receptor is due to a combination of proteolysis and phosphorylation, then resensitization appears to be due to protein synthesis and dephosphorylation (Fig. 3). In platelets, where little protein synthesis occurs, dephosphorylation would be the sole available means for resensitization. Might other mechanisms also be involved, particularly in a nucleated cell? Several others have been described, including receptor internalization and alterations in the rate of receptor synthesis (Sibley *et al*, 1987; Hausdorff *et al*, 1990; Bouvier *et al*, 1989). Using immuno-fluorescence microscopy and the monoclonal antibodies directed against the TRP[42/55] sequence, we have examined the fate of the thrombin receptor on HEL cells and CHRF-288 cells. On resting cells, antibody binding is easily detectable and occurs predominantly on the cell surface. Following the addition of thrombin, the antibody binding sites rapidly disappear from the cell surface and re-appear within the cell. By 10 minutes, this process is complete. Identical results are obtained when the cells are activated with SFLLRN rather than thrombin (unpublished observations). These results suggest, therefore, that internalization is part of the process of thrombin receptor down-regulation and that proteolysis is not required for this to occur.

Proteolysis of the receptor may not be limited to the bond between Arg[41] and Ser[42]. Coller, *et al*. (Coller *et al*, 1991) have shown that plasma peptidases can cleave receptor-derived tethered ligand peptides such as TRP[42/55] into inactive fragments. The process was relatively slow compared to the time frame of receptor activation and desensitization, but was comparable to the rate of resensitization. "Secondary" proteolysis of the receptor could explain the apparent failure of the thrombin-cleaved receptor to reactivate itself once dephosphorylation has occurred - as would be expected to occur if the N-terminus were still able to interact with the remainder of the receptor. Its failure to occur suggests that either critical portions of the N-terminus are no longer available or that additional changes have occurred during the period of phosphorylation and internalization. The recovery of TRP[42/55] responsiveness argues against any sort of global defect in receptor function, but based upon studies with variants of TRP[42/55], removal of Ser[42] would be expected to abolish receptor activity. Whether or not this actually occurs will be established when the receptor can be physically isolated in sufficient quantity for N-terminal sequencing to be performed.

Unresolved issues: Far from answering all of the questions about platelet activation *via* thrombin receptors, the observations that have been reported to date leave a number of issues unresolved. The first is whether there are other types of thrombin receptors on platelets in addition to the one that has been cloned. Although such a possibility has not yet been completely ruled out, the ability of receptor-derived peptides such as SFLLRNPNDKYEPF to mimic many, if not all, of the effects of thrombin on platelets suggests that only a single class of receptors may be present. Efforts to identify a second thrombin receptor are now underway in a number of laboratories and will hope fully settle this issue in the near future.

A second question is the relationship between the cloned thrombin receptor and platelet glycoprotein Ib. Attempts have been made to combine observations about the cloned receptor with the results of earlier studies on the binding of radioiodinated thrombin to the platelet surface. This has produced a proposal that the high affinity thrombin binding site on platelets is located on GP Ib, while the moderate affinity binding site is the cloned receptor (Greco & Jamieson, 1991). This also remains to be established.

A third issue is whether thrombin normally remains bound to its receptor longer than the time required to cleave it and, if so, whether this has functional consequences. Might prolonged binding extend the duration of the thrombin-initiated signal? We found that the addition of hirudin after thrombin shortened the duration of thrombin-induced increases in cytosolic Ca^{++} in HEL cells, even when the hirudin was added at the peak of the Ca^{++} response (Brass et al, 1991). Mitogenesis, which admittedly is not an issue for platelets, seems to require the thrombin to be present for at least several hours. These observations suggest a number of possibilities. One is that the continued presence of thrombin may help to maintain receptors in their active state. Alternatively, some responses may require activation of a greater percentage of thrombin receptors than others and may, therefore, require more time to become self-sustaining. Neither of these possibilities would seem to adequately account for the observation that thrombin receptors become internalized within a few minutes of thrombin addition.

Another issue is whether the "thrombin" receptor is really a more generalized protease receptor. Platelets can be activated by trypsin, cathepsin G and plasmin, as well as thrombin. In theory, any protease capable of cleaving the receptor N-terminus would activate the receptor - although the site of cleavage would have to be between Arg^{41} and Ser^{42} (Vassallo et al, 1992b). Trypsin is known to mimic many of the effects of thrombin on platelets and HEL cells, including causing desensitization of the HEL cell thrombin receptor (Brass, 1992). The anti-receptor antibodies that inhibit platelet activation by thrombin also inhibit platelet activation by trypsin (Vassallo et al, 1992a; Brass et al, 1992). Further work using these approaches should help to determine whether the thrombin receptor is responsible for platelet responses to other proteases as well.

Finally, none of the data discussed thus far have addressed the issue of protein tyrosine phosphorylation in platelets. Several studies have shown that platelets contain $pp60^{src}$ and src-related kinases and that aggregation in response to thrombin is associated with the phosphorylation on tyrosine of multiple platelet proteins (Golden & Brugge, 1989; Iiri et al, 1989). Most of these have not been identified, nor has it been determined whether tyrosine kinase activation in platelets is a cause or a consequence of aggregation. It remains to be determined, therefore, whether there is a direct link between platelet thrombin receptors and tyrosine kinases or whether the association is indirect. Of note, however, it has been shown that receptor-derived peptides such as $TRP^{42/55}$ stimulate tyrosine phosphorylation in platelets (Vassallo et al, 1992b). It seems unlikely, therefore, that kinase activation by thrombin involves an entirely separate mechanism from that involving its cloned receptor.

G PROTEINS

Guanine nucleotide-binding proteins or G proteins are regulatory proteins which mediate the interaction between cell surface receptors, such as those for thrombin, and intracellular effectors such as phospholipases and ion channels. G proteins have a characteristic $\alpha\beta\gamma$ heterotrimeric structure in which the α subunit contains the site for guanine nucleotide binding and, in most cases, is responsible for the interaction with receptors and effectors. ADP-ribosylation of the a subunit by cholera toxin or pertussis toxin alters G protein function - either by mimicking the effects of an agonist-occupied receptor or by inhibiting agonist-evoked responses. An example of the former is the effect of cholera toxin on Gs, the G protein which stimulates cAMP formation. ADP-ribosylation of Gsα causes continuous activation of adenylylcyclase. In contrast, pertussis toxin typically inhibits responses to agonists. An example of this effect is the ability of pertussis toxin to block the Gi-mediated suppression of cAMP formation caused by agonists such as thrombin and epinephrine. G protein β and γ subunits perform at least two roles. First, they form an isoprenylated heterodimer (Gßγ) which may help to anchor the G protein to the plasma membrane. In addition, they have regulatory effects on adenylylcyclase and phospholipase A_2 that are independent of the a subunit (Gilman, 1987; Jelsema & Axelrod, 1987; Burch et al1986; Tang & Gilman, 1991).

Historically, transducin, Gs and Gi were the first G proteins described in both structural and functional terms. Transducin regulates cGMP phosphodiesterase in the retina and mediates the conversion of light to neural impulses. Gs and Gi regulate cellular cAMP levels by, respectively, stimulating and inhibiting adenylylcyclase. Recent cloning studies have shown that most forms of Gα exist as families of several closely-related proteins (Strathmann & Simon, 1991). There are, for example, at least three different forms of Giα, $Giα_1$, $Giα_2$ and $Giα_3$. These three proteins are 85-95% homologous with each other at the amino acid level. All three are predicted to be substrates for pertussis toxin. The significance of the differences between the members of the Giα family is unknown. However, the preservation of the differences in their primary sequence across tissue and species lines suggests that they may play distinct roles. In addition to adenylylcyclase and cGMP phosphodiesterase, G proteins have been implicated in the regulation of K^+ and Ca^{++} channels, the hydrolysis of phosphatidylcholine and the activation of phospholipases A_2 and C. Reviews of G protein structure and function will be found in references (Gilman, 1987; Weiss et al, 1988).

G proteins in platelets: In platelets, as in other cells, G proteins play a major role in signal transduction by mediating the interaction between agonist receptors and the enzymes which produce second messengers (Fig. 1). The G proteins that have been identified in platelets are listed in Table 2. Based upon their functional characteristics, Western blotting with peptide-directed antibodies, ADP-ribosylation by bacterial toxins, Northern blots and nucleotide sequencing, platelets contain both Gs and Gi. Gsα appears to be present in at least its 45 kDa form. Giα is present in all three of its currently known forms, $Giα_1$, $Giα_2$ and $Giα_3$ (Williams et al, 1990). Gs and Gi are thought to play the same role in the regulation of cAMP formation in platelets that they do in other cells. Agents which increase cAMP levels in platelets, such as PGI$_2$, do so via Gs. Agonists such as thrombin and epinephrine, which suppress cAMP formation are thought to do so through Gi.

In addition to adenylylcyclase, at least two other platelet enzymes are thought to be regulated by G proteins: phospholipase C and phospholipase A_2. Regulation of phospholipase C by a G protein is a general phenomenon that has been demonstrated in a wide variety of cells and presumably involves the b form of phospholipase C, although this has not been established. The G protein that stimulates phospholipase C is sometimes referred to as "Gp". However, recent studies suggest that Gp may not be the same in all cells, or even a single G protein. In some cells, particularly those of hematopoietic origin, agonist or hormone-induced phosphoinositide hydrolysis can be blocked by preincubating the cells with pertussis toxin. In other cells, pertussis toxin has no effect. Studies in platelets have been complicated by the inability of pertussis toxin to cross the platelet plasma membrane. However, on balance, it appears that pertussis toxin can inhibit thrombin-induced phosphoinositide hydrolysis in platelets. These observations suggest that platelets contain at least one form of Gp that is a substrate for pertussis toxin. Since all of the pertussis toxin substrates in platelets that have been identified to date are members of Giα family, this raises the possibility that Gi_1, Gi_2 or Gi_3 regulate phospholipase C as well as adenylylcyclase. This has not, however, been proven.

Platelets are also thought to contain a second G protein capable of stimulating

Table 2. G protein α subunits in platelets

	kDa	Toxin	Phosphorylated	Enzyme	Function	Ref.
$G_{s\alpha}$	45	cholera	-	adenylylcyclase	↑ cAMP	90
$G_{i\alpha2} \gg G_{i\alpha3} > G_{i\alpha1}$	40-41	pertussis	no	adenylylcyclase, phospholipase C ?	↓ cAMP, ↑ IP$_3$/DAG ?	66
$G_{z\alpha}$	41	neither	yes	?	?	48,87, 91
$G_{q\alpha}$	42	neither	?	phospholipase C	↑ IP$_3$/DAG	70

phospholipase C, one that is not a pertussis toxin substrate. The initial evidence for this conclusion was indirect: under conditions in which thrombin-induced phosphoinositide hydrolysis was inhibited by pertussis toxin, the ability of the TxA$_2$ analog, U46619, to activate phospholipase C was unimpaired (Brass *et al*, 1987; Brass *et al*, 1988). Subsequently, two new forms of Gα were described (Strathmann & Simon, 1990). These proteins, Gqα and G$_{11}\alpha$, are more closely related to each other than to other forms of Gα. Neither possesses the site at which ADP-ribosylation by pertussis toxin normally occurs and neither is thought to be a substrate for the toxin. Both proteins are widely distributed and antibodies directed against peptides present in each suggest that at least Gqα is present in platelets (Shenker *et al*, 1991). Reconstitution studies with phospholipase Cß and purified or recombinant Gqα/G$_{11}\alpha$ show that these proteins can cause a 5-fold or greater increase in phosphoinositide hydrolysis by the enzyme (Smrcka *et al*, 1991; Taylor & Exton, 1991; Taylor *et al*, 1991; Blank *et al*, 1991). It is assumed, therefore, that one or both of these proteins mediates pertussis toxin-resistant activation of phospholipase C in most cells. Although this has not been specifically established in platelets, Shenker, *et al.* (Shenker *et al*, 1991) have shown that an antibody directed against Gqα/G$_{11}\alpha$ can inhibit thromboxane receptor-stimulated GTPase activity.

Phospholipase A$_2$ is the second phospholipid-hydrolyzing enzyme in platelets whose activity may be regulated by G protein(s). There are two potential mechanisms by which this could be accomplished. One is the indirect activation of phospholipase A$_2$ *via* the increase in cytosolic Ca^{++} caused by Gp- or Gq-dependent phosphoinositide hydrolysis (Fig. 1). A second mechanism is the direct activation of phospholipase A$_2$ by a G protein. Current evidence suggests that phospholipase A$_2$ is primarily located in the platelet cytosol and that arachidonate release and metabolism occur in the dense tubular system (Kramer *et al*, 1986; Yoshimoto *et al*, 1977; Carey *et al*, 1982; Laposata *et al*, 1987). However, this need not preclude a direct interaction between phospholipase A$_2$ and a G protein since evidence in cells other than platelets suggests that Giα at least is widely distributed within the cell and not confined to the plasma membrane (Ercolani *et al*, 1990; Lewis *et al*, 1991).

In cells other than platelets phospholipase A$_2$ appears to be activated by Gßγ, rather than Gα (Kim *et al*, 1989). In platelets GTP and GTPγS have been shown to cause ^{3}H-arachidonate release. The extent of release was unaffected by an inhibitor of diacylglycerol lipase, suggesting that the arachidonate was derived through the action of phospholipase A$_2$ and not from the diacylglycerol produced by phosphoinositide hydrolysis. Thrombin-induced release of ^{3}H-arachidonate from permeabilized platelets appeared to be GTP-dependent and was inhibited by pertussis toxin, suggesting that the G protein involved is a substrate for pertussis toxin (Nakashima *et al*, 1987; Kajiyama *et al*, 1989; Silk *et al*, 1989; Murayama *et al*, 1990). A broadly-reactive anti-Gα antiserum also inhibited arachidonate release. Antisera against G$\beta\gamma$ had no effect on the response to thrombin, but inhibited arachidonate release by histamine (Murayama *et al*, 1990). A study in which chimeric forms of Gα were expressed in Chinese hamster ovary cells implicated Giα_2 in phospholipase A$_2$ regulation (Gupta *et al*, 1990). Therefore, although it appears likely that G proteins are involved in the regulation of phospholipase A$_2$, there remains a great deal of uncertainty about the mechanism involved.

In addition to the G proteins discussed thus far, platelets also contain several newly-described G proteins. The first is these is Gzα, which has proven to be identical to a protein whose DNA has been cloned from brain and retinal cDNA libraries (Carlson *et al*, 1989; Gagnon *et al*, 1991). Like Gqα and G$_{11}\alpha$, Gzα lacks the cysteine residue normally required for ADP-ribosylation and has been shown to not be a substrate for pertussis toxin. Like other forms of Gα, Gzα is able to interact with G$\beta\gamma$ and hydrolyze GTP to GDP, although at a rate slower than other G proteins. Gz has one property, however, that is so far unique among platelet G proteins and is of considerable interest. When platelets are activated by thrombin or TxA$_2$ analogs, both of which activate protein kinase C *via* phosphoinositide hydrolysis, or by phorbol esters, which directly activate protein kinase C, Gzα is phosphorylated (Carlson *et al*, 1989). This phosphorylation occurs with a stoichiometry of 1 mole per mole, can be reproduced with recombinant Gzα and purified protein kinase C, and takes place at a serine residue near the N-terminus of the protein (Lounsbury *et al*, 1991). Under the same conditions, the forms of Giα present in platelets are not phosphorylated (Carlson *et al*, 1989).

In addition to Gzα, transcripts encoding 3 additional forms of Gα have recently been described, G$_{12}\alpha$, G$_{13}\alpha$ and G$_{16}\alpha$ (Strathmann & Simon, 1991; Amatruda *et al*, 1991). Based upon their amino acid sequences, none of these are predicted to be pertussis toxin substrates. G$_{12}\alpha$ and G$_{13}\alpha$ are <45% homologous to other forms of Gα and form a separate family of subunits (Strathmann & Simon, 1991). G$_{16}\alpha$ is thought to be related to Gqα and G$_{11}\alpha$ (Amatruda *et al*, 1991). By RNA analysis after PCR amplification, G$_{12}\alpha$ and G$_{13}\alpha$ are widely, but not universally, distributed (Strathmann & Simon, 1991), while G$_{16}\alpha$ is found predominantly in hematopoietic cells (Amatruda *et al*, 1991). Preliminary studies with peptide-directed antisera suggest that G$_{12}\alpha$ and G$_{13}\alpha$ may be present in platelets. G$_{16}\alpha$, however, was not detected.

Finally, platelets contain at least one additional class of GTP-binding proteins. At 20-30 kDa, these proteins are smaller than the α subunits of the "classical" G proteins. As yet little is known about their function, but several have been shown to be homologous to oncogene products such as *ras* and at least one appears to be phosphorylated by cAMP-dependent protein kinase. It remains to be determined whether these proteins participate in signal transduction or play a totally different role such as facilitating membrane fusion and exocytosis (reviewed in reference (Manning & Brass, 1991)).

Unresolved issues: Although the list of the G proteins present in platelets has grown steadily over the past several years, a number of questions remain unanswered, as is indicated by the number of question marks in Table 2. One is the still-unresolved issue of the identity of Gp, the pertussis toxin-sensitive G protein that activates phospholipase C in response to thrombin. The only forms of pertussis toxin-sensitive forms of Gα that have been identified in platelets to date are members of the Giα family. However, in contrast to the results obtained with Gq, efforts to stimulate phospholipase C with Gi in reconstituted systems have not been dramatically successful. It is possible that platelets contain another G protein that can activate phospholipase C and is a substrate for pertussis toxin. One approach to answering this question would be to isolate platelet thrombin receptors along with the G proteins with which it is normally associated. Efforts to do so are currently underway in our laboratories.

A second issue is the role of Gz in platelets and the effects of phosphorylation on that role. As has already been discussed, the tissue distribution of Gzα is limited. Gzα is present in megakaryocytes, but absent from non-megakaryocytic hematopoietic cells that we have examined and cannot be detected in the megakaryoblastic HEL cell line (Gagnon *et al*, 1991). It is possible that Gz plays a role during megakaryocyte development, but this remains to be established.

A final question is the role of a G protein in the activation of phospholipase A_2. Even setting aside the issue of the identity of such a G protein, the evidence that phospholipase A_2 activation in platelets involves a G protein has never been as clear as the evidence for phospholipase C. It does, however, appear that guanine nucleotides can cause arachidonate mobilization in saponin-permeabilized platelets. Which form of phospholipase A_2 is involved, where the activation occurs and what G protein mediates the activation are all important questions that remain unanswered.

ACKNOWLEDGMENTS

These studies were supported in part by funds from the National Institutes of Health (HL40387, HL45181, CA39712 and GM34781), the American Heart Association and the W.W. Smith Charitable Trust.

REFERENCES

Amatruda, T.T. III., Steele, D.A., Slepak, V.Z., and Simon, M.I., 1991, Gα16, a G protein α subunit specifically expressed in hematopoietic cells, *Proc Natl Acad Sci USA*, 88:5587.

Benovic, J.L., Regan, J.W., Matsui, H., Mayor, F., Cotecchia, S., Leeb-Lundberg, L.M.F, Caron, M.G., and Lefkowitz, R.J., 1987, Agonist-dependent phosphorylation of the α₂-adrenergic receptor by the ß-adrenergic receptor kinase, *J Biol Chem*, 262:17251.

Blank, J.L.,Ross, A.H., and Exton, J.H., 1991, Purification and characterization of two G-proteins that activate the ß1 isozyme of phosphoinositide-specific phospholipase C. Identification as members of the Gq class, *J Biol Chem*, 266:18206.

Bouvier, M., Collins, S., O'Dowd, B.F., Campbell, P.T., Kobilka, B.K., MacGregor, C., Irons, G.P., Caron, M.G., and Lefkowitz, R.J., 1989, Two distinct pathways for cAMP-mediated down-regulation of the ß₂-adrenergic receptor:phosphorylation of the receptor and regulation of its mRNA level, *J Biol Chem*, 264:16786.

Brass, L.F., Shaller, C.C., and Belmonte, E.J., 1987, Inositol 1,4,5-triphosphate-induced granule secretion in platelets. Evidence that the activation of phospholipase C mediated by platelet thromboxane receptors involves a guanine nucleotide binding protein-dependent mechanism distinct from that of thrombin, *J Clin Invest*, 79:1269.

Brass, L.F.,Woolkalis, M.J., and Manning, D.R., 1988, Interactions in platelets between G proteins and the agonists that stimulate phospholipase C and inhibit adenylyl cyclase, *J Biol Chem*, 263:5348.

Brass, L.F., Manning, D.R., Williams, A., Woolkalis, M.J., and Poncz, M., 1991, Receptor and G protein-mediated responses to thrombin in HEL cells, *J Biol Chem*, 266:958.

Brass, L.F., 1992, Homologous desensitization of HEL cell thrombin receptors:distinguishable roles for proteolysis and phosphorylation, *J Biol Chem*, 267:6044.

Brass, L.F., Vassallo, R.R.,Jr., Belmonte, E., Ahuja, M., Cichowski, K., and Hoxie, J.A., 1992, Structure and function of the human platelet thrombin receptor: studies using monoclonal antibodies against a defined epitope within the receptor N-terminus, *J Biol Chem*, 267: 13795.

Brass, L.F., and Woolkalis, M.J., 1992, Dual regulation of cAMP formation by thrombin in HEL cell, a leukemic cell line with megakaryocytic properties, *Biochem J*, 281:73.

Burch, R.M., Luini, A., and Axelrod, J., 1986, Phospholipase A2 and phospholipase C are activated by distinct GTP-binding proteins in response to alpha 1-adrenergic stimulation in FRTL5 thyroid cells, *Proc Natl Acad Sci USA*, 83:7201.

Carey, F., Menashi, S., and Crawford, N., 1982, Localization of cyclo-oxygenase and thromboxane synthetase in human platelet intracellular membranes, *Biochem J*, 264:847.

Carlson, K., Brass, L.F., and Manning, D.R., 1989, Thrombin and phorbol esters cause the selective phosphorylation of a G protein other than Gi in human platelets, *J Biol Chem*, 264:13298.

Cichowski, K., McCormick, F., and Brugge, J.S., 1992, p21^{ras}GAP association with fyn, lyn and yes in thrombin-activated platelets, *J Biol Chem*, 267:5025.

Coller, B.S., Springer, K.T., Scudder, L.E., and Norton, K.J., 1991, Studies of peptides derived from a platelet receptor, *Blood*, 78(Suppl.1):394a.

Coughlin, S.R., and Lingappa, V.R., 1989 Expression of a thrombin receptor in Xenopus oocytes, *Clin Res* 37:379A.

De Marco, L., Mazzucato, M., Masotti, A., Fenton, J.W. II., and Ruggeri, Z.M., 1991), Function of glycoprotein Ibα in platelet activation induced by α-thrombin. *J Biol Chem*, 266:23776.

Dennis, E.A., Rhee, S.G., Billah, M.M., and Hannun, Y.A., 1991, Role of phospholipases in generating lipid second messengers in signal transduction, *FASEB J*, 5:2068.

Ercolani, L., Stow, J.L., Boyle, J.F., Holtman, E.J., Lin, H., Grove, J.R., and Ausiello, D.A., 1990, Membrane localization of the pertussis toxin-sensitive G protein subunits αi₂ and αi₃ and expression of a metallothionein-αi₂ fusion gene in LLC-PK1 cells, *Proc Natl Acad Sci USA*, 87:4635.

Fischer, T.H., Gatling, M.N., Lacal, J-C., and White, G.C.II., 1990, Rap1B, a cAMP-dependent protein kinase substrate, associates with the platelet cytoskeleton, *J Biol Chem*, 265:19405.

Fox, J.E.B., Reynolds, C.C., and Johnson, M.M., 1987, Identification of glycoprotein Ibb as one of the major proteins phosphorylated during expsoure of intact platelets to agents that activate cAMP-dependent protein kinase, *J Biol Chem*, 262:12627.

Gagnon, A.W., Manning, D.R., Catani, L., Gewirtz, A., Poncz, M., and Brass, L.F., 1991, Identification of Gzα as a pertussis toxin-insensitive G protein in human platelets and megakaryocytes, *Blood*, 78:1247.

Gilman, A.G., 1987, G proteins: transducers of receptor-generated signals, *Ann Rev Biochem*, 56:615.

Golden, A., and Brugge, J.S., 1989, Thrombin treatment induces rapid changes in tyrosine phosphorylation in platelets, *Proc Natl Acad Sci USA*, 86:901.

Greco, N.J., and Jamieson, G.A., 1991, High and moderate affinity pathways for α-thrombin-induced platelet activation, *Proc Soc Exp Biol Med*, 198:792.

Gupta, S.K., Diez, E., Heasley, L.E., Osawa, S., and Johnson, G.L., 1990, A G protein mutant that inhibits thrombin and purinergic receptor activation of phospholipase A₂, *Science*, 249:662.

Hallam, T.J., Daniel, J.L., Kendrick Jones, J., and Rink, T.J., 1985, Relationship between cytoplasmic free calcium and myosin light chain phosphorylation in intact platelets, *Biochem J*, 232:373.

Harmon, J.T., and Jamieson, G.A., 1986, Activation of platelets by alpha-thrombin is a receptor-mediated event, *J Biol Chem*, 261:15928.

Harmon, J.T., and Jamieson, G.A., 1988, Platelet activation by thrombin in the absence of the high affinity thrombin receptor, *Biochemistry*, 27:2151.

Haslam, R.J., Lynham, J.A., and Fox, J.E.B., 1979, Effects of collagen, ionophore A23187 and prostaglandin E1 on the phosphorylation of specific proteins in blood platelets, *Biochem J*, 178:397.

Hathway, D.R., and Adelstein, R.S., 1979, Human platelet myosin light chain kinase requires the calcium-binding protein calmodulin for activity, *Proc Natl Acad Sci USA*, 76:1653.

Hausdorff, W.P., Caron, M.G., and Lefkowitz, R.J., 1990, Turning off the signal: desensitization of beta-adrenergic receptor function, *FASEB J*, 4:2881.

Hirata, M., Hayashi, Y., Ushikubi, F., Nakanishi, S., and Narumiya, S., 1991, Cloning and expression of cDNA for a human thromboxane A2 receptor, *Nature*, 349:617.

Honda, Z., Nakamura, M., Miki, I., Minami, M., Watanabe, T., Seyama, Y., Okado, H., Toh, H., Ito, K., Miyamoto, T., and Shimizu, T., 1991, Cloning by functional expression of platelet-activating factor receptor from guinea-pig lung, *Nature*, 349:342.

Huang, E.M., and Detwiler, T.C., 1987, Thrombin-induced phosphoinositide hydrolysis in platelets. Receptor occupancy and desensitization, *Biochem J*, 242:11.

Huang, R., Sorisky, A., Church, W.R., Simons, E.R., and Rittenhouse, S.E., 1991, "Thrombin" receptor-directed ligand accounts for activation by thrombin of platelet phospholipase C and accumulation of 3-phosphorylated phosphoinositides, *J Biol Chem*, 266:18435.

Hung, D.T., Thien-Khai, H.V., Nelken, N.A., and Coughlin, S.R., 1992, Thrombin-induced proteolytic events in non-platelet cells are mediated by the unique proteolytic mechanism established for the cloned platelet thrombin receptor, *J Cell Biol*, 116:827.

Iiri, T., Tohkin, M., Morishima, N., Ohoka, Y., Ui, M., and Katada, T., 1989, Chemotactic peptide receptor-supported ADP-ribosylation of a pertussis toxin substrate GTP-binding protein by cholera toxin in neutrophil-type HL-60 cells, *J Biol Chem*, 264:21394.

Jaffe, E.A., Grulich, J., Weksler, B.B., Hampel, G., and Watanabe, K., 1987, Correlation between thrombin-induced prostacyclin production and inositol trisphosphate and cytosolic free calcium levels in cultured human endothelial cells, *J Biol Chem*, 262:8557.

Jamieson, G.A., and Okumura, T., 1978, Reduced thrombin binding and aggregation in Bernard-Soulier platelets, *J Clin Invest*, 61:861.

Jelsema, C.L., and Axelrod, J., 1987, Stimulation of phospholipase A2 activity in bovine rod outer segments by the beta gamma subunits of transducin and its inhibition by the alpha subunit, *Proc Natl Acad Sci USA*, 84:3623.

Jones, L.G., McDonough, P.M., and Brown, J.H., 1989, Thrombin and trypsin act at the same site to stimulate phosphoinositide hydrolysis and calcium mobilization, *Mol Pharmacol*, 36:142.

Kajijama, Y., Murayama, T., and Nomura, Y., 1989, Pertussis toxin-sensitive GTP-binding proteins may regulate phospholipase A_2 in response to thrombin in rabbit platelets, *Arch Biochem Biophys*, 274:200.

Kaser-Glanzmann, R., Gerber, E., and Luscher, E.F., 1979, Regulation of the intracellular calcium level in human blood platelets: cAMP dependent phosphorylation of a 22,000 Dalton component in isolated Ca2+-accumulating vesicles, *Biochim Biophys Acta*, 558:34.

Kelleher, D.J., Pessin, J.E., Ruoho, A.E., and Johnson, G.L., 1984, Phorbol ester induces desensitization of adenylate cyclase and phosphorylation of the ß-drenergic receptor in turkey erythrocytes, *Proc Natl Acad Sci USA*, 81:4316.

Kemp, B.E., and Pearson, R.B., 1990, Protein kinase recognition sequence motifs, *Trends in Biochemical Sciences*, 15:342.

Kim, D., Lewis, D.L., Graziadei, L., Neer, E.J., Bar-Sagi, D., and Clapham, D.E., 1989, G-protein ßgamma-subunits activate the cardiac muscarinic K^+-channel via phospholipase A_2, *Nature*, 337:557.

Kobilka, B.K., Matsui, H., Kobilka, T.S., Yang Feng, T.L., Francke, U., Caron, M.G., Lefkowitz, R.J., and Regan, J.W., 1987, Cloning, sequencing, and expression of the gene coding for the human platelet alpha 2-adrenergic receptor, *Science*, 238:650.

Kramer, R., Checani, G., Deykin, A., Pritzker, C., and Deykin, D., 1986, Solubilization and properties of Ca^{2+}-dependent human platelet phospholipase A2, *Biochim Biophys Acta*, 878:394.

Laposata, M., Krueger, C.M., and Saffitz, J.E., 1987, Selective uptake of [3H]arachidonic acid into the dense tubular system of human platelets, *Blood*, 70:832.

Leeb-Lundberg, L.M.F., Cotecchia, S., Lomasney, J.W., DeBernardis, J.F., Lefkowitz, R.J., and Caron, M.G., 1985, Phorbol esters promote α1-adrenergic receptor phosphorylation and receptor uncoupling from inositol phospholipid metabolism, *Proc Natl Acad Sci USA*, 82:5651.

Lewis, J.M., Woolkalis, M.J., Gerton, G.L., Smith, R.M., Jarett, L., and Manning, D.R., 1991, Subcellular distribution of the α subunit(s) of G_i: visualization by immunofluorescent and immunogold labeling, *Cell Regulation*, 2:1097.

Liu, L-W., Vu, T-K.H., Esmon, C.T., and Coughlin, S.R., 1991, The region of the thrombin receptor resembling hirudin binds to thrombin and alters enzyme specificity, *J Biol Chem*, 266:16977.

Lounsbury, K.M., Brass, L.F., and Manning, D.R., 1990, Phosphorylation of Gzα in human platelets: proximity to the amino-terminus of the subunit, *J Cell Biol*, 111:334a.

Lounsbury, K.M., Casey, P.J., Brass, L.F., and Manning, D.R., 1991, Phosphorylation of G_z in human platelets: selectivity and site of modification, *J Biol Chem*, 266:22051.

Manning, D.R., and Brass, L.F., 1991, The role of GTP-binding proteins in platelet activation, *Thromb Haemost*, 66:393.

Matsui, Y., Kikuchi, A., Kawata, M., Kondo, J., Teranishi, Y., and Takai, Y., 1990, Molecular cloning of smg p21B and identification of smg p21 purified from bovine brain and human platelets as smg p21B, *Biochem Biophys Res Commun*, 166:1010.

Murayama, T., Kajiyama, Y., and Nomura, Y., 1990, Histamine-stimulated and GTP-binding proteins-mediated phospholipase A_2 activation in rabbit platelets, *J Biol Chem*, 265:4290.

Nakashima, S., Hattori, H., Shirato, L., Takenaka, A., and Nozawa, Y., 1987, Differential sensitivity of arachidonic acid release and 1,2-diacylglycerol formation to pertussis toxin, GDPßS and NaF in saponin-permeabilized human platelets; possible evidence for distinct GTP-binding proteins involving phospholipase C and A2 activation, *Biochem Biophys Res Commun*, 148:971.

Ngaiza, J.R., and Jaffe, E.A., 1991, A 14 amino acid peptide derived from the amino terminus of the cleaved thrombin receptor elevates intracellular calcium and stimulates prostacyclin production in human endothelial cells, *Biochem Biophys Res Commun*, 179:1656.

Onorato, J.J., Palczewski, K., Regan, J.W., Caron, M.G., Lefkowitz, R.J., and Benovic, J.L., 1991, Role of acidic amino acids in peptide substrates of the beta-adrenergic receptor kinase and rhodopsin kinase, *Biochemistry*, 30:5118.

Paris, S., Magnaldo, I., and Pouysségur, J., 1988, Homologous desensitization of thrombin-induced phosphoinositide breakdown in hamster lung fibroblasts, *J Biol Chem*, 263:11250.

Pipili-Synetos, E., Gershengorn, M.C., and Jaffe, E.A., 1990, Expression of functional thrombin receptors in Xenopus oocytes injected with human endothelial cell mRNA, *Biochem Biophys Res Commun*, 171:913.

Rasmussen, U.B., Vouret-Craviari, V., Jallat, S., Schlesinger, Y., Pagès, G., Pavirani, A., Lecocq, J-P., Pouysségur, J., and Van Obberghen-Schilling, E., 1991, cDNA Cloning and expression of a hamster α-thrombin receptor coupled to Ca^{2+} mobilization, *FEBS Letters*, 288:123.

Raymond, J.R., 1991, Protein kinase C induces phosphorylation and desensitization of the human 5-HT$_{1A}$ receptor, *J Biol Chem*, 266:14747.

Ryu, S.H., Kim, U-H., Wahl, M.I., Brown, A.B., Carpenter, G., Huang, K-P., and Rhee, S.G., 1990, Feedback regulation of phospholipase C-ß by protein kinase C, *J Biol Chem*, 265:17941.

Salter, R.D., Howell, D.N., and Cresswell, P., 1985, Genes regulating HLA class I antigen expression in T-B lymphoblast hybrids, *Immunogenetics*, 21:235.

Seiler, S.M., Michel, I.M., and Fenton, J.W.II., 1992, Involvement of the "tethered-ligand" receptor in thrombin inhibition of adenylate cyclase, *Biochem Biophys Res Commun*, 182:1296.

Shenker, A., Goldsmith, P., Unson, C.G., and Spiegel, A.M., 1991, The G protein coupled to the thomboxane A$_2$ receptor in human platelets is a member of the novel Gq family, *J Biol Chem*, 266:9309.

Sibley, D.R., Benovic, J.L., Caron, M.G., and Lefkowitz, R.J., 1987, Regulation of transmembrane signaling by receptor phosphorylation, *Cell*, 48:913.

Silk, S.T., Clejan, S., and Witkom, K., 1989, Evidence of GTP-binding protein regulation of phospholipase A$_2$ activity in isolated human platelet membranes, *J Biol Chem*, 264:21466.

Simmoteit, R., Schulzki, H-D., Palm, D., Mollner, S., and Pfeuffer, T., 1991, Chemical and functional analysis of components of adenylyl cyclase from human platelets treated with phorbolesters, *FEBS Lett*, 285:99.

Smith, S.K., and Limbird, L.E., 1982, Evidence that human platelet alpha-adrenergic receptors coupled to inhibition of adenylate cyclase are not associated with the subunit of adenylate cyclase ADP-ribosylated by cholera toxin, *J Biol Chem*, 257:10471.

Smrcka, A.V., Hepler, J.R., Brown, K.O., and Sternweis, P.C., 1991, Regulation of polyphosphoinositide -specific phospholipase C activity by purified Gq, *Science*, 251:804.

Strathmann, M., and Simon, M.I., 1990, G protein diversity: a distinct class of α subunits is present in vertebrates and invertebrates, *Proc Natl Acad Sci USA*, 87:9113.

Strathmann, M.P., and Simon, M.I., 1991, Gα12 and Gα13 subunits define a fourth class of G protein α subunits, *Proc Natl Acad Sci USA*, 88:5582.

Tang, W-J., and Gilman, A.G., 1991, Type-specific regulation of adenylyl cyclase by G protein ßgamma subunits, *Science*, 254:1500.

Taylor, S.J., Chae, H.Z., Rhee, S.G., and Exton, J.H., 1991, Activation of the ß1 isozyme of phospholipase C by α subunits of the Gq class of G proteins. *Nature*, 350:516.

Taylor, S.J., and Exton, J.H., 1991, Two α subunits of the Gq class of G proteins stimulate phosphoinositide phospholipase C-ß1 activity, *FEBS Letts*, 286:214.

Van Obberghen-Schilling, E., Chambard, J.C., Lory, P., Nargeot, J., and Pouysségur, J., 1990, Functional expression of Ca^{2+}-mobilizing α-thrombin receptors in mRNA-injected *Xenopus* oocytes, *FEBS Letts*, 262:330.

Vassallo, R.R. Jr., Hoxie, J.A., and Brass, L.F., 1992a, Antibodies to the N-terminus of the thrombin receptor inhibit platelet activation by thrombin, *Clin Res*, in press.

Vassallo, R.R. Jr., Kieber-Emmons, T., Cichowski, K., and Brass, L.F., 1992b, Structure/function relationships in the activation of platelet thrombin receptors by receptor-derived peptides, *J Biol Chem*, 267:6081.

Vouret-Craviari, V., Van Obberghen-Schilling, E., Rasmussen, U.B., Pavirani, A., Lecocq, J-P., and Pouysségur, J., 1992, Synthetic α-thrombin receptor peptides activate G protein-coupled signaling pathways but are unable to induce mitogenesis, *Mol Biol Cell*, 3:95.

Vu, T-K.H., Hung, D.T., Wheaton, V.I., and Coughlin, S.R., 1991a, Molecular cloning of a functional thrombin receptor reveals a novel proteolytic mechanism of receptor activation, *Cell*, 64:1057.

Vu, T-K.H., Wheaton, V.I., Hung, D.T., Charo, I., and Coughlin, S.R., 1991b, Domains specifying thrombin-receptor interaction, *Nature*, 353:674.

Wallach, D., Davies, P.J.A., and Pastan, I., 1978, Cyclic AMP-dependent phosphorylation of filamin in mammalian smooth muscle, *J Biol Chem*, 253:4739.

Wardell, M.R., Reynolds, C.C., Berndt, M.C., Wallace, R.W., and Fox, J.E.B., 1989, Platelet glycoprotein Ibß is phosphorylated on serine 166 by cyclic AMP-dependent protein kinase, *J Biol Chem*, 264:15656.

Weiss, E.R., Kelleher, D.J., Woon, C.W., Soparkar, S., Osawa, S., Heasley, L.E., and Johnson, G.L., 1988, Receptor activation of G proteins, *FASEB J*, 2:2841.

White, T.E., Lacal, J-C., Reep, B., Fischer, T.H., Lapetina, E.G., and White, G.C. II., 1990, Thrombolamban, the 22-kDa platelet substrate of cyclic AMP-dependent protein kinase, is immunologically homologous with the Ras family of GTP-binding proteins, *Proc Natl Acad Sci USA*, 87:758.

Williams, A., Woolkalis, M.J., Poncz, M., Manning, D.R., Gewirtz, A., and Brass, L.F., 1990, Identification of the pertussis toxin-sensitive G proteins in platelets, megakaryocytes and HEL cells, *Blood*, 76:721.

Yoshimoto, T., Yamamoto, S., Okuma, M., and Hayaishi, O., 1977, Solubilization and resolution of thromboxane synthesizing system from microsomes of bovine blood platelets, *J Biol Chem*, 252:5871.

REGULATION OF PHOSPHOINOSITIDE-SPECIFIC PHOSPHOLIPASE C

ACTIVITY IN HUMAN PLATELETS

Yoshinori Nozawa, Yoshiko Banno and Koh-ichi Nagata

Department of Biochemistry
Gifu University School of Medicine
Tsukasamachi-40, Gifu 500
Japan

INTRODUCTION

The importance of phosphoinositide turnover in signal transduction has been well documented and the role of phosphoinositide-specific phospholipase C (PI-PLC) has stimulated much progress in defining the PI-PLC isozymes. PI-PLC is known to comprise at least nine isozymes which can be separated into three structurally related classes (PLC-ß, PLC-γ, PLC-δ) (Rhee et al. 1989; Kriz et al. 1990). However, the mechanism is not completely understood by which the isozymes are activated upon receptor stimulation to generate the second messengers inositol 1,4,5-trisphosphate (IP$_3$), and 1,2-diacylglycerol (DG). Recent studies reveal that multiple PLC isozymes exist within a single cell (Rhee et al. 1991), suggesting that their respective physiological roles and regulation may be distinct. Receptors which activate PI-PLC after ligand binding act via two established mechanisms. Several lines of evidence indicate that GTP-binding protein(s), putatively called Gp, mediate receptor-coupled phosphoinositide hydrolysis. There are pertussis toxin (PT)-sensitive and insensitive forms of Gp. Although the PT-sensitive Gp has not yet been specified, recently one of the PT-insensitive G proteins, Gq, which couples to PLC-ß1 has been isolated (Taylor et al. 1991). A second type of receptor-mediated activation of PI-PLC is catalyzed by receptors with intrinsic protein tyrosine kinase activity such as EGF and PDGF receptors, where the PLC-γ1 isozyme is phosphorylated on tyrosine residues (Rhee et al. 1991). Thus it is conceivable that different PLC isozymes could be coupled to different receptors and involved in expression of different functions.

The platelet is one of the most suitable model systems to investigate the regulation of PI-PLC isozymes in conjunction with responses such as shape change, secretion and aggregation. In human platelets, multiple forms of PI-PLC have been identified in cytosol and membrane fractions (Nozawa and Banno, 1991), although their full identification and characterization has not yet been reported. Many studies indicate involvement of G proteins in receptor signalling through PI-PLC in human platelets (Manning and Brass,

Mechanisms of Platelet Activation and Control, Edited by
K.S. Authi *et al.*, Plenum Press, New York, 1993

1991). Thrombin stimulates PI-PLC activity, in part, via a PT-sensitive Gp, while thromboxane A_2 (TxA_2) is coupled to Gq (Shenker et al. 1991). Low molecular weight GTP-binding protein(s) may not be involved in activation of human platelet PI-PLC. Recent studies have demonstrated that the cytoskeleton may play an important role in the regulation of PI-PLC activity (Goldschmidt-Clermont and Janmey, 1991).

Purification and Identification of Human Platelet Phosphoinositide-specific Phospholipase C Isoforms

There is substantial evidence to indicate a multiplicity of PI-PLC in mammalian tissues. Various types of PI-PLCs from platelets have been studied in several laboratories. However, because of the different assay conditions used (pH, substrate, detergent, Ca^{2+} concentration), satisfactory identification of the isozymes has not yet been achieved. Hakata et al. (1982) first isolated a soluble PLC of 143 kDa from bovine platelets, which had an optimum activity at pH 7.0. On the other hand, Low et al. (1986) identified multiple PLC isoforms in human platelets and suggested that they may be formed following aggregation or truncation by calpain. Two polymeric PLC isoforms (trimer and dimer of 146 kDa polypeptide) were purified by Moriyama et al. (1990). These enzymes appear to be similar in molecular mass (140-150 kDa) and pH optimum (at 7.0 in the presence of 0.1% deoxycholate). Furthermore, Manne (1987) suggested the presence of a poly-phosphoinositide-specific PLC that did not hydrolyze PI. We have isolated PLCs (61-63 kDa) with preferential activity towards PIP_2 from the cholate extract of membranes pretreated with 2M KCl (Banno et al. 1988). Similar low Mr PLCs were isolated from cytosol (57 kDa) and membrane fractions (67 and 64 kDa) by Baldassare et al. (1989). Although these observations indicate multiple forms of PLC, full characterization of these enzymes has not been achieved.

As shown in Figure 1, when human platelet cytosol was subjected to Fast Q-Sepharose column chromatography, several PLC activity peaks were resolved as assayed with PI at pH 5.5 or 7.0 and with PIP_2 at pH 6.5 (Fig.1, upper). The PIP_2-hydrolysing activity fraction (Fr.II) obtained from the Q-Sepharose column was further separated into three activity peaks by heparin-Sepharose column chromatography (Fig.1, lower). By these two column chromatographies, five PLC activity peaks (Ia, Ib, IIa, IIb, IIc) were resolved. These five PLC isoforms were analyzed by immuno-blotting with antibodies raised against rat brain PLC isozymes (PLC-β_1, -γ_1, -δ_1). The anti-PLC-β_1 antibody reacted with the peak IIc (150 kDa) and anti-PLC-γ_1 antibody with the peak Ib (145 kDa). Moreover, the anti-PLC-δ_1 antibody cross-reacted with the peak Ia (85 kDa). The peak IIa and IIb fractions were not recognizable by these antibodies. Previously, the presence of PLC-γ_2 was observed in the peak Ib fraction by positive reaction with an appropriate antibody and by purification to homogeneity (Banno et al. 1990). It had an apparent molecular mass of 145 kDa and similar enzymatic properties to those of bovine spleen PLC-γ_2 (Homma et al. 1990). The PLC-β_1 type was purified from peak IIc. The purified PLC-β_1 was truncated form with higher activity for PIP_2 hydrolysis (Banno et al. 1992). These results indicate that human platelet cytosol possesses at least four types of PLC isozymes (PLC-β_1, γ_1, γ_2, -δ_1). In other laboratories, two PI-PLC forms with 146 kDa (Moriyama et al. 1990) and low Mr PI-PLC-II (57 kDa) were purified to homogeneity (Baldassare et al. 1989). However, these enzymes were not cross-reactive with the antibodies against PLC isozymes described above, suggesting that they may correspond to the PLC-IIa and IIb fractions. This is supported by their higher activities for PI and PIP_2 hydrolysis at neutral pH in the presence of 0.1% deoxycholate.

Twenty per cent of human platelet PLC activity was detected in the membrane fraction as assayed with PIP_2 as substrate. The prepared extraction of membranes with 2M KCl solubilized a considerable proportion of membrane-associated PIP_2-PLC activity. We purified two membrane-bound PIP_2-PLCs (mPLC-I, mPLC-II) with 61-63 kDa from the

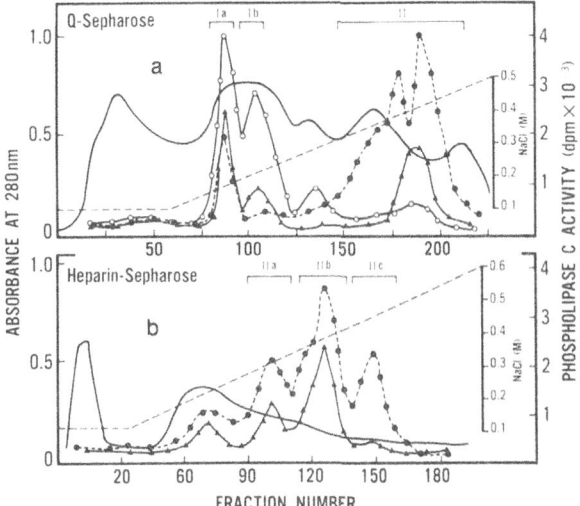

Figure 1   Q-Sepharose and heparin-Sepharose column chromatography of human platelet cytosolic PLC isoforms. The human platelet cytosolic fraction was applied onto Q-Sepharose column and the peak fractions (Ia, Ib and II) were separately pooled (upper panel). The fraction II was further applied onto heparin-Sepharose column and eluted with linear NaCl gradient (0.1-0.7M). Three activity fractions (IIa, IIb and IIc) were pooled (lower pane). Dashed lines, NaCl concentration; solid lines, absorbance at 280 nm; $\bigcirc$, PLC activity for PI at pH 5.5; $\blacktriangle$, PLC activity for PI at pH 7.0; $\bullet$, PLC activity for PIP$_2$ at pH 6.5.

cholate extract of the membrane fraction that had been pre-extracted with 2M KCl (Banno et al. 1988). Similar enzymes (63 and 69 kDa) were also purified by Baldassare et al. (1989). When the membrane fraction was directly extracted with 1% sodium cholate and subjected to column chromatographies on Fast Q-Sepharose, heparin-Sepharose and Ultrogel AcA-44 gel columns, three main activity peaks were separated. A major activity peak with molecular mass of 150 kDa was obtained in a pure state and was recognized by anti-PLC-ß$_1$ antibody. The minor activity peak with 61 kDa (mPLC-II) was separated from the PLC-ß$_1$ peak upon Mono Q column chromatography and did not cross-react with anti-PLC-ß$_1$ antibody, suggesting that mPLC-II was not a proteolytic product of PLC-ß$_1$.

Several low Mr types of PLC have also been purified from other tissues. Membrane-bound PLC species with 54-66 kDa have been co-purified with PLC-ß$_1$ in rabbit brain membrane (Carter et al. 1990) and a PLC of 18 kDa has been purified from human spleen membrane, which was dissociated from PLCs (60-70 kDa) by high salt concentration (Roy et al. 1991). Although these observations provide evidence which supports the presence of low Mr forms of PLCs, cDNA cloning will be required for their full identification.

GTP-BINDING PROTEINS IN HUMAN PLATELETS

Heterotrimeric Type

Signal-transducing G proteins are largely classified into two groups, heterotrimeric G proteins and monomeric low Mr GTP-binding proteins such as ras p21. G proteins are composed of α, ß, γ subunits and play important roles in agonist-mediated transmembrane signalling between cell surface receptors and intracellular effectors such as adenyl cyclase, cyclic GMP phosphodiesterase, phospholipase C and A_2 (Casey and Gilman, 1989; Kaziro et al. 1991). The α-subunits contain a GTP/GDP-binding site and possess GTP-hydrolysis activity. Differences in the α-subunits serve to distinguish the various G protein oligomers.

Several kinds of G proteins are known to be present in platelets (see reviews, Nagata and Nozawa, 1990; Manning and Brass, 1991). The regulation of adenyl cyclase in response to extracellular stimuli in platelets resembles that in other cells. Gs, cholera toxin substrate, and Gi, PT substrate, transduce stimulatory and inhibitory signals for adenyl cyclase, respectively. A major PT substrate in the human platelet, Gi_2, has been purified and identified (Nagata et al. 1988). Gi_3 and Gi_1 are also present as minor components. Gi_2 was observed to mediate inhibition of adenyl cyclase by adrenaline (Simonds et al. 1989). Recently, PT-insensitive G proteins, Gz(Gx) and Gq, have been identified. Gz was reported to be present in platelets (Gagnon et al. 1991). We have examined the amount of Gq in human platelet membrane as compared to brain membrane by using the antiserum raised against $G_L2\alpha$ (α-subunit of bovine liver G11) (Nakamura et al. 1991). As shown in Figure 2, the antiserum recognized the 42 kDa-protein in human platelet membrane extract and brain membrane extract.

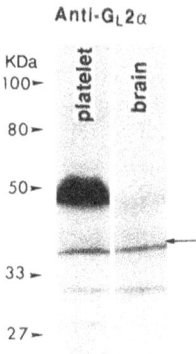

Anti-$G_L2\alpha$

Figure 2   Immunoblot with anti-$G_L2\alpha$ antibody in human platelet and rat brain membranes. The platelet membrane and rat brain membrane were extracted with 20 mM Tris-HCl buffer containing 1% Triton X-100 and the extracts (100 μg protein) were subjected to SDS-PAGE, then transferred to nitrocellulose membrane and probed with the antibody against $G_L2\alpha$.

Low Molecular Weight Types

There is a superfamily of GTP-binding proteins with Mr ranging 20,000 to 30,000. More than 30 genes of such GTP-binding proteins have been identified. Although the functions of many of them remain to be clarified, they are believed to be involved in controlling essential cellular events such as growth, differentiation, cytoskeletal organization, intracellular vesicle transport, and secretion.

In platelets, at least 15 low Mr GTP-binding proteins have been reported to be present and several of them have been isolated, characterized biochemically, and identified based on proteins or cDNA sequence homologies with other molecules (Manning and Brass, 1991). The sequence of the GTP-binding proteins is most highly conserved in the domains involving GTP/GDP-binding and GTP-hydrolysis. The major low Mr GTP-binding protein in human platelet membranes was purified and identified as *rap*1B (*smg* 21B) protein (Nagata and Nozawa, 1990; Takai et al. 1992). *ras* p21 and 24 kDa-proteins

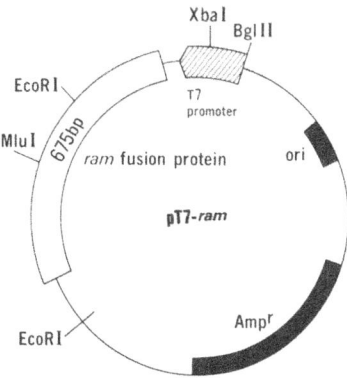

Figure 3   Construction of the *ram*-expression vector. The *ram* protein is produced as a fused protein under the control of T7 promoter. ori, the origin of replication; Amp^r, ampicillin-resistant gene.

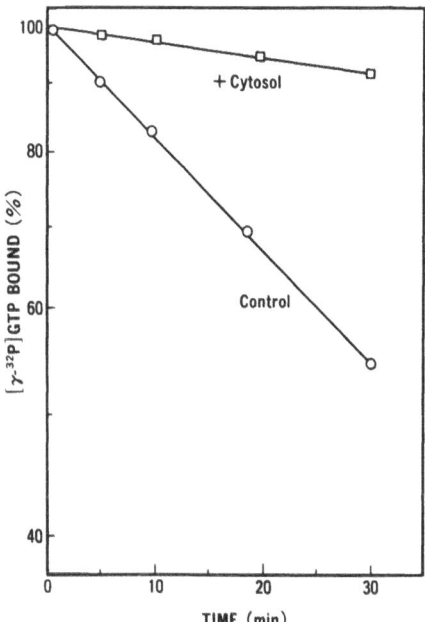

Figure 4   Time-course of inhibition of GTP-hydrolysis of *ram*p25. The [γ-32P]GTP-bound *ram*p25 was incubated in the absence (○) and presence (□) of 0.2 mg/ml of human platelet cytosol.

were also purified (Ohmori et al. 1989). Moreover, G25K, *ral* and *rac*1 were identified in human platelet membranes (Manning and Brass, 1991). Although the physiological functions of these proteins are not established, *rap*1B protein was observed to be phosphorylated in intact platelets by cAMP-elevating prostaglandin (PG)E₁ or dibutyryl cAMP, indicating that this protein is a substrate of the cAMP-dependent protein kinase (A-kinase) and is a component of the cAMP-dependent signal transduction pathway (Kawata et al 1989). Lapetina et al. (1989) have proposed that translocation of *rap*1B from membrane to cytosol occurs in response to PGE₁ and dibutyryl cAMP.

c25KG is a GTP-binding protein with Mr of 25,000 first purified from human platelet cytosolic fraction and shows both GTP/GDP-binding and GTPase activities like other low Mr GTP-binding proteins (Nagata et al. 1989). The cDNA of this protein has not been isolated, but a homologous cDNA, designated as ram encoding a GTP-binding protein of 221 amino acids with a calculated Mr of 25,068, was cloned from rat megakaryocyte library (Nagata et al. 1990). The homology between ram protein (ram p25) and c25KG is 80% and their putative effector domains are almost identical. When expressed in *Escherichia coli* as in Fig.3 and purified to near homogeneity, ram p25 possesses GTP/GDP-binding and GTP-hydrolysis activities. The notable feature of ram p25 and c25KG is that they share about 60% amino acid homology to yeast $SEC4$ protein, which acts at a late step of secretory pathway, and their putative effector domains are very similar to that of $SEC4$ protein. These results suggest that ram p25 and c25KG would have important roles in the secretion of platelets.

There are regulatory proteins for low Mr GTP-binding proteins, which include GTPase activating protein (GAP), GDP dissociation stimulator (GDS) and GDP dissociation inhibitor (GDI) (see review, Takai et al. 1992). In platelets, two GAPs for rap1B protein are reported, although the modes of their interactions have not been clarified (Takai et al. 1992). As for ram p25, we were unable to detect its GAP activity but observed GTPase inhibiting activity in human platelet cytosol as shown in Figure 4. It is thus conceivable that the function of ram p25 would be regulated by this factor *in vivo*.

REGULATION OF PHOSPHOINOSITIDE-SPECIFIC PHOSPHOLIPASE C ACTIVITY

PI-PLC activity is known to be influenced by various conditions *in vitro* and *in vivo*. PI-PLCs are Ca^{2+}-dependent requiring micromolar Ca^{2+} for PIP_2 hydrolysis and millimolar Ca^{2+} for PI hydrolysis *in vitro*. Therefore, in the cell (10^{-7} M - 10^{-6}M Ca^{2+}) PI-PLC may act preferentially on PIP_2. Recent studies demonstrate two different systems which induce receptor-mediated stimulation of PI-PLC; coupling to Gp and cross-talk with protein tyrosine kinase (PTK). On the other hand, protein kinase C (C-kinase) and A-kinase exert negative modulation on PI-PLC activation. Despite abundant findings, the regulatory mechanisms of receptor-mediated PI-PLC activity in human platelets are poorly understood.

GTP-Binding Protein

Several lines of evidence indicate that a G protein(s) (Gp) is involved in receptor-coupled phosphoinositide hydrolysis in platelets. Haslam and Davidson (1984) first demonstrated that addition of GTP and GTPγS to electrically permeabilised human platelets enhanced thrombin-induced DG formation, suggesting G protein participation in thrombin receptor-coupled PI-PLC activation. Similar experiments with saponin-permeabilised platelets were performed by Brass et al. (1986) who observed that thrombin-induced DG formation and ^{45}Ca release were inhibited by preincubation with PT. This may imply that a Gi-like protein is coupled to thrombin-induced PI-PLC activation. In addition to thrombin, thromboxane also stimulates phosphoinositide hydrolysis. However, the TxA_2 analog, U46619-induced phosphoinositide hydrolysis was unaffected by PT (Brass et al. 1988). It appears that platelets contain two different G proteins (PT-sensitive and PT-insensitive) that mediate activation of PI-PLCs. The TxA_2 receptor in human platelets was demonstrated to couple to a PT-insensitive G protein, Gq (Shenker et al. 1991). Indeed, it has been shown in reconstitution experiments that a PT-insensitive G protein (Gq class) stimulated PLC-ß$_1$, but not PLC-γ$_1$ and PLC-δ$_1$ (Taylor et al. 1991). We have shown the presence of PLC-ß$_1$ and Gq class in the human platelet membrane (unpublished data).

These results strongly suggest the pathway of TxA$_2$ receptor-coupled Gq·PLC-ß$_1$ in human platelets.

On the other hand, there are some reports indicating the presence of soluble Gp which may be involved in PI-PLC activation. Majerus's group (1986) has investigated effects of guanine nucleotides and NaF on PIP$_2$ hydrolysis by supernatant, and they postulated a 41-kDa PT substrate as a candidate for Gp. In other experiments, Baldassare et al (1988) reported that a low *Mr* GTP-binding protein (29 kDa) was associated with human platelet cytosolic PI-PLC. We also have obtained a similar finding which could be explained by a direct effect of GTP on cytosolic PLC rather than by involvement of Gp. Furthermore, we have indicated evidence that the GTPγS binding activity fraction could be separated from a soluble PI-PLC activity by using hydroxyapatite column chromatography, and also that the purified protein with an apparent GTPγS binding activity was not due to GTP-binding proteins as present in other tissues (Banno et al. 1992a).

Interaction with Cytoskeleton

Recent studies demonstrate that stimulation of platelets by thrombin causes the association of signalling enzymes responsible for lipid phosphorylation and hydrolysis (DG kinase, PI3 kinase, PI4 kinase and PI-PLC) with the cytoskeleton and that this may occur at cytoskeleton anchoring points to the membranes (Grondin et al. 1991; Zhang et al. 1992). Actin is a major component of the cytoskeleton system in the platelet, and actin-binding protein(s) is involved in the association of GPIa, a collagen receptor with actin filaments. This GPIa-actin filament linkage may play a role in signal transduction in collagen-stimulated platelets.

We have recently found that the co-purified protein with a PI-PLC (IIa) from human platelet cytosol was a complex of gelsolin and actin (molar ratio 1:1) by Western blotting analysis (Banno et al. 1992a). The presence of the gelsolin complex was also observed in other PLC isoforms (Fig.5). Its content was relatively high in peak Ia (PLC-δ$_1$) and the peak Ib (PLC-γ$_1$) and lower IIa and IIb. It was shown that the PLCγ$_1$ and the gelsolin complex were co-immunoprecipitated, suggesting that the PLCγ$_1$ was physically associated with the gelsolin complex.

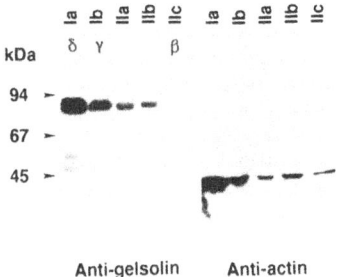

Figure 5   Immunoblot of platelet soluble PI-PLC isoforms with anti-actin and anti-gelsolin antibodies. The five PI-PLC fractions (Ia, Ib, IIa, IIb and IIc) described in Figure 1 were immunoblotted with anti-actin and anti-gelsolin antibodies.

Grondin et al. (1991) have demonstrated that upon thrombin stimulation of human platelets PI-PLC, DG kinase and PI kinases were translocated to cytoskeleton (pellet obtained by centrifugation (12,000 g, 10 s) of Triton X-100 extract). Although our findings do not provide direct evidence that PLCγ$_1$ is associated with the gelsolin complex in the platelet, it can be assumed that PLCγ$_1$ is associated somehow with the cytoskeleton. Platelet activation with thrombin causes marked changes accompanied with actin polymerization and gelsolin complex formation (Lind et al. 1987). The stimulation of platelets with thrombin causes quantitative conversion of profilin and gelsolin to complexes with actin. One may thus speculate that translocation of PI-PLC to the membrane cytoskeleton would occur for this activation.

PIP_2 is known to promote dissociation of actin:gelsolin complex *in vitro* by its high binding affinity for gelsolin, thereby leading to actin polymerization (Janmey and Stossel, 1989). Profilin, another actin-binding protein, was shown to inhibit PIP_2 hydrolysis by a cytosolic PLC due to its preferential binding to substrate PIP_2 (Goldschmidt-Clermont and Janmey, 1991). Our finding that the gelsolin complex was co-purified with cytosolic PLC isoforms leads us to investigate whether the gelsolin complex affects PIP_2 hydrolysis by PLC isoforms. Although the complex *per se* had little effect on PIP_2 hydrolysing activities of PLC isoforms, free gelsolin separated from the complex affects, to various extents, the PLC activities. The activities of PLC-Ia (δ_1 type), PLC-Ib (γ_1 type) and PLC-IIa (unidentified type) were markedly inhibited by free gelsolin, but PLC-IIc (β_1 type) and PLC-IIb activities were less affected. These results indicated differences in the inhibitory potency of free gelsolin among platelet cytosolic PLC isoforms *in vitro*. Such inhibition is due to the decrease in the available substrate PIP_2 for PI-PLCs by its strong interaction with gelsolin. Similar observations were also obtained with profilin in PI-PLC isozymes *in vitro* (Goldschmidt-Clermont and Janmey 1991). Profilin inhibited PIP_2 hydrolysis by PLCγ_1 but not by PLC-β_1. Moreover, it was demonstrated that tyrosine-phosphorylated PLC-γ_1 can hydrolyze PIP_2 even in the presence of profilin.

Protein Phosphorylation

Several studies provided evidence for the involvement of receptor PTKs in PLC-γ_1 activation (Rhee et al. 1991). Phosphorylation of PLC-γ_1 by the receptor PTK regulates both the membrane association and catalytic activity of the enzyme in fibroblasts. Furthermore, recent data with lymphocytes indicate that the activation of PLC-γ_1 by tyrosine phosphoryation is not restricted to receptor PTKs but can also be achieved through non-receptor PTKs (Secrist et al. 1991). Moreover, stimulation of rat basophilic leukemia cells with oligomeric IgE elicits a rapid and transient tyrosine phosphorylation of PLC-γ_1, probably mediated by non-receptor PTKs, *src* and *fyn* products (Park et al. 1991). Therefore, it is interesting to know whether PLC-γ_1 is associated with tyrosine phosphorylation in platelets. The level of pp60[c-src] in the human platelet is very high and is comparable to the level observed in transformed cells (Golden and Brugge. 1989). Since physiological agonists, such as thrombin and collagen, cause the elevation of phosphotyrosine content, it would be possible that PLC-γ_1 is activated in agonist-stimulated human platelets via tyrosine phosphorylation. We observed upon stimulation by thrombin the increases in tyrosine phosphorylation on 130, 105, 85, 78 kDa protein bands. However, unlike NIH-3T3 fibroblasts stimulated with PDGF, no tyrosine phosphorylation of PLC-γ_1 was observed in thrombin-stimulated human platelets under our conditions (Fig.6). A PTK inhibitor, genistein, inhibited the TxA_2-induced events such as PI-PLC activation, aggregation and secretion (Nakashima et al. 1990). However, these results, together with reduced tyrosine phosphorylation, were also obtained with a genistein analogue which has no ability to inhibit PTK. The apparent inhibition of PIP_2 hydrolysis by genistein was interpreted by the interference of the thromboxane binding to receptors. At present, we have no strong evidence that PLC-γ_1 is stimulated by PTKs in thrombin-stimulated human platelets.

On the other hand, there is considerable evidence to suggest that receptor-mediated phosphoinositide hydrolysis is affected by A-kinase and C-kinase activation, exerting a negative feedback regulation of PI-PLC in platelets (see reviews, Lapetina, 1990; Nozawa et al. 1991). The pretreatment of human platelets with a tumour-promoting phorbol ester (PMA) and permeable DG suppressed thrombin-induced formation of inositol phosphates

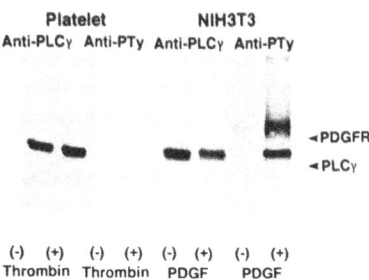

Platelet NIH3T3
Anti-PLCγ Anti-PTy Anti-PLCγ Anti-PTy

◄PDGFR

◄PLCγ

(-) (+) (-) (+) (-) (+) (-) (+)
Thrombin Thrombin PDGF PDGF

Figure 6   Immunoblot of immunoprecipitated PLC-γ_1 of platelet and NIH3T3 fibroblast with anti-phosphotyrosine antibody. Human platelets and NIH3T3 fibroblasts were stimulated with 1 U thrombin for 5 min and 1 μM PDGF for 30 min, respectively. Cells were then lysed with 1% Triton X-100 and the extracts were subjected to immunoprecipitation with anti-PLC-γ_1 antibody. The precipitated proteins were separated on SDS-PAGE, then transferred to nitrocellulose membrane and probed with anti-PLC-γ_1 antibody and anti-phosphotyrosine antibody.

and Ca^{2+} mobilization (Watson and Lapetina, 1985). Such inhibitory effects were prevented by the pretreatment with C-kinase inhibitor H-7 (Tohmatsu et al. 1987). The mechanism underlying inhibition remains to be disclosed, but it is likely that the possible targets of C-kinase and A-kinase actions are PLC-ß$_1$ and PLC-γ_1 (Rhee et al. 1991). In cells containing PLC-ß$_1$, PLC-γ_1 and PLC-δ$_1$, the activation of C-kinase selectively phosphorylates PLC-ß$_1$ and PLC-γ_1, whereas elevation of intracellular concentration of cyclic AMP causes phosphorylation of PLC-γ_1 by A-kinase but not PLC-ß$_1$ and PLC-δ$_1$. However, direct inhibition of the PLCs is not likely because the purified enzymes are phosphorylated by kinases *in vitro* but not changed in their activities. Although we have demonstrated that PIP$_2$ hydrolytic activity was enhanced by GTPγS in isolated human platelet membranes, such GTPγS effect was not observed in membranes from dibutyryl cAMP-pretreated platelets (Yada et al. 1990). Therefore, it can be postulated that the inhibition of PI-PLC is caused by impairment of Gp coupling to PI-PLC. Lapetina (1990) has proposed a view that membrane-bound *rap*1B is phosphorylated by A-kinase in thrombin-stimulated human platelets and then transfers from membrane to cytosol, thereby resulting in PI-PLC inhibition.

ACKNOWLEDGEMENTS

This work was in part supported by a Grant-in-Aid from the Ministry of Education, Science and Culture of Japan. We gratefully thank Dr. T. Takenawa (Tokyo Metropolitan Institute of Gerontology, Tokyo) and Dr. S.G. Rhee (National Heart, Lung and Blood Institute, Bethesda) for kindly providing anti-PLC isozyme antibodies and Dr. K. Haga (Tokyo University, Tokyo) for a kind supply of anti-$G_L 2\alpha$ antibody.

REFERENCES

Baldassare, J.J., Knipp, M.A., Henderson, P.A., and Fisher, G.J., 1988, GTPγ S-stimulated hydrolysis of phosphatidylinositol-4,5-bisphosphate by soluble phospholipase C from human platelets requires soluble GTP-binding protein, *Biochem. Biophys. Res. Commun.* 154:351.

Baldassare, J.J., Henderson, P.A., and Fisher G.J., 1989, Isolation and characterization of one soluble and two membrane-associated forms of phosphoinositide-specific phospholipase C from human platelets, *Biochemistry* 28:6010.

Banno, Y., Yada, Y., and Nozawa, Y., 1988, Purification and characterization of membrane-bound phospholipase C specific for phosphoinositides from human platelets, *J. Biol. Chem.* 263:11459.

Banno, Y., Yu, A., Nakashima, T., Homma, Y., Takenawa, T., and Nozawa, Y.,1990, Characterization of a cytosolic phosphoinositide-phospholipase C (γ_2-type) from human platelets, *Biochem. Biophys. Res. Commun.* 167:396.

Banno, Y., Nakashima, T., Kumada, T., Ebisawa, K., Nonomura, Y., and Nozawa, Y., 1992a, Effects of gelsolin on human platelet cytosolic phosphoinositide-phospholipase C isozymes, *J. Biol. Chem.* 267:6488.

Banno, Y., Suzuki, T., and Nozawa, Y., 1992b, Isolation of a polyphosphoinositide-phospholipase C (type ß) rom cytosolic and membrane fractions of human platelets, *Platelets* 2:69.

Brass, L.F., Laposata, M., Banga, H.S., and Rittenhouse, S.E., 1986, Regulation of the phosphoinositide hydrolysis pathway in thrombin-stimulated platelets by a pertussis toxin-sensitive guanine nucleotide-binding protein: Evaluation of its contribution to platelet activation and comparisons with the adenylate cyclase inhibitory protein, Gi, *J. Biol. Chem.* 261:16838.

Brass, L.F., Woolkalis, M.J., and Manning, D.R., 1988, Interaction in platelets between G protein and agonists that stimulate phospholipase C and inhibit adenyl cyclase, *J. Biol. Chem.* 263:5348.

Carter, H.R., Wallace, M.A., and Fain, J.N., 1990, Purification and characterization of PLC-ßm, a muscarinic cholinergic regulated phospholipase C from rabbit brain membrane, *Biochim. Biophys. Acta* 1054:119.

Casey, P.J., and Gilman, A.G., 1988, G protein involvement in receptor-effector coupling, *J. Biol. Chem.* 263:2577.

Deckmyn, H.M., Tu, S.M., and Majerus, P.W., 1986, Guanine nucleotides stimulate soluble phosphoinositide-specific phospholipase C in the absence of membranes, *J. Biol. Chem.* 261:16553.

Gagnon, A.W., Manning, D.R., Catani, L., Gewirtz, A., Poncz, M., and Brass, L.F., 1991, Identification of Gz as a pertussis toxin-insensitive G protein in human platelets and megakaryocytes, *Blood* 78:1247.

Golden, A., and Brugge, J.S., 1989, Thrombin treatment induces rapid changes in tyrosine phosphorylation in platelets, *Proc. Natl. Acad. Sci. USA* 86:901.

Goldschmidt-Clermont, P., and Janmey, P.A., 1991, Profilin, a weak CAP for actin and *ras*, *Cell* 66:419.

Grondin, P., Plantavid, M., Sultan, C., Breton, M., Mauco, G., and Chap, H., 1991, Interaction of pp60[c-src],phospholipase C, inositol-lipid, and diacylglycerol kinases with the cytoskeletons of thrombin-stimulated platelets, *J. Biol. Chem.* 266:15705.

Hakata, H., Kambayashi, J., and Kosai, G., 1982, Purification and characterization of phosphatidylinositol-specific phospholipase C from bovine platelets, *J. Biochem.* 92:929.

Haslam, R.J., and Davidson, M.M.L., 1984, Receptor-induced diacylglycerol formation in permeabilized platelet: Possible role for GTP-binding protein, *J. Recept. Res.* 4:605.

Homma, Y., Emori, Y., Shibasaki, F., Suzuki, K., and Takenawa, T., 1990, Isolation and characterization of a γ-type phosphoinositide-specific phospholipase C (PLC-γ_2), *Biochem. J.* 269:13.

Janmey, P.A., and Stossel, T.P., 1989, Gelsolin-polyphosphoinositide interaction: Full expression of gelsolin-inhibiting function by polyphosphoinositides in vesicular form and inactivation by dilution, aggregation, or masking of the inositol head group, *J. Biol. Chem.* 264:4825.

Kawata, M., Kikuchi, A., Hoshijima, M., Yamamoto, K., Hashimoto, E., Yamamura, H., and Takai, Y., 1989, Phosphorylation of *smg* p21, a *ras*-like GTP-binding protein, by cyclic AMP-dependent protein kinase in a cell-free system and in response to prostaglandin E1 in intact human platelets, *J. Biol. Chem.* 264:15688.

Kaziro, Y., Itoh, H., Kozasa, T., Nakafuku, M., and Satoh, T., 1991, Structure and function of signal-transducing GTP-binding proteins, *Annu. Rev. Biochem.* 60:349.

Kriz, R., Lim, L.L., Sultzman, L., Ellis, C., Heldin, C.H., Pawson, C.H., Pawson, T., and Knopf, J., 1990, Phospholipase C isozymes: Structure and functional similarities, *Ciba Foundation Symposium*, 150:112.

Lapetina, E.G., Lacal, J.C., Reep, B.R., and Vedia, L.M., 1989, A *ras*-related protein is phosphorylated and translocated by agonists that increase cAMP levels in human platelets, *Proc. Natl. Acad. Sci. U.S.A.* 86:3131.

Lapetina, E.G., 1990, The signal transduction induced by thrombin in human platelets, *FEBS Lett.* 268:400.

Lind, S.E., Janmey, P.A., Chaponnier, C., Herbert, T.J., and Stossel, T.P., 1987, Reversible binding of actin to gelsolin and profilin in human platelet extracts, *J. Cell Biol.* 105:833.

Low, M.G., Carroll, R.C., and Cox, A.C., 1986, Characterization of multiple forms of phosphoinositide-specific phospholipase C purified from human platelets, *Biochem. J.* 237:139.

Manne, V., 1987, Identification of polyphosphoinositide-specific phospholipase C and its resolution from phosphoinositide-specific phospholipase C from human platelet extract, *Oncogene* 2:49.

Manning, D.R., and Brass, L.F., 1991, The role of GTP-binding proteins in platelet activation, *Thromb. Haemost.* 66:393.

Moriyama, T., Narita, H., Oki, M., Matsuura, T., and Kito, M., 1990, Purification of polymeric phospholipase Cs from human platelets, *J. Biochem.* 108:414.

Nagata, K., Katada, T., Tohkin, M., Itoh, H., Kaziro, Y., Ui, M., and Nozawa, Y., 1988, GTP-binding protein in human platelet membranes serving as the specific substrate of islet-activating protein, pertussis toxin, *FEBS Lett.* 237:113.

Nagata, K., Itoh, H., Katada, T., Takenaka, T., Ui, M., Kaziro, Y., and Nozawa, Y., 1989, Purification, identification and characterization of two GTP-binding proteins with molecular weight of 25,000 and 21,000 in human platelet cytosol: One is the *rap1/smg21/Krev1* and the other is a novel GTP binding protein. *J. Biol. Chem.* 264:17000.

Nagata, K., and Nozawa, Y., 1990, GTP-binding proteins in human platelets, *Platelets* 1:67.

Nagata, K., Satoh, T., Itoh, H., Kozasa, T., Okano, Y., Doi, T., Kaziro, Y., and Nozawa, Y., 1990, The *ram*: A novel low molecular weight GTP-binding protein cDNA from a rat megakaryocyte library, *FEBS Lett.* 275:29.

Nakamura, F., Oga, K., Shiozaki, K., Kameyama, K., Ohara, K., Haga, T., and Nukada, T., 1991, Identification of two novel GTP-binding protein α-subunits that lack apparent ADP-ribosylation sites for pertussis toxin, *J. Biol. Chem.* 266:12681.

Nakashima, S., Kioke, T., and Nozawa, Y., 1990, Genistein, a protein tyrosine kinase inhibitor, inhibits thromboxane A$_2$-mediated human platelet responses, *Mol. Pharmacol.* 39:475.

Nozawa, Y., Nakashima, S., and Nagata, K., 1991, Phospholipid-mediated signalling in receptor activation of human platelets, *Biochim. Biophys. Acta* 1082:219.

Nozawa, Y., and Banno, Y., 1991, Phosphatidylinositol-specific phospholipase C from human platelets, *Methods Enzymol.* 197:518.

Ohmori, T., Kikuchi, A., Yamamoto, K., Kim, S., and Takai, Y., 1989, Small molecular weight GTP-binding proteins in human platelet membranes: Purification and characterization of a novel GTP-binding protein with a molecular weight of 22,000, *J. Biol. Chem.* 264:1877.

Park, D.J., Min, H.K., and Rhee, S.G., 1991, IgE-induced tyrosine phosphorylation of phospholipase C-γ in rat basophilic leukemia cells, *J. Biol. Chem.* 266:24237.

Rhee, S.G., Suh, P.H., and Lee, S.Y., 1989, Studies of inositol phospholipid-specific phospholipase C, *Science* 244:546.

Rhee, S.G., Kim, H., Suh, P.H., and Choi, W., 1991, Multiple forms of phosphoinositide-specific phospholipase C and different modes of activation, *Biochem. Soc. Trans.* 19:337.

Roy, G., Villar, L.M., Lazaro, I., Gonzalez, M., Bootello, A., and Gonzalez-Porque, P., 1991, Purification and properties of membrane and cytosolic phosphatidylinositol-specific phospholipase C from human spleen, *J. Biol. Chem.* 266:11495.

Secrist, J.P., Karnitz, L., and Abraham R.T., 1991, T-cell antigen receptor ligation induces tyrosine phosphorylation of phospholipase C-γ$_1$, *J. Biol. Chem.* 266:12135.

Shenker, A., Goldsmith, P., Unson, C.G., and Spiegel, A.M., 1991, The G protein coupled to the thromboxane A$_2$ receptor in human platelets is a member of the novel Gq family, *J. Biol. Chem.* 266:9309.

Simonds, M.F., Goldsmith, P.K., Codina, J., Unson, C.G., and Spiegel, A.M., 1989, Gi$_2$ mediates α$_2$-adrenergic inhibition of adenyl cyclase in platelet membranes: *In situ* identification with Gα C-terminal antibodies, *Proc. Natl. Acad. Sci. U.S.A.* 86:7809.

Takai, Y., Kaibuchi, K., Kikuchi, A., and Kawata, M., 1992, Small GTP-binding proteins, *Int. Rev. Cytol.* 133:187.

Taylor, S.J., Chae, H.Z., Rhee, S.G., and Exton, J.H., 1991, Activation of the ß$_1$ isozymes of phospholipase C by α subunits of the Gq class of G proteins, *Nature* 350:516.

Tohmatsu, T., Hattori, H., Nagao, S., Ohki, K., and Nozawa, Y., 1986, Reversal by protein kinase C inhibitor of suppressive actions of phorbol-12-myristate-13-acetate on polyphosphoinositide metabolism and cytosolic Ca^{2+} mobilization in thrombin-stimulated human platelets, *Biochem. Biophys. Res. Commun.* 134:868.

Watson, S.P., and Lapetina, E.G., 1985, 1,2-Diacylglycerol and phorbol ester inhibit agonist-induced formation of inositol phosphates in human platelets: Possible implications for negative feedback regulation of inositol phospholipid hydrolysis, *Proc. Natl. Acad. Sci. U.S.A.* 82:2623.

Yada, Y., Okano, Y., and Nozawa, Y., 1990, Inhibition by cyclic AMP of guanine nucleotide-induced activation of phosphoinositide-specific phospholipase C in human platelets, *Biochem. Biophys. Res. Commun.* 172:256.

Zhang, J., Fry, M.J., Waterfield, M.D., Jaken, S., Liao, L., Fox, J.E.B., and Rittenhouse, S.E., 1992, Activated phosphoinositide 3-kinase associates with membrane skeleton in thrombin-exposed platelets, *J. Biol. Chem.* 267:4686.

RAP1B AND PLATELET FUNCTION

Eduardo G. Lapetina and Francis X. Farrell

Division of Cell Biology, Burroughs Wellcome Co.
3030 Cornwallis Road
Research Triangle Park
North Carolina 27709

INTRODUCTION

Platelets contain multiple low molecular weight GTP-binding proteins which share strong sequence similarity to ras including rap, rac, ral, and rho (Nagata et al., 1989; Polakis et al., 1989). In addition, platelets contain regulatory molecules which both control the hydrolysis of GTP bound to the protein and/or promote the exchange of GDP for GTP (Hart et al., 1991). Surprisingly, platelets do not contain the ras molecule in a significant amount, yet possess high levels of the ras regulatory molecule rasGAP. RasGAP has been shown to bind rap1a with high affinity without increasing its GTPase activity (Frech et al., 1990). It is generally accepted that in addition to rasGAP acting as a GTPase activating protein, it may function as the downstream target molecule of ras (Hall, 1990a). For this reason, the role of rasGAP and rap1 in platelets is intriguing given that platelets also contain the GTPase activating protein specific for rap1. Recent data has proposed that this complex interaction may play a controling role in platelet signal transduction.

Rap Proteins

Rap proteins were identified by screening cDNA libraries with probes homologous to regions of ras (Pizon et al., 1988). Rap proteins like ras are low molecular weight GTP-binding proteins. They share 50% homology to ras and are expressed in a wide variety of tissues. More importantly, rap proteins are found in high abundance in platelets. On the other hand, ras is not found in abundance as judged by western analysis. Two rap familes designated rap1 and rap2 have been identified to date. Within each family are two members denoted as a and b, i.e., rap1a, rap1b, rap2a, and rap2b (Ohmstede et al., 1990). The two members within each family are 90% homologous. The rap proteins are members of a larger family of proteins homologous to ras including ral, rho, rac, and rab which has been termed the ras-related protein family (Hall, 1990b; Downward, 1990a).

More important than rap's 50% homology to ras is the fact that rap proteins share the greatest homology to ras in the regions important for ras transformation including the nucleotide binding region, the putative effector region, and a membrane attachment site. As shown in figure 1, rap1 and rap2 are almost identical to Kirsten ras in these regions. The membrane attachment site is the much studied CAAX motif (Cysteine-aliphatic-aliphatic -last amino acid) found at the C-terminal of ras and rap (Hancock et al., 1989). This sequence motif has also been identified in other ras-related proteins, nuclear lamins, the gamma subunits of the heterotrimeric G-proteins, cGMP phosphodiesterase, and several yeast mating factors. An important difference to note is that all rap proteins contain a threonine at position 61 instead of glutamine.

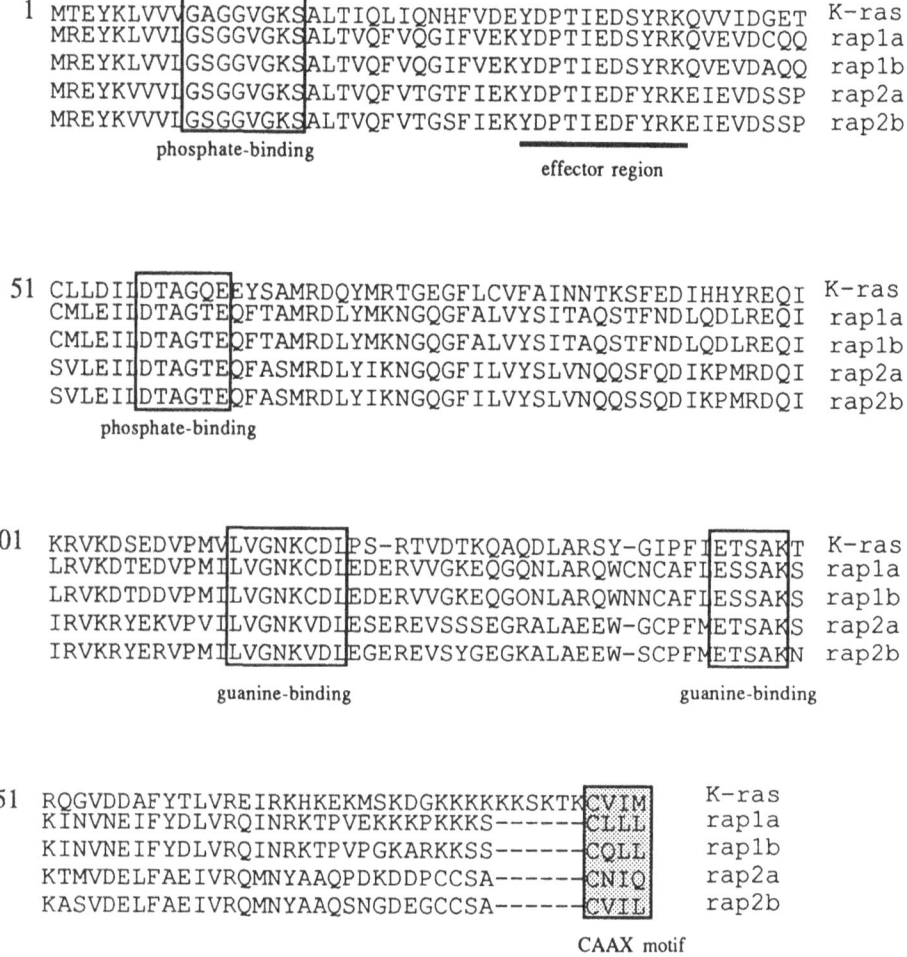

Figure 1. Aligment of rap proteins with Kirsten ras. Phosphate and guanine binding regions are indicated by boxes. The effector region is shown by the solid line and CAAX is denoted by the shaded box.

Rap proteins like all GTP-binding proteins cycle between an active GTP-bound and inactive GDP-bound state. The hydrolysis of GTP to GDP can occur by the intrinsic low GTPase activity of the protein itself. For example, the intrinsic GTPase activities of rap2b and ras have rate constants of (0.2×10^{-2} min^{-1}) and (2.6×10^{-2} min^{-1}), respectively (Molina et al., 1990). Proteins have been identifed which stimulate this intrinsic GTPase activity (Halenbeck et al., 1990). In addition to proteins which stimulate the intrinsic GTPase activity (GAP), other proteins have been described which affect the exchange of GDP for

GTP in the molecule (Downward et al., 1990b). Proteins which both promote and inhibit nucleotide exchange have been described (Araki et al., 1990).

GTPase Activating Proteins

The GAP molecule which stimulates the GTPase activity of ras is the most characterized to date (McCormick, 1989; Vogel et al., 1988). Several other GAP molecules have been described including rap1GAP, rhoGAP, and racGAP (Burstein et al., 1991; Polakis et al., 1991). It is postulated that GAP molecules specific for other ras superfamily members will be identified. The cDNA sequences for rasGAP and rap1GAP have been cloned and sequenced (Vogel et al., 1988; Rubinfeld et al., 1991). The deduced amino acid sequence of rap1GAP is not homologous to any known protein sequences including rasGAP.

Besides its GTPase activating activity towards ras, the ras GAP protein has been shown to play a complex role in cellular signaling. The rasGAP protein is a 120 kDa protein expressed in a wide variety of cells (Trahey et al., 1988). It has been shown that the GTPase activity resides in the C-terminal portion of the molecule specifically amino acids 714 to 1047 (Gideon et al., 1992). Interestingly, the non-catalytic region of rasGAP contains regions homologous with the Src kinase family (Koch et al., 1991). These regions have been designated as Src homology 2 (SH2) and Src homology 3 (SH3) domains. The SH2 domain(s) have also been identified in phospholipases C-γ1 and -2, the v-Crk oncoprotein, the p85 subunit of phosphatidylinositol 3-kinase (PI3K) and tensin (Seidel-Dugan et al., 1992). The SH3 region has been found in several cytoskeletal or membrane associated proteins including an actin-binding protein and two neutrophil NADPH oxidase-associated proteins (Koch et al., 1991). The SH2 domains bind phosphotyrosine residues on several growth factor receptors including platelet-derived growth factor receptor (Anderson et al., 1990). It is thought that SH2 domains will exhibit specificity to distinct phosphotyrosine residues.

Another mechanism by which rasGAP exerts a signaling potential is through its interaction with two tightly associated proteins, p62 and p190 (Wong et al., 1992; Settleman et al., 1992). These two tyrosine phosphorylated proteins have been recently cloned and shown to bind both DNA and RNA. It is postulated that these associated proteins may transmit signals impinging on the cell surface to the nucleus through rasGAP. The p190 protein has been shown to contain strong sequence similarity to both rhoGAP and the glucocorticoid repressor protein.

The role of rap proteins on platelet function

As stated above, platelets contain rap1 in high abundance. Furthermore, when platelets were treated with the prostacyclin analog iloprost, an agonist which increases cAMP levels we observed a shift of the protein to an apparent molecular weight of 24 kDa (Lapetina et al., 1989).

The shift of rap1b to an apparent molecular weight of 24 kDa upon increasing cAMP levels was also seen in human erythroleukemia (HEL) cells (Lazarowski et al., 1989). HEL cells possess a number of platelet-like features including granules that morphologically resemble the platelet alpha granule. The shift was observed when HEL cells were treated with prostacyclin, or theophylline and dibutyryl cAMP. On the other hand, no change in mobility was observed when HEL cells were treated with thrombin, phorbol dibutyrate, or the Ca^{2+} ionophore A23187 (Lazarowski et al., 1989). This result suggests that rap1b is not a substrate of protein kinase C.

Interestingly, this shift in molecular weight was accompanied by its translocation from the particulate to the cytosolic fraction (Siess et al., 1990). We concluded that this shift in molecular weight was due to phosphorylation by a cAMP dependent protein kinase. Further studies revealed that the ratio of phosphate to protein using recombinant rap1b and protein kinase A is 1:1 and identified serine 179 as the site of phosphorylation.

One of the first insights of a possible function of rap1b was observed in HEL cells (Lazarowski et al., 1990). We observed that the increase in phosphorylation of rap1b by the addition of iloprost is concomitant with a decrease in the phopholipase C-induced formation of inositol phosphates. This result suggested that rap1b might have a functional role in the regulation of phospholipase C-γ1 and the metabolism of the inositol phospholipids (Lapetina, 1990). This model has been postulated given that thrombin leads to the activation of phospholipase C-γ1, which hydrolizes phosphatidylinositol 4,5-bisphosphate to produce two second messengers, 1,2-diacylglycerol and 1,4,5-trisphosphate. Phosphorylation of rap1b by the action of the cyclic AMP-dependent protein kinase seems to uncouple the thrombin receptor from phospholipase C-γ1.

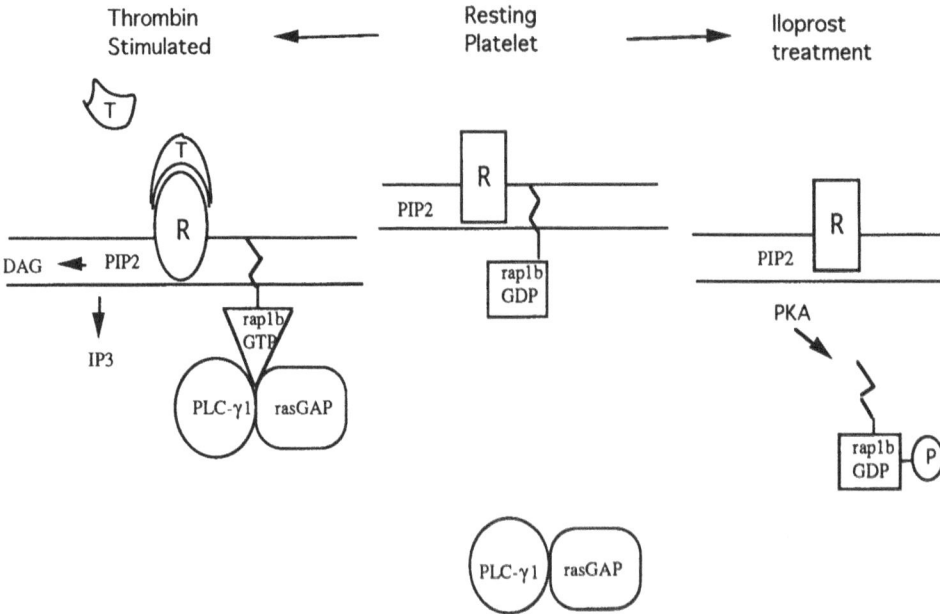

Figure 2. Proposed model of rap1b in platelet function after thrombin or iloprost treatment. In resting platelets, rap1b is found associated with the membrane by virtue of its isoprenoid moiety. Thrombin treatment increases the GTP bound state of rap1b producing a subsequent association with cytosolic rasGAP/PLC-γ1. When resting platelets are treated with iloprost, a compound which increases cAMP levels, rap1b becomes phosphorylated by a cAMP dependent kinase (PKA). This phosphorylation is accompanied by a translocation of the protein from the membrane to the cytosol. R, thrombin receptor; T, thrombin; PIP2, phosphatidylinositol bisphosphate; DAG, diacylglycerol; IP3, inositol trisphosphate.

Thrombin induces an association of Phospholipase C-γ1 and ras GTPase activating protein with rap1b

As stated above thrombin causes rapid activation of phospholipase C-γ1 in human platelets. Phospholipase C plays a role in growth factor signal transduction and has been shown to associate with the activated PDGF receptor, PI3-kinase, and ras GTPase activating protein (Anderson et al., 1990). In both resting and thrombin-stimulated platelets we observed a significant amount of phospholipase C-γ1 in anti rasGAP immunoprecipitates

(Torti and Lapetina, 1992). In addition, we observed that the phospholipase C-γ1 associated with rasGAP is phosphorylated. Further experiments showed that the phosphorylation of platelet PLC-γ1 was not on tyrosine residues and was thrombin independent, i.e., the level of phosphorylation was the same in control and thrombin stimulated platelets. These results suggest that the phosphorylation of PLC-γ1 is the signal for its association with rasGAP. That thrombin does not increase phosphorylation on tyrosine residues of PLC-γ1 yet increases rasGAP tyrosine phosphorylation suggests that the interaction is through SH2 groups of PLC-1 and phosphotyrosine residues on rasGAP. Moreover, since the tyrosine phosphorylated proteins p190 and p62 are constitutively associated with rasGAP, their phosphotyrosine groups may also serve as a bridge between the SH2 domains of the two proteins (Moran et al., 1990).

Previously, it had been shown that rasGAP binds to rap1b with high affinity yet does not increase the intrinsic GTPase activity of the protein (Frech et al., 1990). We sought to investigate if rap1b and rasGAP are associated in platelets owing to the fact that platelets contain a very small amount of ras. We observed that rap1b forms a complex with rasGAP and phospholipase C-γ1 in thrombin stimulated platelets (Torti and Lapetina, 1992). We concluded that rasGAP and phospholipase C-γ1 are physically associated in quiesent platelets. Upon thrombin treatment, which increases the GTP bound state of rap1b, a tertiary complex of rasGAP/phospholipase C-γ1 is formed with rap1b. We propose that rap1b in platelets may serve as a signal to recruit phospholipase C-γ1 to the cell membrane where it can act upon membrane-bound inositol phospholipids, see figure 2. Moreover, the data suggests that rap1b may play an important role in antagonistic signaling pathways when platelets are stimulated with iloprost or thrombin.

Erythropoietin induces ras activation and rasGAP tyrosine phosphorylation in HEL cells

Although HEL cells possess many platelet characteristics, they differ in that they express constitutive levels of p21ras. Previous results have shown that the hematopoietic growth factor, erythropoietin induces a rapid tyrosine phosphorylation of several membrane-associated proteins (Quelle and Wojchowski, 1991). This result demonstrates that tyrosine kinase activation is an early event associated with erythropoietin interaction with its receptor. Since ras is involved in transducing signals from normal and oncogenic tyrosine kinases, we sought to determine if erythropoietin exerts a direct effect on p21ras.

In erythropoietin stimulated HEL cells, an increase in the activated GTP bound•ras state is observed (Torti et al., 1992). This erythropoietin induced activation is very rapid and was strongly correlated with the erythropoietin-induced tyrosine phosphorylation. In addition, the p21ras•GTP complex is inhibited by the tyrosine kinase inhibitor, genistein.

Concomitant with the erythropoietin induced activation of ras, a decrease in rasGAP activity was observed (Torti et al., 1992). Analogous to ras activation, this erythropoietin induced inhibition of rasGAP was rapid and transient. Moreover, when lysates from erythropoietin stimulated HEL cells were blotted and probed with both antiphosphotyrosine and rasGAP antibodies, a time dependent tyrosine phosphorylation of rasGAP was observed (Torti et al., 1992). Phosphorylation of rasGAP was maximal at 1 minute. This result suggested that the erythropoietin induced inhibition of rasGAP was due to tyrosine phosphorylation.

CONCLUSION

Platelets contain several low molecular weight GTP-binding proteins including a high abundance of rap proteins. Surprisingly, platelets are mostly devoid of p21ras, yet contain a significant amount of rasGAP. Rap1 proteins have been shown to bind rasGAP with high affinity without a subsequent activation of GTPase activity. Upon thrombin stimulation of platelets a rap1b/rasGAP complex is observed. In platelets, rasGAP is constitutively associated with PLC-γ1. Therefore, thrombin by activation of rap1b into its active GTP bound state produces a rap1b/rasGAP/PLC-γ1 complex at the membrane. This produces an increase in hydrolysis of membrane-bound inositol phospholipids by the action of PLC-γ1. An opposite effect is observed upon iloprost treatment of platelets. Iloprost causes phosphorylation of rap1b by cAMP-dependent protein kinase. This phosphorylation induces dissociation of rap1b from the membrane. This dissociation event most likely causes a lower percentage of PLC-γ1 to be associated at the membrane producing a inhibition of PLC-γ1 upon iloprost treatment.

REFERENCES

Anderson, D., Koch, C. A., Grey, L., Ellis, C., Moran, M. F., and Pawson, T. 1990, Binding of SH2 domains of phospholipase C-γ1, GAP, and Src to activated growth factor receptors, *Science* 250:979.

Araki, S., Kikuchi, A., Hata, Y., Isomura, M., and Takai, Y. 1990, Regulation of reversible binding of smg p25A, a *ras* p21-like GTP-binding protein, to synaptic plasma membranes and vesicles by its specific regulatory protein, GDP dissociation inhibitor, *J. Biol. Chem.* 265:13007.

Burstein, E. S., Linko-Stentz, K., Lu, Z., and Macara, I. G. 1991, Regulation of the GTPase activity of the *ras*-like protein p25^{rab3a}, *J. Biol. Chem.* 266:2689.

Downward, J. 1990a, The ras superfamily of small GTP-binding proteins, *Trends Biochem. Sci.* 15:469.

Downward, J., Riehl, R., Wu, L., and Weinberg, R. A. 1990b, Identification of a nucleotide exchange-promoting activity for p21^{ras}, *Proc. Natl. Acad. Sci. USA.* 87:5998.

Frech, M., John, J., Pizon, V., Chardin, P., Tavitian, A., Clark, R., McCormick, F., and Wittinghofer, A. 1990, Inhibition of GTPase activating protein stimulation of Ras-p21 GTPase by the Krev-1 gene product, *Science* 249:169.

Gideon, P., John, J., Frech, M., Lautwein, A., Clark, R., Scheffler, J. E., and Wittinghofer, A. 1992, Mutational and kinetic analyses of the GTPase-activating protein (GAP)-p21 Interaction: The C-terminal domain of GAP is not sufficient for full activity, *Mol. Cell Biol.* 12:2050.

Halenback, R., Crosier, W. J., Clark, R., McCormick, F., and Koths, K. 1990, Purification, characterization and western blot analysis of human GTPase-activating protein from native and recombinant sources, *J. Biol. Chem.* 265:21922.

Hall, A. 1990a, ras and GAP-who's controlling whom? *Cell* 61:921.

Hall, A. 1990b, The cellular functions of small GTP-binding proteins, *Science* 249:635.

Hancock, J. F., Magee, A. I., Childs, J.E., and Marshall, C. J., 1989, All ras proteins are polyisoprenylated but only some are palmitoylated, *Cell* 57:1167.

Hart, M. J., Shinjo, K., Hall. A., Evans, T., and Cerione, R, A. 1991, Identification of the human platelet GTPase activating protein for the CDC42Hs protein, *J. Biol. Chem.* 266:20840.

Koch, C. A., Anderson, D., Moran, M. F., Ellis, C., and Pawson, T. 1991, SH2 and SH3 domains: elements that control interactions of cytoplasmic signaling proteins, *Science* 252:668.

Lapetina, E. G., Lacal, J. C., Reep, B. R. and Molina, L. yV. 1989, A ras-related protein is phosphorylated and translocated by agonists that increase cAMP levels in human platelets, *Proc. Natl. Acad. Sci. USA* 86:3131.

Lapetina, E. G. 1990, The signal transduction induced by thrombin in human platelets, *FEBS Letts.* 286:400.

Lazarowski, E. R., Lacal, J. C., and Lapetina, E. G. 1989, Agonist-induced phosphorylation of an immunologically ras-related protein in human erythroleukemia cells, *Biochem. Biophys. Res. Comm.* 161:972.

Lazarowski, E. R., Winegar, D. A., Nolan, R. D., Oberdisse, E., and Lapetina, E. G. 1990, Effect of protein kinase A on inositide metabolism and rap1 G-protein in human erythroleukemia cells, *J. Biol. Chem.* 265:13118.

McCormick, F. 1989, *ras* GTPase activating protein: signal transmitter and signal terminator, *Cell* 56:5.

Molina, L. yV., Ohmstede, C.-A., and Lapetina, E. G. 1990, Properties of the exchange rate of guanine nucleotides to the novel rap2b protein. *Biochem. Biophys. Res. Comm.* 171:319.

Moran, M. F., Koch, C. A., Anderson, D., Ellis, C., England, L., Martin, G. S., and Pawson, T.

1990, Src homology region 2 domains direct protein-protein interactions in signal transduction, *Proc. Natl. Acad. Sci. USA* 87:8622.

Nagata, K.-I., Nagao, S., and Nozawa, Y. 1989, Low Mr GTP-binding proteins in human platelets; cyclic AMP-dependent protein kinase phosphorylates m22KG(I) in membrane but not c21KG in cytosol, *Biochem. Biophys. Res.Comm.* 160:235.

Ohmstede, C.-A., Farrell, F. X., Reep, B. R., Clemetson, K. J., and Lapetina, E. G. (1990). Rap-2B: A ras-related GTP-binding protein from platelets, *Proc. Natl. Acad. Sci USA.* 87:6527.

Pizon, V., Chardin, P., Lerosey, I., Olofsson, B. and Tavitian, A. 1988, Human cDNAs rap1 and rap2 homologous to the Drosophila gene Dras3 encode proteins closely related to ras in the "effector" region, *Oncogene* 3:201.

Polakis, P. G., Weber, R. F., Nevins, B., Didsbury, J. R., Evans, T., and Synderman, R. 1989, Identification of the *ral* and *rac*1 gene products, low molecular mass GTP-binding proteins from human platelets, *J. Biol. Chem.* 264:16383.

Polakis, P. G., Rubinfeld, B., Evans T., and McCormick, F. (1991). Purification of plasma membrane-associated GTPase-activating protein specific for rap1/Krev-1 from HL-60 cells, *Proc. Natl. Acad. Sci. USA* 88:239.

Quelle, F. W. and Wojchowski D. M. 1991, Proliferative action of erythropoietin is associated with rapid protein tyrosine phosphorylation in responsive B6SUt.EP cells, *J. Biol. Chem.* 266:609.

Rubinfeld, B., Munemitsu, S., Clark, R., Conroy, L., Watt, K., Crosier, W. J., McCormick, F. and Polakis, P. 1991, Molecular cloning of a GTPase activating protein specific for the *Krev*-1 protein p21^{rap1}, *Cell* 65:1033.

Seidel-Dugan, C., Meyer, B. E., Thomas, S. M. and Brugge, J. S. 1992, Effects of SH2 and SH3 deletions on the functional activities of wild-type and transforming variants of c-Src, *Mol. Cell. Biol.* 12:1835.

Siess, W., Winegar, D. A. and Lapetina, E. G. 1990, Rap1b is phosphorylated by protein kinase A in intact human platelets, *Biochem. Biophys. Res. Comm.* 170:944.

Settleman, J., Narasimhan, V., Foster, L. C. and Weinberg, R. A. 1992, Molecular cloning of cDNAs encoding the GAP-associated protein p190: Implications for a signaling pathway from ras to the nucleus, *Cell* 69:539.

Torti, M., Bencke Marti, K., Altschuler, D., Yamamoto, K. and Lapetina, E. G. 1992, Erythropoietin induces p21ras activation and p120GAP tyrosine phosphorylation in human erythroleukemia cells, *J. Biol. Chem.* 267:8293.

Torti, M. and Lapetina, E. G. 1992, The role of rap1b and p21ras GTPase-activating protein in the regulation of phospholipase C-γ1 in human platelets, *Proc. Natl. Acad. Sci. USA.* 891:7796.

Trahey, M., Wong, G., Halenbeck, R., Rubinfeld, B., Martin, G. A., Lander, M., Long, C. M., Crosier, W. J., Watt, K., Koths, K. and McCormick, F. 1988, Molecular cloning of two types of GAP complementary DNA from human placenta, *Science* 242:1697.

Vogel, U. S., Dixon, R. A. F., Schaber, M. D., Diehl, R. E., Marshall, M. S., Scolnick, E. M., Sigal, I. S. and Gibbs, J. B. 1988, Cloning of bovine GAP and its interaction with oncogenic *ras* p21, *Nature* 335:90.

Wong, G., Muller, O., Clark, R., Conroy, L., Moran, M. F., Polakis, P. and McCormick, F, 1992, Molecular cloning and nucleic acid binding properties of the GAP-associated tyrosine phosphoprotein p62, *Cell* 69:551.

CALCIUM SIGNALLING AND PHOSPHOINOSITIDE METABOLISM IN PLATELETS: SUBSECOND EVENTS REVEALED BY QUENCHED-FLOW TECHNIQUES

Adrian R.L. Gear and Sanghamitra Raha

Department of Biochemistry
University of Virginia
Charlottesville, VA 22908, U.S.A.

INTRODUCTION

Thrombin and adenosine diphosphate (ADP) are perhaps the two most important activators of platelet function (Stormorken, 1986). Both agents elicit a rapid "shape change" where platelet discoid shape is lost and pseudopodia are extended, followed by aggregation. Thrombin also induces massive granule secretion under a variety of conditions, while ADP is much less potent (Mustard et al., 1975). There is therefore considerable functional similarity for both agents and much research has been directed to understanding the biochemical basis for how they cause the initial events in signal transduction, leading to "shape change", activation of the glycoprotein IIb/IIIa receptor with binding of fibrinogen, and resulting platelet aggregation.

A critical question for these biochemical studies has been the linkage between receptor occupancy and the major increase in intraplatelet calcium caused by ADP and thrombin, from about 0.1 μM to 1 μM $[Ca^{2+}]_i$ or more (Hallam et al., 1984; Hallam and Rink, 1985; Davies et al., 1990). With the exciting developments in inositol phospholipid and calcium metabolism (Michell, 1975) and the role of 1,4,5-inositol trisphosphate ($InsP_3$) in releasing intracellular calcium (Streb et al., 1983), it was natural that researchers evaluated whether thrombin and ADP both caused release of 1,4,5-$InsP_3$. Since about 1985, published research has failed to reveal a consistent pattern. Fisher et al. (1985) found that ADP did not hydrolyse phosphatidylinositol-4,5-bisphosphate ($PtdInsP_2$) or liberate soluble inositol phosphates at 10 sec or later after platelet activation. In sharp contrast, thrombin induced significant changes, while both agonists caused similar elevation of $[Ca^{2+}]_i$ to greater than 1 μM by 10 sec, as sensed by the dye Quin-2. Therefore, Fisher et al. (1985) suggested that the increase in platelet $[Ca^{2+}]_i$ was not causally linked to hydrolysis of $PtdInsP_2$. A subsequent   report by Sweatt et al. (1986) also suggests that ADP does not directly stimulate formation of inositol phosphates, while thrombin does.

In line with this is a recent careful study of human platelets by Vickers et al. (1990) who found that ADP failed to change the chemical amount of PtdInsP$_2$ or increase [^{3}H] inositol-labelled InsP$_3$ at 10 sec after stimulation. They concluded that ADP did not stimulate phospholipase C (PLC).

Other research supports an ADP-induced activation of phospholipase C, since increases in phosphatidic acid occur in rat and rabbit platelet (Lloyd et al., 1973; Vickers et al., 1982; Vickers et al., 1986; Feliste et al., 1988). One study by Daniel et al. (1986) on human platelets directly reveals an ADP-induced liberation of inositol phosphates and shows that [^{32}P]-labelled 1,4,5-InsP$_3$ increased 1.7 fold at 5 sec and was unaffected by the presence or absence of external calcium. However, the increase in [Ca^{2+}]$_i$ in their platelet preparation was weak, only reaching about 150 nM in the absence of external calcium and about 290 nM in its presence, from a baseline of about 70 nM. These data need to be related to several studies which strongly suggest that ADP stimulates <u>interconversion</u> between the major phosphatidylinositol species (PtdIns, PtdInsP, PtdInsP$_2$) rather than activation of phospholipase C, to yield 1,4,5-InsP$_3$ (reviewed by Vickers et al., 1990). Thus, ADP can stimulate PtdIns turnover, but may not directly cause 1,4,5-InsP$_3$ formation, in distinction to thrombin where many reports have confirmed liberation of 1,4,5-InsP$_3$, degradation of PtdInsP$_2$ and formation of diglyceride (Rittenhouse-Simmons, 1979; Mauco et al., 1979; Bell and Majerus, 1980; Agranoff et al., 1983).

An important point in all these studies on phosphatidylinositol metabolism and calcium mobilization by ADP and thrombin concerns critical temporal relationships with the actual kinetics of platelet function. The earliest time point usually studied has been 10 sec after agonist addition, such as in Vickers et al. (1990). Sweatt et al. (1986) even measured inositol phosphate production after 5 min of platelet activation. However, a notable attempt to follow earlier kinetics was that of Daniel et al. (1987) who reported one and two-second time points for thrombin-stimulated human platelets.

During the last ten years new information on the kinetics of platelet function clearly shows that the "shape change" may be nearly complete within 2 sec after ADP or thrombin stimulation (Gear, 1984), and that aggregation is over 50% complete within 5 sec (Gear, 1982), when assayed as the loss of single particles. Even secretion of serotonin caused by thrombin can be maximal by 5 sec (Gear and Burke, 1982). Consequently, the kinetics of biochemical events responsible for the rapid functional phases must be even faster, especially for calcium signalling and phosphatidylinositol metabolism.

Two laboratories have been concerned in establishing the earliest calcium kinetics after platelet activation by ADP or thrombin. Our own work (Jones and Gear, 1988; Jones and Gear, 1990) has involved a continuous-flow approach with Indo-1 loaded platelets. That of Sage and Rink's laboratory used a stopped-flow technique with Fura-2 loaded platelets (Sage and Rink, 1986; Sage and Rink, 1987; Sage et al., 1990). These rapid-mixing techniques have enabled determination of accurate calcium kinetics during the important 0 to 2 sec time period, something not achieved with cuvette stirring and earlier use of Quin-2 or Fura-2 loaded platelets. The major conclusion of these groups is that both ADP and thrombin increase [Ca^{2+}]$_i$ from about 0.1 μM to 1 to 5 μM within 1 to 2 sec, a process barely affected by the absence of extracellular calcium. Consequently, the metabolic signals responsible for the agonist-induced [Ca^{2+}]$_i$ increase must have been generated within one second of cell activation. This question represents the specific focus for the research to be described here. In addition, our earlier discussion on whether ADP and thrombin both stimulate phospholipase C, is a critical related question. In other words, do both ADP and thrombin stimulate release of 1,4,5-InsP$_3$ with kinetics consistent for causing the rapid ($<$ 2 sec) increase in [Ca^{2+}]$_i$ detected by the continuous or stopped-flow approaches? The mechanisms of ADP-induced signal transduction have been a mystery for many years (Gachet and Cazenave, 1991) and the questions posed for this study may begin to lead to answers.

METHODS

A number of procedures have been employed to address the specific aims outlined above. Some of these procedures have been published and the new ones are described in full in a separate manuscript submitted for publication. A summary will be given here of the newer techniques.

Platelet Isolation, [³H]-Inositol Loading and Electroporation

Human platelets were isolated by differential centrifugation of venous blood anti-coagulated with acid-citrate-dextrose (ACD). Three brief centrifugations at 350 g of 3, 3 and 5 min respectively are used (Haver and Gear, 1981). To obtain washed platelets for electroporation, the platelet rich plasma (PRP) was centrifuged in the presence of indomethacin, apyrase and PGI_2 to minimise platelet activation (Jones and Gear, 1988). The resultant pellet was suspended in a potassium-rich medium containing ATP and is similar to that described by Authi et al. (1989). One critical change we made to this medium was to include Ca^{2+} at 0.2 μM free concentration, knowing the association constant for ATP at pH 7.4. The inclusion of calcium was essential to recover platelet reactivity to ADP, in contrast to thrombin which was much less sensitive to the presence or absence of calcium during electroporation and subsequent resealing. Based on trial experiments, we used 8 discharges at 20 KV/cm in a 1.3 ml cell possessing stainless-steel electrodes.

The electroporated platelets were then exposed to 40 $\mu Ci/ml$ of [³H]-D-*myo*inositol and resealed for 90 min at 37°C, before centrifuging and final suspension in an Eagles-Hepes buffer containing 0.5 mg/ml fibrinogen and 1.5 mM $CaCl_2$ to enable platelet aggregation. Full resealing of the cells was seen after about 15 min at 37°C, as monitored by resistive-volume decrease from about 12 to 14 fl down to the normal cell volume of 6 to 7 fl.

Quenched-Flow Aggregometry and Inositol Phosphate Extraction and Analysis

The basic experimental approach using electroporated platelets, the quenched-flow system (Gear, 1982) and subsequent HPLC analysis is illustrated in Fig 1. The critical part of the system is the reaction loop which usually has an internal diameter of 0.25 mm and flow rates are typically 6.7 μl sec $^{-1}$, giving shear stresses of about 50 dyn cm^{-2}, or less at lower pumping speeds. Platelets loaded with [³H]-inositol as described above were then pumped through the quenched-flow system to stop the state of platelet activation as short as 130 msec after stimulation with ADP or thrombin, and for up to 5 sec. This period was chosen to cover the time during which the large increase in $[Ca^{2+}]_i$ occurs (Jones and Gear, 1988). Our specific aim was to use $HClO_4$ as quenching acid in the quenched-flow apparatus and then extract and analyse the [³H]-labelled inositol phosphates by a flow-through radiation detector coupled to an HPLC system. Preliminary experiments determined that 40 μCi [³H]-inositol per ml of electroporated platelets was sufficient for detection of the various inositol phosphate isomers in 250 μl of washed, labelled cells. Control counts of 1,4,5-InsP₃ for non-activated cells were 155 $\pm$ 75 dpm/10⁹ platelets and we typically obtained a ten-fold increase after stimulation with ADP and thrombin (data shown below). Identification of 1,4,5-InsP₃ and 1,3,4,5-InsP₄ was by comparison with the retention times off the HPLC column using radiolabeled standards supplied by American Radiolabeled Chemicals, Inc. The standards were run both in the morning and late afternoon to ensure correct identification of the peaks. For ease of presentation, data are usually expressed as dpm per 10⁶ platelets. Each experimental day involved 2 static controls and 6 quenched-flow runs to cover five reaction times ranging from 0.13 to 1 sec,

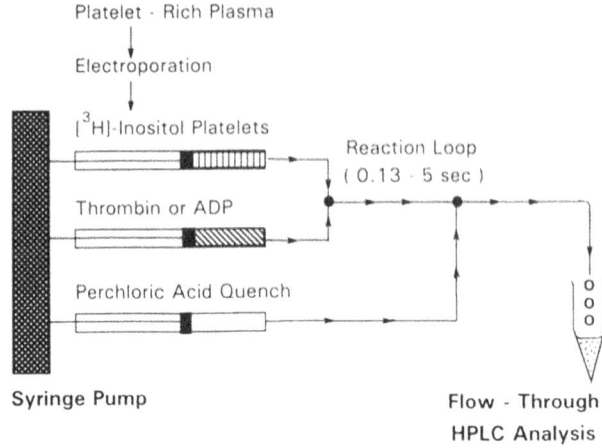

Figure 1   Quenched-flow approach to earliest events in [³H]-inositol phosphate synthesis in ADP and thrombin-activated platelets.

or from 1.5 to 3.5 sec, for the shear controls, ADP and thrombin situations. Values at each time point are usually means from 8 separate platelet preparations, although some values were derives from fewer numbers due to unavoidable losses.

Platelet aggregation kinetics during the first few seconds after agonist activation were assessed as the loss of single particles, following reaction in the quenched-flow system. The procedures were as described earlier (Gear, 1982), with data being expressed as the percent of single platelets aggregating at each reaction time, or as a rate in percent per second. These rates were 18 and 21 per cent per second for ADP and thrombin, respectively, values close to those seen in PRP (Gear, 1982).

Intracellular Platelet Calcium Analysis

This was done by pumping platelets loaded with Indo-1 through a continuous-flow system very similar in general principle to that described above in Fig 1. The "reaction loop" in this case was a transparent teflon tube which passed through a microcuvette placed in an SLM 4800 spectrofluorimeter (Urbana IL). The procedural details have been published and the data presented here were calculated from the original experiments (Jones and Gear, 1988). It is important to emphasize that the shear forces generated in this continuous-flow system as well as in the quenched-flow one used for following the changes in [³H]-inositol phosphate metabolism were the same. We normally generated shear forces of between 15 to 50 dyn cm^{-2} (Gear, 1982), and which are comparable to those seen in the microcirculation. As will be seen below, the ability to expose platelets to controlled shear forces was critical for observing liberation of inositol phosphates.

Data Evaluation and Presentation

Unless otherwise indicated, the data are expressed as means plus and minus the standard error of the mean, with numbers of individuals and preparations of platelets also provided. Tests of statistical difference between different time points were evaluated by Student's paired or unpaired t tests, when appropriate.

RESULTS

Comparison of ADP and Thrombin-Induced Calcium Increase within Two Seconds of Platelet Activation

We have found that ADP and thrombin induce essentially the same rapid increase in cytosolic $[Ca^{2+}]_i$ sensed by the dye Indo-1 and using the continuous-flow approach (Jones and Gear, 1988). These data are summarized in Fig 2, which also provides results on shear forces control. ADP (10 μM) and thrombin (5 U/ml) cause virtually identical increased increases which peaked at about 1 to 2 μM by 2 second. For ADP, there was then a slow decline, while a steady increase up to 5 μM or more by 8 sec was seen with thrombin (data not shown). The initial phase up to 2 sec was unaffected by the removal of extraplatelet $[Ca^{2+}]_i$ by EGTA (Jones and Gear, 1988), strongly suggesting that intraplatelet release of calcium was solely responsible (data not shown). Also, evident in Fig 2 is that shear forces acting alone caused only a very small increase over the resting platelet level of $[Ca^{2+}]_i$ of 104 $\pm$ 17 nM (n = 8).

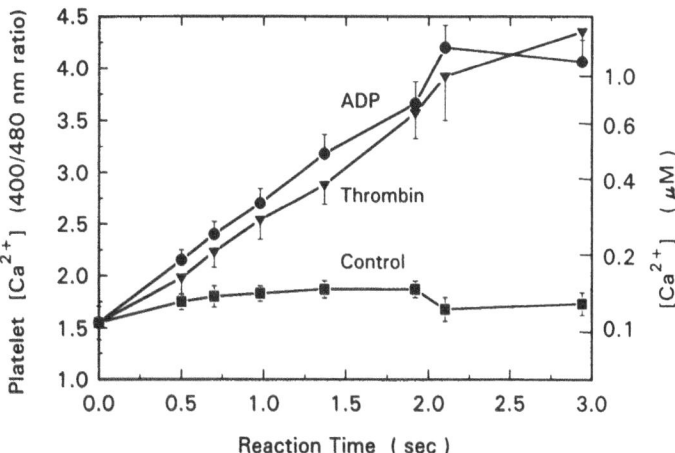

Figure 2   Continuous-flow Analysis of Early Calcium Changes in ADP and Thrombin-Activated Platelets. Values are expressed as the ratio of Indo-1 fluorescence at 400 to 480 nm, as well as $[Ca^{2+}]_i$ in μM.

Kinetics of 1,4,5-InsP$_3$ Formation During Shear, ADP and Thrombin Activation

The use of electroporated platelets to load enough [^{3}H]-inositol for subsequent detection proved essential for use of the flow-through HPLC radiation detector. Our initial experiments were began at 0.5 sec after activation with either thrombin in ADP. This time was chosen to precede the large increase in $[Ca^{2+}]_i$ detected by following the Indo-1 sensed changes (Fig 2). However, the first results revealed some preparations had already generated significant amounts of 1,4,5-InsP$_3$ and 1,3,4,5-InsP$_4$ at 0.5 sec. Therefore, subsequent runs were begun at 130 msec with additional times at 200 and 350 msec before continuing with 500, 750 and 1000 msec. Subsequently, we chose to analyse every 0.5 sec until 4 sec, with a final point at 5 sec after stimulation with ADP, thrombin or shear forces acting alone.

Our data are presented in Fig 3, 4 and 5 for shear forces acting alone, ADP (10 μM) and thrombin (1 U/ml), respectively. As outlined in the Introduction, our aim was to examine potential temporal correlations of changes in inositol-phosphates with those in cytoplasmic $[Ca^{2+}]_i$ as sensed by Indo-1. An initial surprise common to all three situations (shear forces, ADP or thrombin) was the finding of a very rapid 'burst' in synthesis of 1,4,5-InsP$_3$ before 0.5 sec, well before any significant increase in $[Ca^{2+}]_i$ occurred. Thrombin stimulated a peak at about 130 msec, with most platelet preparations reacting at this time, but some did peak at 200 msec. With ADP as agonist, the peak was a little later, nearer 200 msec, while a few preparations were maximal at 130 msec. The statistical

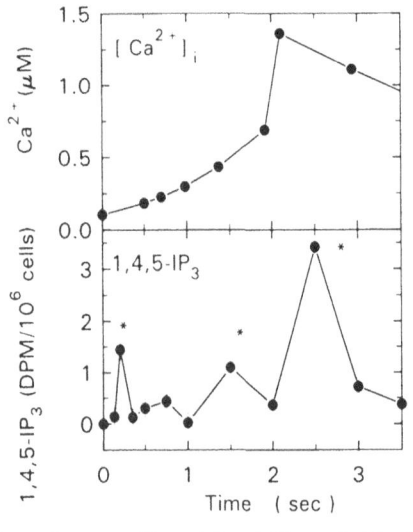

Figure 3   Shear-Induced Change in Platelet Calcium and 1,4,5-InsP$_3$.

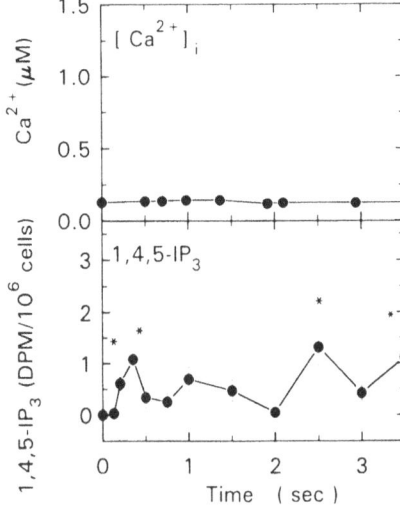

Figure 4   ADP-Induced Changes Calcium and 1,4,5-InsP$_3$.

significance of the peak versus the data points before or after them is indicated by an asterisk *, with p values of 0.05 or less. The data reveal that ADP and thrombin stimulated different patterns of 1,4,5-InsP$_3$ synthesis. Both agents caused a similar initial burst near 130 to 200 msec, but after this early rise, thrombin evoked a much greater second burst between 350 msec to 1 sec. This increase was highly significant, while none was noted with ADP as agonist. A direct comparison of the two agents is shown in Fig 6. After 1 sec, the pattern given by ADP was also distinct from that seen with thrombin. Transient increases in 1,4,5-InsP$_3$ were seen, but with different time courses. An important finding in this series of experiments is seen in Fig 3 where shear forces alone stimulated highly-significant increases in the inositol phosphate, but <u>failed</u> to raise intraplatelet calcium levels.

Early Synthesis of 1,3,4,5-InsP$_4$ During Shear, ADP and Thrombin Activation

Some ten minutes after elution of the inositol trisphosphates emerge a group of three inositol tetrakisphosphates. The data in Fig 7 reveal that both agonists caused an initial burst in 1,3,4,5-InsP$_4$ at 130 to 200 msec, very similar to that seen with the liberation of [^{3}H]-1,4,5-InsP$_3$ (Fig 6). Most intriguingly, the amounts formed as dpm [^{3}H] as well as the kinetics were almost identical for both compounds. It was impossible to detect whether 1,4,5-InsP$_3$ was liberated before 1,3,4,5-InsP$_4$. Following this initial transient increase, thrombin caused a highly-significant (P < .001) second peak at 500 msec, while ADP did not. This distinction is reminiscent of 1,4,5-InsP$_3$ (Fig 6). Subsequently, both agents elicited several peaks up to 3.5 sec of reaction in the quenched-flow system. Shear forces acting alone (Fig 8) caused a significant transient increase over 1000 dpm at 500 msec, but none later. Overall, formation of 1,3,4,5-InsP$_4$ until 1 sec was similar to that seen with either ADP or thrombin, yet no significant increase in [Ca^{2+}]$_i$ was detected by Indo-1 with shear forces alone. The data are shown here in the form of bar graphs for ready comparison of the three situations, and between the early 'burst' period of 0 to 350 msec and then up until 1 sec.

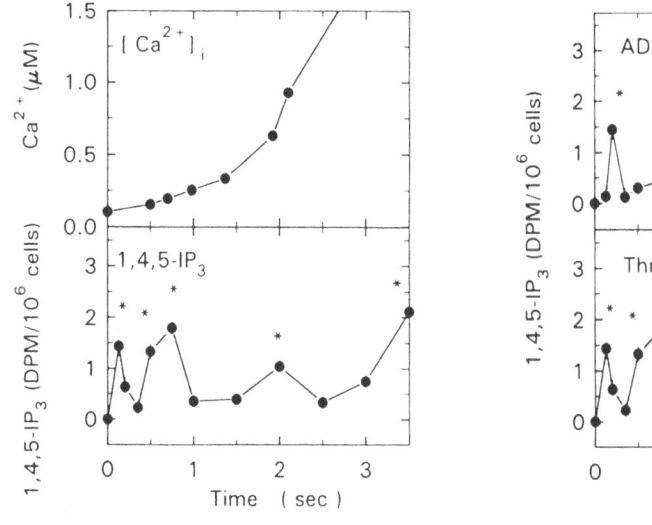

Figure 5   Thrombin-Induced Changes in Platelet Calcium and 1,4,5-InsP$_3$.

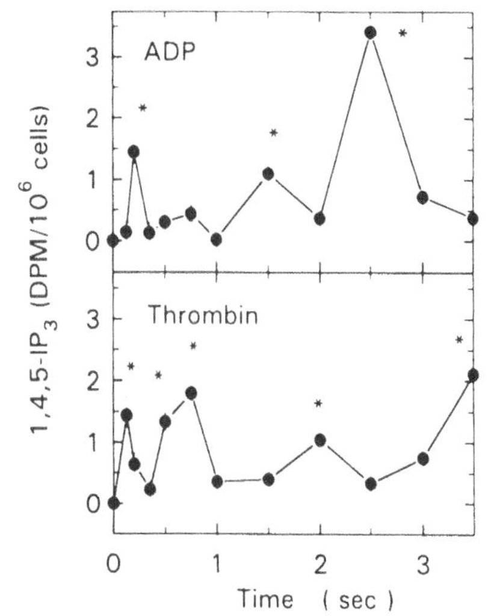

Figure 6   Comparison of ADP and Early Kinetics of 1,4,5-InsP$_3$ Formation.

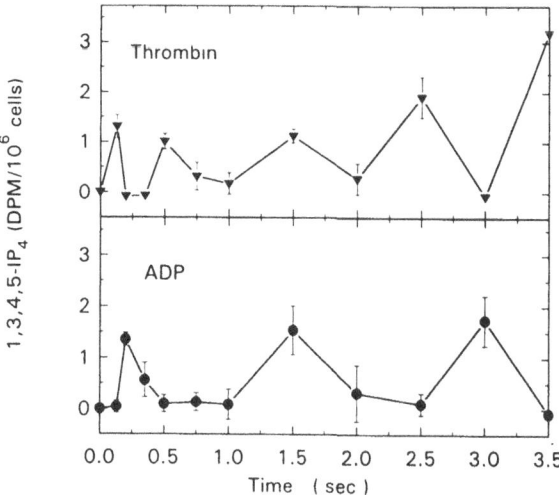

Figure 7   Comparison of ADP and Thrombin: Early Kinetics of 1,3,4,5-InsP$_4$ Formation.

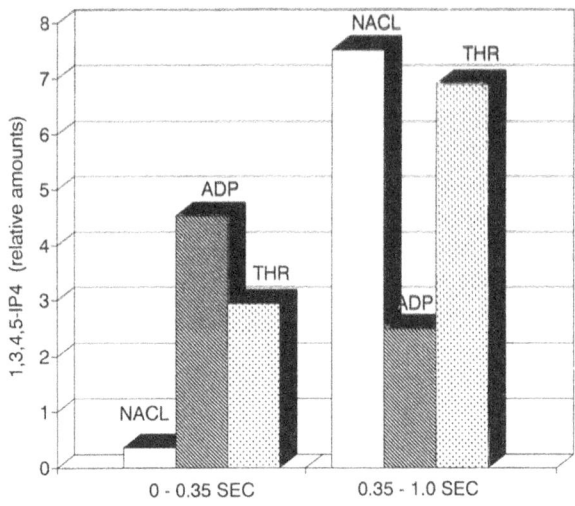

Figure 8   Overall Liberation of 1,3,4,5-InsP₄ by Shear Forces, ADP or Thrombin. Amounts produced from 0 to 0.35 sec and from 0.35 to 1.0 sec were integrated, respectively.

DISCUSSION

The data obtained with the quenched-flow system have revealed for the first time that platelets can liberate significant amounts of both 1,4,5-InsP₃ and 1,3,4,5-InsP₄ in a rapid transient 'burst', whether the stimulus is simply shear stress, or ADP, or thrombin (Fig 3-5). This burst before 500 msec was very similar in total amount for the three stimuli, yet only the two agonists ADP and thrombin induced a significant increase in cytoplasmic $[Ca^{2+}]_i$ (Fig 2). The data also show that apart from the similar initial peak near 130 to 200 msec for ADP and thrombin, the pattern and subsequent amounts of the 1,4,5-InsP₃ liberated were very different between 0.35 to 2 sec. Thus, our major question for this research remains a mystery; that is, why do both agonists cause virtually identical elevations in cytoplasmic $[Ca^{2+}]_i$ from about 0.1 to near 1 μM by 2 sec (Jones and Gear, 1988), a process dependent only on intracellular release? There is therefore no clear linkage between the kinetics or amounts of 1,4,5-InsP₃ formed and the increase in $[Ca^{2+}]_i$. The absence of any direct link is also emphasized by our finding that shear forces acting alone released the same amount of 1,4,5-InsP₃ as did ADP for up to 2 sec (Fig 3,4), yet only ADP increased $[Ca^{2+}]_i$.

It should be emphasized that we chose to express our results as "raw" data rather than by subtracting flowing control values from the ADP or thrombin situations. This is because the platelet experiences a combination of both shear force and agonist stimulation, and only this <u>combination</u> caused an increase in $[Ca^{2+}]_l$ (Fig.3,4). When flowing control values are subtracted from corresponding agonist-stimulated ones, highly significant peaks are seen for 1,4,5-InsP, at 0.13 and 0.13 and 0.2 sec for thrombin and ADP respectively (p < .001, data not shown). However, apparent negative values then occur at 0.35 sec (p < .02), which do not reflect the real situation experienced by the cell.

A second observation from our results is that both 1,4,5-InsP₃ and 1,3,4,5-InsP₄ were liberated in pulses of about 1 sec frequency. There must therefore be extremely rapid mechanisms for the liberation and subsequent removal of these inositol phosphates. This suggests exquisite regulation of not just phospholipase C, but also the various kinases and phosphatases involved in inositol phosphate metabolism (Downes and MacPhee, 1991).

Our observation of a highly-transient synthesis of the inositol phosphates, with half lives of less than a second parallels that of Breer et al. (1990) on cockroach olfactory signal transduction. They noted a single spike in 1,4,5-InsP$_3$ at about 50 msec after stimulation of antennal cell preparations by odorants. This research group used a quenched-flow system similar to ours, with perchloric acid as quenching agent, and found baseline levels were restored by 100 to 500 msec. Another research group (Harootunian et al., 1991) considered the linkage between transient increases in $[Ca^{2+}]_i$ in fibroblasts and 1,4,5-InsP$_3$ levels. Their evidence for inositol phosphate fluctuations was weak, with no statistics presented and a limited number of time points. A further attempt to demonstrate a direct linkage between $[Ca^{2+}]_i$ oscillations and oscillations in 1,4,5-InsP$_3$, was not successful in a study of pancreatic acinar cells (Wakui et al., 1989), since application of a non-hydrolysable InsP$_3$ analogue via a patch pipette caused fluctuations in $[Ca^{2+}]_i$ similar to those elicited by 1,4,5-InsP$_3$.

This discussion, together with our new results on rapid transients in platelet 1,4,5-InsP$_3$ as well as in 1,3,4,5-InsP$_4$, indicate no simple linkage between increase in the inositol phosphates and cytoplasmic $[Ca^{2+}]_i$. Several explanations are possible. The very rapid initial burst in 1,4,5-InsP$_3$ at 130 to 200 msec observed in control, ADP and thrombin stimulated platelets may release a small amount of internal calcium not sensed by Indo-1. It is conceivable that a local, calcium-induced calcium release may then occur before the large increase is sensed between 0.5 to 2 second in ADP and thrombin-stimulated cells. The important work of Finch et al. (1991) on the ability of varying $[Ca^{2+}]_i$ levels to modulate 1,4,5-InsP$_3$-induced calcium release is consistent with this possibility. Alternatively, other second messengers may exist which are more directly linked to intracellular $[Ca^{2+}]_i$ release. This possibility is supported by our finding that shear forces acting in the absence of ADP or thrombin generated substantial amounts of 1,4,5-InsP$_3$ during the first second, yet no increase in $[Ca^{2+}]_i$ was recorded (Fig 3). Therefore, it is likely that the presence of ADP or thrombin caused an unidentified second messenger to be liberated. A recent report by Murphy et al. (1991) also suggests PLC-independent mechanisms as being responsible. These workers noted poor correlations between the PAF-induced increase in rabbit-platelet $[Ca^{2+}]_i$, diglyceride and 1,4,5-InsP$_3$.

A final observation of considerable interest is our data on the parallel liberation of 1,4,5-InsP$_3$ and 1,3,4,5-InsP$_4$ at 130 to 200 msec, with either ADP or thrombin (Fig 4,5). The kinetics and amounts of the [^{3}H] compounds formed were nearly identical. This suggests that formation of the tetrakisphosphate from the trisphosphate via action of a 3-kinase (Downes and MacPhee, 1991) is unlikely, following initial breakdown of PtInsP$_2$ by PLC. It raises the possibility that a parent PtInsP$_3$ may be the substrate for a PLC which leads directly to liberation of 1,3,4,5-InsP$_4$. However, there is no evidence yet for such a mechanism and the function of the initial rapid synthesis of the tetrakisphosphate is also not at all clear.

In conclusion, we have found that both ADP and thrombin caused an extremely rapid 'burst' of inositol phosphate synthesis in platelets. We concur that ADP may indeed stimulate PLC, as suggested by Daniel et al. (1986), but it is also very clear that thrombin stimulated a much greater net liberation of 1,4,5-InsP$_3$ during the first critical second of platelet activation. There is still the enigma that both agonists elicited nearly identical increases in platelet $[Ca^{2+}]_i$ up to 2 sec. Also, shear forces acting alone failed to raise $[Ca^{2+}]_i$, even though comparable amounts of 1,4,5-InsP$_3$ were formed with ADP (Fig 3,4). This is demonstrated more clearly in Fig.8 where the total production of 1,4,5-InsP$_3$ caused by shear forces acting alone for 1 sec was almost the same as elicited by ADP in combination with shear forces. Only in the latter situation was $[Ca^{2+}]_i$ significantly elevated (Fig.4). Therefore, there is little correlation between overall inositol phosphate liberation and $[Ca^{2+}]_i$. These data strongly indicate that the signalling mechanisms for ADP and thrombin need to be pursued further.

Acknowledgements

This work was supported by the NIH (HL-27014). The critical help of Dr. Glen Jones is gratefully acknowledged for the calcium studies. Professor Neville Crawford and Dr. Kalwant Authi are especially thanked for valuable discussions and encouragement about the electroporation procedure.

REFERENCES

Agranoff, B.W., Murthy, P. and Seguin, E.B., 1983, Thrombin-induced phosphodiesteratic cleavage of phosphatidylinositol biphosphate in human platelets, *J Biol Chem*, 258:2076.

Authi, K.S., Hughes, K. and Crawford, N., 1989, High incorporation of [^{3}H] inositol into phosphoinositides of human platelets during reversible electropermeabilization, *FEBS Lett*, 254:52.

Bell, R.L. and Majerus, P.W., 1980, Thrombin-induced hydrolysis of phosphatidylinositol in human platelets, *J Biol Chem*, 255:1790.

Breer, H., Boekhoff, I. and Tareilus, E., 1990, Rapid kinetics of second messenger formation in olfactory transduction, *Nature*, 345:65.

Daniel, J.J., Dangelmaier, C.A. and Smith, J.B., 1987, Formation and metabolism of inositol 1,4,5-trisphosphate in human platelets, *Biochem J*, 246:109.

Daniel, J.L., Dangelmaier, C.A., Selak, M. and Smith, J.B., 1986, ADP stimulates IP$_3$ formation in human platelets, *FEBS Lett*, 206:299.

Davies, T.A., Weil, G.J. and Simons, E.R., 1990, Simultaneous flow cytometric measurements of thrombin-induced cytosolic pH and Ca^{2+} fluxes in human platelets, *J Biol Chem*, 265:11,522.

Downes, C.P. and MacPhee, C.H., 1991, Myo-inositol metabolites as cellular signals, *Eur J Biochem*, 193:1.

Feliste, R., Simon, M.F., Chap, H., Douste-Blazy, L., Defreyn, G. and Maffrand, J.P., 1988, Effect of PCR 4099 on ADP-induced calcium movements and phosphatidic acid production in rat platelets, *Biochem Pharmacol*, 37:2559.

Finch, E.A., Turner, T.J., and Golden, S.M., 1991, Calcium as a coagonist of inositol 1,4,5-trisphosphate-induced calcium release, *Science*, 252:443.

Fisher, G.J., Bakshian, S. and Baldassare, J.J., 1985, Activation of human platelets to ADP causes a rapid rise in cytosolic free calcium without hydrolysis of phosphatidylinositol-4,5-bisphosphate, *Biochem Biophys Res Commun*, 129:958.

Gachet, C. and Cazenave, J.P., 1991, ADP-induced blood platelet activation: a review, *Nouv Rev Fr Hematol*, 33:347.

Gear, A.R.L., 1982, Rapid reactions of platelets studies by a quenched-flow approach: aggregation kinetics, *J Lab Clin Med*, 100:866.

Gear, A.R.L. and Burke, D., 1982, Thrombin-induced secretion of serotonin from platelets can occur in seconds, *Blood*, 60:1231.

Gear, A.R.L., 1984, Rapid platelet morphological changes visualized by scanning-electron microscopy: Kinetics derived from a quenched-flow approach, *Br J Haematol*, 56:387.

Hallam, T.J., Sanchez, A. and Rink, T.J., 1984, Stimulus-response coupling in human platelets. Changes evoked by platelet-activating factor in cytoplasmic free calcium monitored with the fluorescent calcium indicator quin2, *Biochem J*, 218:819.

Hallam, T.J. and Rink, T.J., 1985, Responses to adenosine diphosphate in human platelets loaded with the fluorescent calcium indicator quin2, *J Physiol. (Lond)*, 368:131.

Harootunian, A.T., Kao, J.P.Y., Parenjape, S. and Tsien, R.Y., 1991, Generation of calcium oscillations in fibroblasts by positive feedback between calcium and IP$_3$, *Science*, 251:75.

Haver, V.M. and Gear, A.R.L., 1981, Functional fractionation of platelets, *J Lab Clin Med*, 97:187.

Jones, G.D. and Gear, A.R.L., 1988, Sub-second calcium dynamics in ADP and thrombin-stimulated platelets: A continuous-flow approach using Indo-1, *Blood*, 71:1539.

Jones, G.D. and Gear, A.R.L., 1990, Rapid blood platelet activation: Continuous and quenched-flow versus stopped-flow approaches, *Biochem J*, 265:305.

Lloyd, J.V., Nishizawa, E.E., Joist, J.H. and Mustard, J.F., 1973, Effect of ADP-induced aggregation on ^{32}PO$_4$ incorporation into phosphatidic acid and the phosphoinositides of rabbit platelets, *Br J Haematol*, 24:589.

Mauco, G., Chap, H. and Douste-Blazy, L., 1979, Characterization and properties of a phosphatidyl-inositol phosphodiesterase (phospholipase C) from platelet cytosol, *Febs Lett*, 100:367.

Michell, R.H., 1975, Inositol phospholipids and cell surface receptor function, *Biochim Biophys Acta*, 415:81.

Murphy, C.T., Elmore, M., Kellie, S. and Westwick, J., 1991, The relationship between cytosolic Ca^{2+}, sn-1,2-diacylglycerol and inositol 1,4,5-trisphosphate elevation in platelet-activating-factor-stimulated rabbit platelets, *Biochem J*, 278:255.

Mustard, J.F., Perry, D.W., Kinlough-Rathbone, R.L. and Packham, M.A., 1975, Factors responsible for ADP-induced release reaction of human platelets, *Am J Physiol*, 228:1757.

Rittenhouse-Simmons, S.E., 1979, Production of diglyceride from phosphatidylinositol in activated human platelets. *J Clin Invest*, 63:680.

Sage, S.O. and Rink, T.J., 1986, Kinetic differences between thrombin-induced and ADP-induced calcium influx and release from internal stores in fura-2-loaded human platelets. *Biochem Biophys Res Commun*, 136:1124.

Sage, S.O. and Rink, T.J., 1987, The kinetics of changes in intracellular calcium concentration in fura-2-loaded human platelets, *J Biol Chem*, 262:13364.

Sage, S.O., Reast, R. and Rink, T.J., 1990, ADP evokes biphasic Ca^{2+} influx in fura-2-loaded human platelets., *Biochem J*, 265:675.

Stormorken, H., 1986, Platelets in hemostasis and thrombosis, in: "Platelet responses and metabolism", Vol. 1, pp. 4-29, "Responses", H. Holmsen, ed. CRC Press, Inc., Boca Raton, Florida.

Streb, H., Irvine, R.F., Berridge, M.J. and Schulz, I., 1983, Release of Ca^{2+} from a nonmitochondrial intracellular store in pancreatic acinar cells by inositol-1,4,5-trisphosphate, *Nature*, 306:67.

Sweatt, J.D., Blair, I.A., Cragoe, E.J. and Limbird, L.E., 1986, Inhibitors of Na^+/H^+ exchange block epinephrine and ADP-induced stimulation of human platelet phospholipase C by blockade of arachidonic acid release at a prior step, *J Biol Chem*, 261:8660.

Vickers, J.D., Kinlough-Rathbone, R.L. and Mustard, J.F., 1982, Changes in phosphatidyl-4,5-bisphosphate 10 seconds after stimulation of washed rabbit platelets with ADP, *Blood*, 60:1247.

Vickers, J.D., Kinlough-Rathbone, R.L. and Mustard, J.F., 1986, The decrease in phosphatidylinositol-4,5-bisphosphate in ADP-stimulated washed rabbit platelets is not primarily due to phospholipase C activation, *Biochem J*, 237:327.

Vickers, J.D., Kinlough-Rathbone, R.L., Packham, M.A. and Mustard, J. F., 1990, Inositol phospholipid metabolism in human platelets stimulated by ADP, *Eur J Biochem*, 193:521.

Wakui, M., Potter, B.V.L. and Petersen, O.H., 1989, Pulsatile intracellular calcium release does not depend on fluctuations in inositol trisphosphate concentration, *Nature*, 339:317.

CALCIUM INFLUX MECHANISMS AND SIGNAL

ORGANISATION IN HUMAN PLATELETS

Stewart O. Sage, Paul Sargeant, Johan W.M. Heemskerk* and
Martyn P. Mahaut-Smith

The Physiological Laboratory
University of Cambridge
Downing Street
Cambridge CB3 9ET
U.K.

*Dept. of Biochemistry
University of Limburg
P.O. Box 616
6200 MD Maastricht
The Netherlands

INTRODUCTION

One of the key intracellular events in platelet activation is a rise in the cytosolic calcium concentration ($[Ca^{2+}]_i$). Ca^{2+} from two sources contributes to the signal: release from intracellular stores in the dense tubular system, and entry across the plasma membrane.

As in many other cell types, the link between receptor occupation and the release of Ca^{2+} from intracellular stores in platelets seems to be the formation of the diffusible messenger, inositol 1,4,5-trisphosphate (Ins 1,4,5-P_3). Ins 1,4,5-P3 is generated in response to most platelet agonists (Rink & Sage, 1990) and this messenger releases Ca^{2+} from permeabilised platelets and platelet microsomes (Brass & Joseph, 1985; Authi & Crawford, 1985). The only doubt surrounds the release of stored Ca^{2+} by ADP. Unlike other agonists, ADP evokes little (Daniel, Dangelmaier, Selak & Smith, 1986) or no (Fisher, Bakshian & Baldassare, 1985) Ins 1,4,5-P_3 formation, which may indicate an alternative transduction pathway.

The mechanisms by which agonists evoke Ca^{2+} entry in platelets are less well understood. It is generally accepted that these cells lack voltage-operated calcium channels (Rink & Sage, 1990) and that one or more forms of receptor mediated Ca^{2+} entry are activated by agonists. These mechanisms and the evidence for them form the major part of this chapter.

Mechanisms of Platelet Activation and Control, Edited by
K.S. Authi *et al.*, Plenum Press, New York, 1993

Until recently, most studies of Ca^{2+} signalling in platelets were conducted using suspensions of cells loaded with fluorescent Ca^{2+} indicator dyes. The signals recorded were thus averages of the $[Ca^{2+}]_i$ in thousands of cells. Recently, digital video imaging has been used to study details of Ca^{2+} signalling at the single platelet level. Agonists appear to evoke spikes or oscillations in $[Ca^{2+}]_i$ in platelets, as in many other cells. A discussion of single platelet work concludes this chapter.

CALCIUM INFLUX

Receptor-Mediated Calcium Entry

Receptor-mediated calcium entry has been proposed to take several forms. Receptor-operated channels are opened directly by agonist-receptor binding, either by conformational events within subunits of the same protein, or by close coupling via a GTP-binding protein. Second messenger-operated channels are opened by a diffusible messenger, itself resulting from agonist-receptor binding. Finally, Ca^{2+} entry may be regulated by the state of filling of the intracellular calcium store (Putney, 1990). There is some evidence to support the presence of all three classes of receptor-mediated calcium entry in human platelets.

Clues From Stopped-Flow Fluorimetry

Important clues as to the types of mechanism employed by agonists in generating Ca^{2+} entry have been gained from stopped- flow studies using fura-2-loaded platelets (Sage & Rink, 1987; Sage, Merritt, Hallam & Rink, 1989; Sage, Reast & Rink, 1990). These experiments allow the rapid mixing of cells with agonist (<30 ms) and the continous recording of the fluorescence signal on a 10 ms timescale (Sage & Rink, 1987). The stopped-flow procedure does not activate or damage the cells (Sage & Rink, 1990) and has the advantage over continuous flow methods (Jones & Gear, 1990) of collecting data over the critical first few hundreds of ms of activation.

Using the stopped-flow technique, ADP-evoked Ca^{2+} entry was shown to commence without measurable delay (i.e. within the 30 ms mixing period) (Fig. 1a). The rapid onset of the ADP response suggested that ADP might be activating a receptor-operated channel (Sage & Rink, 1987). The same entry pathway was shown to be permeable to Mn^{2+} (Sage et al, 1989), Ba^{2+} (Mahaut-Smith, Sage & Rink, 1990), and Na+ (Sage, Rink & Mahaut-Smith, 1991), but blocked by Ni^{2+}.

Stopped-flow records of the ADP-evoked response at 37°C indicated that the rise in $[Ca^{2+}]_i$ was biphasic, with a first, fast phase, commencing without measurable delay, and a later acceleration in the rate of rise after about 200 ms. The biphasic nature of the response was more easily studied at 17°C, when the first phase continued to be evoked without measurable delay, but the later phase was delayed in onset from agonist addition by about 800 ms (fig. 1b) (Sage et al, 1990).

To monitor Ca^{2+} entry independently of release from stores, Mn^{2+} was used as a tracer for Ca^{2+}. Mn^{2+} binds fura-2, quenching its fluorescence. If the fluorescence of the dye is excited at its isobestic wavelength (the wavelength at which fluorecence is insensitive to $[Ca^{2+}]$), the Mn^{2+} quench can be monitored independently of changes in $[Ca^{2+}]_i$ (Sage et al, 1989). Stimulating platelets with ADP in the presence of Ca^{2+} and Mn^{2+} showed that both phases of the rise in $[Ca^{2+}]i$ coincided with Mn^{2+} quench (Fig. 1b). This indicates that both phases of the ADP response are associated with Ca^{2+} entry (Sage et al, 1990). The first, fast phase of ADP-evoked Ca^{2+} entry could be accounted for by a receptor-operated channel. What of the second, delayed phase?

Several pieces of evidence suggested that the delayed ADP-evoked entry might be linked to the emptying of the intracellular Ca^{2+} store. At 17°C, stimulation with ADP in the absence of extracellular Ca^{2+} evoked a rise in $[Ca^{2+}]_i$, due to the release of stored Ca^{2+}, after a delay of 1400 ms. The addition of Mn^{2+} (but not Ca^{2+}), reveals Mn^{2+} entry also delayed by 1400 ms (Sage et al, 1990). The temporal coincidence of the two events suggests a causal link. Further evidence comes from observations of the effects of elevating the levels of cAMP. High concentrations of forskolin, an activator of the catalytic subunit of adenylate cyclase, left the rapid entry of Ca^{2+} (and Mn^{2+}) intact, but completely inhibited the release of Ca^{2+} from internal stores. Under the same conditions the second phase of the ADP-evoked rise in $[Ca^{2+}]_i$ and the delayed phase of Mn^{2+} quench were totally abolished (Sage et al, 1990).

ADP was the only platelet agonist identified as being able to evoke Ca^{2+} entry without measurable delay. Stopped-flow experiments indicated that other agonists, including thrombin, platelet activating factor, vasopressin, and the thromboxane A_2 analogue, U46619, all evoked Ca^{2+} influx with an irreducible delay of at least 200 ms (Fig. 1c) (Sage & Rink, 1987). This suggests that several biochemical steps may lie between receptor occupation and the generation of Ca^{2+} entry by these agonists. These results would, therefore, be compatible with agonists other than ADP generating Ca^{2+} entry via a second messenger-operated channel.

More recent work has endeavoured to test the hypotheses that ADP opens a receptor-operated channel whilst other agonists employ second-messenger operated channels, and that in addition, the emptying of the platelet intracellular Ca^{2+} store activates Ca^{2+} entry.

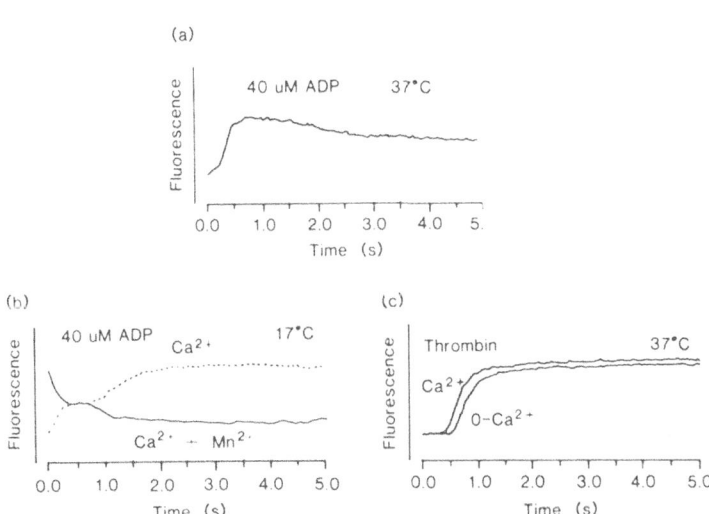

Figure 1   Stopped-flow fluorimetry with fura-2-loaded human platelets. (a) Record from platelets stimulated with 40 μM ADP in the presence of 1 mM external Ca^{2+} at 37°C. Excitation was at 340 nm, with an increase in fluorescence indicating a rise in $[Ca^{2+}]i$. (b) Stimulation with 40 μM ADP at 17°C, revealing a biphasic response. Broken trace shows fluorescence with excitation at 340 nm in presence of 1 mM external Ca^{2+}. Solid trace shows fluorescence with excitation at 360 nm to indicate Mn^{2+} quench (see text). External Mn^{2+} was 200 μM. (c) Records from platelets stimulated with 4 U/ml thrombin at 37°C, with 1 mM external Ca^{2+} or without added Ca^{2+}. Reproduced with kind permission of The Biochemical Society and The Portland Press.

Evidence for an ADP Receptor-Operated Channel

Stopped-flow fluorimetry indicates that ADP activates Ca^{2+} entry without a measurable delay, suggesting that this agonist might activate a receptor-operated channel. To further investigate the route by which ADP evokes rapid Ca^{2+} entry, electro-physiological recordings were employed to directly seek evidence for agonist-evoked channels.

In cell-attached platelet membrane patches, single channel currents were activated when ADP was included in the patch pipette (Fig 2a), but not when added to the bath (Mahaut-Smith, Sage & Rink, 1990). This indicates that the ADP-evoked channel is indeed receptor-operated, and does not require the generation of a diffusible messenger.

The activity of the ADP-evoked channel was essentially independent of membrane potential. The channel was non-selective and cation permeable: inward currents were carried by both Ba^{2+} and Na^+ at the resting potential and replacement of Cl^- with an impermeant anion was without effect. The channel is believed to be permeable to Ca^{2+} on the basis that Ba^{2+} is a good surrogate for Ca^{2+} in other Ca^{2+}-permeable channels and since stopped-flow studies indicate that ADP-evoked Ca^{2+} and Ba^{2+} entry have similar timecourses (Mahaut-Smith et al, 1990). In normal NaCl saline, containing 1 mM Ca^{2+}, the current-voltage relationship of the ADP-evoked current was curvilinear, with the single channel conductance increasing as the membrane was hyperpolarised from rest. It is not understood why this should be so.

To record ADP-evoked channels in cell-attached patches, a high concentration of ADP was added to the patch pipette shank and allowed to diffuse down to the tip during the experiment. This avoided desensitisation of the response before a seal was formed and recording conditions were established. A typical response consisted of 5-20 channel openings followed by desensitisation, usually within one minute of formation of the cell attached patch. This protocol did not, therefore, allow the determination of the kinetics of channel activation or inactivation, or estimation of the channel density.

To study these points, whole-cell patch clamp recordings were made using the nystatin permeabilised patch technique (Horn & Marty, 1988). Nystatin forms pores in the membrane patch which are selective for monovalent cations, giving good electrical access to the cytosol but avoiding dialysis of the cell, which is a problem with conventional whole-cell recording. Superfusion with ADP activated inward currents in platelets clamped at or near the resting potential (Fig. 2b) (Mahaut-Smith, Sage & Rink, 1992). The whole-cell current was activated with a latency as short as 7 ms and desensitised over 1-2 seconds. Single channel openings with a conductance of 15 pS at the resting potential, similar to those observed in cell-attached patches, could be resolved in the decaying phase of the whole-cell current.

The precise selectivity of the ADP-evoked channel is uncertain because outward currents could not be identified in cell-attached or whole-cell recordings, so preventing accurate determination of the reversal potential. The estimated reversal potential was close to 0 mV, indicative of a non-selective channel, and the single channel conductance was similar in nominally Ca^{2+}-free and Ca^{2+}-containing Na^+ salines. These findings suggest that the channel conducts a significant Na^+ current under physiological conditions.

This predicted Na^+ permeability was in agreement with measurements of ADP-evoked changes in cytosolic Na^+ concentration ($[Na^+]_i$) in platelets loaded with the fluorescent Na^+ indicator, SFBI (Sage et al, 1991). Stopped-flow measurements showed that the ADP-evoked rise in $[Na^+]_i$ commenced without measurable delay and peaked within 1 second, a similar timecourse to that of the rise in $[Ca^{2+}]_i$. The measured ADP-evoked rise in $[Na^+]_i$ was from about 6 mM to about 18 mM. Given a single channel conductance of 15 pS, a driving force for Na^+ entry of 100 mV and a surface membrane area of 20 um^2, this provides an estimate for channel density of 33 per platelet, assuming

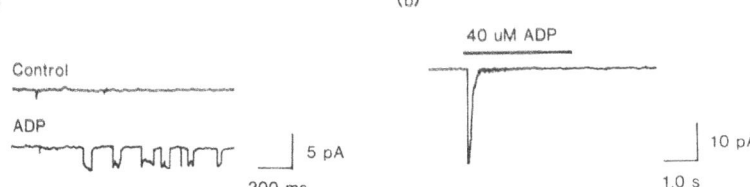

Figure 2   Platelet electrophysiology. (a) Membrane current records from cell attached platelet membrane patch. The control record shows the lack of currents in the absence of agonist, 45 s after seal formation. The lower record shows single channel openings 45 s after sealing to a pipette back-filled with ADP. The pipette potential was 100 mV (patch potential is membrane potential - pipette potential). The external medium contained 145 mM Na^+ and 1 mM Ca^{2+}. (b) Whole-cell current evoked by 40 μM and recorded using a nystatin permeabilised patch. External medium as for (a). The pipette contained 150 mM KCl with 50 μM nystatin. The holding potential was -93 mV. Reproduced by kind permission of The American Society for Biochemistry and Molecular Biology, Inc.

the channels are open for 20% of the time, as observed in cell-attached recordings (Mahaut- Smith et al, 1991). The peak ADP-evoked whole-cell current varied between platelets, with a range of 13 - 31 pA at -73 mV. This corresponded to between 15 and 35 channels per cell, in good agreement with the estimate from Na^+ flux experiments.

Evidence For Store-Regulated Calcium Entry

Various pieces of evidence from stopped-flow work indicated that ADP may activate a component of Ca^{2+} entry which is caused by the discharge of the intracellular Ca^{2+} store (Sage et al, 1990). A similar explanation was offered for the delayed nature of thrombin-evoked Mn^{2+} entry, which lags behind store release (Sage et al, 1989). To test more directly for store-regulated Ca^{2+} entry, use was made of the inhibitors of the endomembrane Ca^{2+}-ATPase, thapsigargin (TG; Jackson, Patterson, Thastrup & Hanley, 1988) and 5-di-(t-butyl)-1,4-benzohydroquinone (tBuBHQ; Kass, Duddy, Moore & Orrenius, 1989). These compounds have been shown to lead to depletion of the intracellular Ca^{2+} stores, by opposing the resequestration of leaked Ca^{2+}, in various cells including platelets (Thastrup, Foder & Scharff, 1987; Brune & Ullrich, 1991). This depletion has been shown to activate Ca^{2+} entry in many cells (Putney, 1990).

The application of TG or tBuBHQ to platelet suspensions in the absence of external Ca^{2+} evoked only small rises in $[Ca^{2+}]_i$ and the subsequent addition of agonists evoked a substantial release of stored Ca^{2+} (Sargeant, Clarkson, Sage & Heemskerk, 1992). This indicated that even when exposed to the Ca^{2+}-ATPase inhibitors for up to two hours, the agonist-releasable Ca^{2+} stores were only partially depleted under the conditions of these experiments. Similar results have been reported by others, who also observed only partial store depletion in platelets in which cyclooxygenase was inhibited (Brune & Ullrich, 1991).

Although TG and tBuBHQ only evoked partial store depletion and a small rise in $[Ca^{2+}]_i$ in the absence of external Ca^{2+}, much larger rises in $[Ca^{2+}]_i$ were observed in the presence of external Ca^{2+}, indicating that store-depletion leads to Ca^{2+} entry (Fig. 3) (Sargeant et al, 1992). Store-activated divalent cation entry was confirmed using the Mn^{2+} quench technique.

The Route For Store-Regulated Ca^{2+} Entry

When store-regulated, or "capacitative" Ca^{2+} entry was first proposed, the model postulated the passage of Ca^{2+} across a restricted area of the cytosol between the plasma

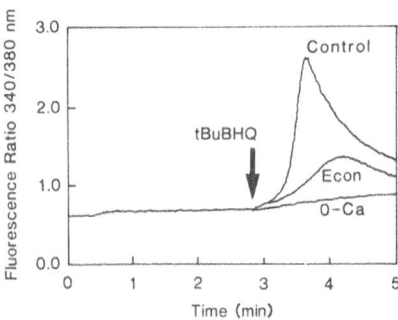

Figure 3   Effects of 50 μM tBuBHQ on fura-2-loaded platelets in the presence of 1 mM external Ca^{2+} (upper two records) or with 1 mM EGTA (lower record). Where indicated, the cells were preincubated with 3 μM econazole for 5 min. Traces show the fura-2 340/380 nm fluorescence ratio, indicating $[Ca^{2+}]_i$.

membrane and the intracellular store (Putney, 1986). A modification was later proposed on the basis of refilling experiments in parotid acinar cells (Merritt & Rink, 1987). In this type of experiment, the cells are stimulated in the absence of external Ca^{2+}, depleting the intracellular Ca^{2+} store. They are then exposed to external Ca^{2+} after the agonist has been washed away or its receptors blocked. Ca^{2+} is then removed again, and the cells re-stimulated to show that the stores have refilled. On finding that during the refilling period $[Ca^{2+}]_i$ was not significantly elevated, it was suggested that there might be a direct route for Ca^{2+} entry from the external space into the intracellular store, with this perhaps being a Ca^{2+} regulated structure like a gap junction (Merritt & Rink, 1987).

Subsequent work has argued against direct store-regulated Ca^{2+} entry, favouring instead initial entry into the cytosol and subsequent sequestration by the store (Putney, 1990). For example, it has been shown that increasing cytosolic Ca^{2+} buffering using the chelator BAPTA, slows the refilling of the intracellular Ca^{2+} stores in parotid acinar cells (Muallem, Khademazad & Sachs, 1990). This result can only be explained if Ca^{2+} enters the store after passage through the cytosol. Similarly, in lacrimal acinar cells, refilling type protocols have been used to show that store-depletion promotes the entry of Ba^{2+} and Sr^{2+}, as well as Ca^{2+}, into the cells (Kwan & Putney, 1990). However, only Ca^{2+} and Sr^{2+}, but not Ba^{2+}, enters the stores, indicating initial entry into the cytosol.

To date there is only limited information concerning the route of store-regulated Ca^{2+} entry in platelets. This is a difficult phenomenon to investigate in these cells, because activation and aggregation make refilling experiments problematic. In the one reported study of this type, in which platelets were activated and washed many times, store depletion was shown to promote the entry of Ba^{2+} and Sr^{2+}, as well as that of Ca^{2+} (Ozaki, Yatomi & Kume, 1992). In contrast to the lacrimal gland results, all three ions were shown to enter the stores. This could be explained by different selectivities of the internal store Ca^{2+}-ATPase or release channels, or by direct entry of divalent ions into the store when depleted.

The Coupling Of Store-Depletion To Increased Ca^{2+} Entry

The mechanism by which the depletion of intracellular Ca^{2+} stores increases plasma membrane Ca^{2+} permeability is not known. A model proposing physical coupling between the plasma membrane and that of the intracellular store has been put forward (Irvine, 1990). This suggests that the Ins 1,4,5-P$_3$ receptor in the membrane of the intracellular store is in contact with another protein in the plasma membrane, and that this second protein is the receptor for inositol 1,3,4,5-tetrakisphosphate (Ins 1,3,4,5-P_4). The two receptors would thus resemble the ryanodine and dihydropyridine receptors believed to be responsible for Ca^{2+} release from the sarcoplasmic reticulum in excitation-contraction coupling in skeletal muscle (Agnew, 1987).

In the non-excitable cell model, Ca^{2+} influx is suggested to be generated when the two receptor proteins dissociate. Dissociation is proposed to be promoted by the binding of Ins 1,4,5-P_3 or Ins 1,3,4,5-P_4 to their receptors, and by a reduction in the [Ca^{2+}] in the lumen of the intracellular store (Irvine, 1990). This model is attractive in that it might explain the rapid coupling between Ca^{2+} release and the second phase of entry apparent in stopped-flow recordings from ADP-stimulated platelets (Sage et al, 1990). As yet, however, there is no experimental evidence to directly support this physical coupling hypothesis in platelets or any other cell type.

It has been suggested that cytosolic and stored Ca^{2+} might antagonistically control tyrosine phosphorylation of specific platelet proteins, which may in turn regulate the Ca^{2+} permeability of the plasma membrane (Vostal, Jackman & Schulman, 1991). The proposal is that elevated [Ca^{2+}]$_i$ activates a tyrosine kinase, so increasing Ca^{2+} entry. A tyrosine phosphatase is then activated when the intracellular Ca^{2+} store is refilled, so reversing this. Such a system could account for store-regulated Ca^{2+} entry, but any contribution of a mechanism of this type to agonist-evoked events remains to be demonstrated.

Another proposal is that the link between store emptying and increased Ca^{2+} entry is microsomal cytochrome P-450 or one of its metabolites (Alvarez, Montero & Garcia-Sancho, 1992). This model is based on the observation that inhibitors of cytochrome P-450 reduce Mn^{2+} entry in cells which have been Ca^{2+} depleted in low Ca^{2+} medium. Alonso and colleagues (1991) have reported this phenomenon in platelets, and also demonstrated the inhibition of agonist-evoked Mn^{2+} entry by the cytochrome P-450 inhibitors, econazole and miconazole, in these cells.

Our own laboratory too has looked at the possible role of cytochrome P-450 in platelets, also using the imidazole antimycotics, econazole and miconazole. Like Alonso et al (1991), we find that these compounds inhibit Mn^{2+} entry in platelets which have been Ca^{2+} depleted in low Ca^{2+} medium (Sargeant et al, 1992). At a concentration of 3 μM, econazole reduced Mn^{2+} entry in Ca^{2+} depleted cells to a similar degree as preincubation with external Ca^{2+}, which refills the intracellular Ca^{2+} stores. The same concentration of econazole significantly inhibited Ca^{2+} (and Mn^{2+}) entry evoked by tBuBHQ (Fig. 3) (Sargeant et al, 1992). However, the imadazole antimycotics were found to be markedly less effective in inhibiting the Ca^{2+} and Mn^{2+} entry evoked by agonists.

For example, econazole only reduced ADP-evoked Mn^{2+} entry by about 40% compared with the control value (determined 20 s after agonist addition, when the rate of Mn^{2+} entry had returned to the pre-stimulus, basal leak rate). This suggests against a key role for cytochrome P-450 in generating agonist-evoked entry as claimed by Alonso et al (1991). An important finding was that even when a reduction in ADP-evoked Mn^{2+} entry was apparent, the Ca^{2+} signal was not measurably affected by econazole (Sage, Sargeant, Merritt, Mahaut-Smith & Rink, 1992). This suggests that both Mn^{2+} permeable and impermeable routes for Ca^{2+} entry may be present in platelets, as previously proposed (Sage et al, 1989). Recently, a calcium current activated by depletion of intracellular Ca^{2+} stores has been reported in mast cells (Hoth & Penner, 1992). This store-dependent influx pathway was found to be highly selective for Ca^{2+}, with little or no permeability to Mn^{2+}, Ba^{2+} or Sr^{2+}, highlighting the possible misinterpretation of data if it is assumed that Mn^{2+} acts as a surrogate for Ca^{2+} through all pathways. Alonso et al (1991) report no [Ca^{2+}]$_i$ data obtained in the presence of external Ca^{2+}, relying on Mn^{2+} quench data alone.

In our hands, the cytochrome inhibitors were also relatively ineffective in inhibiting Mn^{2+} entry evoked by vasopressin, PAF and thrombin, reducing these responses by only 13%, 15% and 7% respectively (Sargeant et al., 1992). Alonso et al (1991) have also suggested that the lipoxygenase inhibitor, nordihydroguaiaretic acid, may inhibit Ca^{2+} entry because it inhibits cytochrome P-450. We find that this compound is without effect on ADP-evoked Ca^{2+} signals at concentrations below which it itself affects $[Ca^{2+}]_i$ (Vindlacheruvu, Rink & Sage, 1991).

Our results with cytochrome inhibitors are strikingly different from those of Alonso et al. (1991), who reported essentially complete inhibition by econazole of Mn^{2+} entry evoked by ADP, thrombin and PAF. A possible explanation of these conflicting findings is the difference in the method of cell preparation used by the two groups. Alonso et al (1991) loaded their cells with fura-2 in a washed suspension, whilst our group loads in citrated plasma (Sargeant et al, 1992). Also, we determine Mn^{2+} entry in the presence of external Ca^{2+}, when the intracellular Ca^{2+} stores should be full. Alonso et al conducted Mn^{2+} experiments in the absence of external Ca^{2+}. Considering this group also load cells in the absence of external Ca^{2+}, significant depletion of the intracellular stores would be expected to occur. This might sensitize any store-regulated, cytochrome-dependent pathway and increase its contribution to agonist-evoked responses compared with that occuring physiologically. Parallel preparation of cells by the two methods indicates that the sensitivity to econazole is increased in double-washed, Ca^{2+} depleted cells (Sargeant & Sage, unpublished observations).

To further test for a possible role for cytochrome P-450, we have looked at the effects of econazole on the early stages of agonist-evoked rises in $[Ca^{2+}]_i$ and Mn^{2+} entry using stopped flow fluorimetry (Sargeant et al, 1992). No effects were discernable over the first 10s of activation, during which time the rises in $[Ca^{2+}]_i$ evoked by ADP, thrombin and PAF all reach their peak.

Taken together, our data obtained with the cytochrome inhibitors suggest against any key role for cytochrome P-450 in generating at least the initial stages of agonist evoked Ca^{2+} signals, including the component of the ADP response that we have suggested may be store-regulated (Sage et al, 1990). However, we could not rule out a more long-term role for the cytochrome in maintaining the state of filling of the intracellular Ca^{2+} stores. A long term role has been proposed for the cytochrome in neutrophils (Montero, Alvarez & Garcia-Sancho, 1991).

Although store-regulated Ca^{2+} entry is believed to occur in platelets (Sage et al, 1990; Sargeant et al, 1992; Ozaki et al, 1992), as in other cells we seem to be some way from understanding how this mechanism operates.

Second Messenger-Operated Channels

Over the last five years or so, several receptor-operated channels have been clearly identified and store-regulated Ca^{2+} entry has been shown to occur in many cell types (Sage, 1992; Putney, 1990). Second messenger-operated channels have proved more elusive. A channel opened by Ins $1,4,5$-P_3 in lymphocytes (Kuno & Gardner, 1987) and one opened by Ca^{2+} in neutrophils (Von Tscharner, Prod'hom, Baggiolini & Reuter, 1986) have been reported, but these types of channel have not found widespread support (Meldolesi, Clementi, Fasolato, Zacchetti & Pozzan, 1991). A recent report of a second-messenger-operated channel in endothelial cells, activated by Ca^{2+} and Ins $1,3,4,5$-P_3 has, however, reopened interest in this type of entry mechanism (Lückhof & Clapham, 1992).

Stopped-flow fluorimetry indicates that Ca^{2+} entry evoked by platelet agonists other than ADP lags behind agonist addition by several hundred ms, but nevertheless slightly preceeds the release of Ca^{2+} from intracellular stores (Sage & Rink, 1987). This suggests that these agonists may open second messenger-operated channels, which could account for a component of Ca^{2+} entry preceeding any activated by discharge of the intracellular store. Possible support comes from the demonstration that Ins $1,4,5$-P_3, a putative intermediary which could open a second messenger-operated channel, releases Ca^{2+} from platelet plasma membrane vesicles (Rengasamy & Feinberg, 1988). However, contamination with internal membrane is difficult to exclude. Direct evidence for membrane currents evoked by agonists other than ADP in intact platelets has yet to be forthcoming.

Lack Of Thrombin-Evoked Membrane Currents

After the succesful detection of ADP-evoked currents in cell-attached and whole-cell patch clamp recordings, we turned to thrombin as an agonist representative of those which evoke delayed Ca^{2+} entry. Surprisingly, a 3s application of thrombin evoked no inward current detectable with the nystatin whole-cell patch technique (Mahaut-Smith et al, 1992). Application of ADP to the same cells evoked a typical inward whole-cell current. Stopped-flow experiments performed under the same conditions confirmed that thrombin-evoked Ca^{2+} and Mn^{2+} entry were detectable 1.1 s and 1.4 s after agonist addition respectively, indicating that the cells were responding within the timescale of the electrophysiological experiments.

Most receptor- and second messenger-operated channels reported to date are relatively non-selective, conducting significant Na^+ as well as Ca^{2+} currents under physiological conditions. The failure to detect any thrombin-evoked current in platelets might be explained by the opening of small conductance, highly selective divalent cation channels, beyond the resolution of our patch-clamp recordings. In fact, the Ca^{2+}-selective current activated by depletion of Ca^{2+} stores in mast cells (Hoth & Penner, 1992) was difficult to resolve at normal intracellular Ca^{2+} buffering capacity, and could only be clearly measured in whole cell patch clamp recordings after Ca^{2+} buffering was increased using EGTA. Given the smaller surface area of a single human platelet compared with a mast cell, a similar Ca^{2+}- selective current would be too small to be detected in our nystatin whole-platelet recordings with only the normal intracellular Ca^{2+} buffering capacity. Another explanation might be the activation of an electroneutral transporter, which allows, for example, Ca^{2+} to enter the cell in exchange for K^+ or H^+, or by co-transport with Cl^-. Supporting a Ca^{2+} selective pathway, we find that the thrombin-evoked rise in $[Na^+]_i$, determined in SBFI- loaded platelets, is slower than the onset of rise in $[Ca^{2+}]_i$ (Mahaut-Smith et al, 1992).

Regardless of the mechanism by which thrombin evokes the initial component of Ca^{2+} entry, stopped-flow measurements indicate that after a short additional lag, Ca^{2+} is released from internal stores (Fig. 1c). This would in turn be expected to evoke Ca^{2+} entry (Sargeant et al, 1992; Ozaki et al, 1992). This raises the question as to whether store-regulated Ca^{2+} entry could explain the whole of the influx generated by thrombin and similar agonists. This might be so, if it is assumed that the lumen of the intracellular store is connected to the extracellular space, and that the flux of Ca^{2+} across the store membrane is electrically silent due to a counter flux of K^+ or if the pathway is highly selective for Ca^{2+}. This would still leave the slightly earlier onset of the rise in $[Ca^{2+}]_i$ seen in stopped-flow experiments in the presence of external Ca^{2+}, compared with its absence, to be explained. These observations indicate that as entry commences before release. However, it is difficult to completely exclude the possibility that removal of external Ca^{2+} slows some component of the signal transduction system, or somewhat depletes the intracellular store, such that detectable internal release is delayed.

Zschauer and coworkers (1988) have reported the detection of a channel in vesicles from thrombin-stimulated platelets incorporated into artificial membranes that was absent in membranes from unstimulated cells. The channel was selective for Ba^{2+} over Na^+. Why this channel should remain open throughout the long membrane preparation procedure and how it is gated is not clear (Rink, 1988). The prolonged opening is suggestive of covalent modification. These reconstitution studies are rather indirect evidence for a thrombin-evoked channel, since it is unclear whether the channel originates from the plasma membrane or contaminating membrane from intracellular organelles. Our work, which failed to find thrombin-evoked currents in intact platelets, could be reconciled with the data of Zschauer et al (1988), if the channel they report has a much smaller conductance at physiological levels of external Ca^{2+}, below the resolution of our patch clamp recordings.

The complex emerging picture of receptor-mediated Ca^{2+} entry in platelets is summarised in Fig. 4.

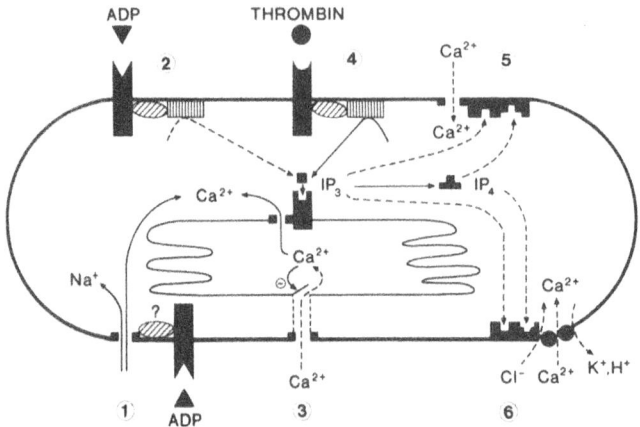

Figure 4   Proposed (————) and possible (- - -) platelet calcium entry mechanisms. ADP rapidly opens a receptor-operated non-selective cation channel, possible coupled by a G-protein (1). ADP also releases Ca^{2+} from an intracellular store (2). This release of stored Ca^{2+} results in further Ca^{2+} entry by an unidentified route (3). The link between the store and the plasma membrane might be physical or chemical. Thrombin (and other agonists) release Ca^{2+} from the intracellular store via the production of Ins $1,4,5$-P_3 (4). This may also lead to secondary Ca^{2+} entry (3). Since thrombin-evoked entry appears to precede release, other pathways may exist. The delayed and electrically undetectable nature of the response could be explained by a Ca^{2+} selective channel (5) or some sort of counter- or co- transporter (6), either of which might be activated by diffusible messengers such as Ins $1,4,5$-P_3 and or Ins $1,3,4,5$-P_4. Reproduced by kind permission of The American Physiological Society and The International Union of Physiological Societies.

TEMPORAL ORGANISATION OF THE PLATELET CALCIUM SIGNAL

Single Platelet Calcium Signals

In recent years, studies of Ca^{2+} signalling in a variety of non-excitable cells have revealed a complexity which was obscured in earlier population measurements. In many cells, agonists have been shown to evoke oscillations or spikes in $[Ca^{2+}]_i$, rather than the smooth changes suggested by population work (Berridge, 1990).

We have now investigated agonist-evoked responses in single, fura-2-loaded human platelets using digital video imaging (Heemskerk, Hoyland, Mason & Sage, 1992). A major difficulty in this approach was the need to immobilize the platelets for the fluorecence imaging technique. In an earlier attempt, ADP-evoked rises in $[Ca^{2+}]_i$ were demonstrated in platelets which had been allowed to attach to glass (Hallam, Poenie & Tsien, 1986). However, the cells were activated by contact with the glass prior to ADP addition, and no clear oscillatory responses were observed.

In our studies, we immobilized platelets on fibrinogen- coated coverslips

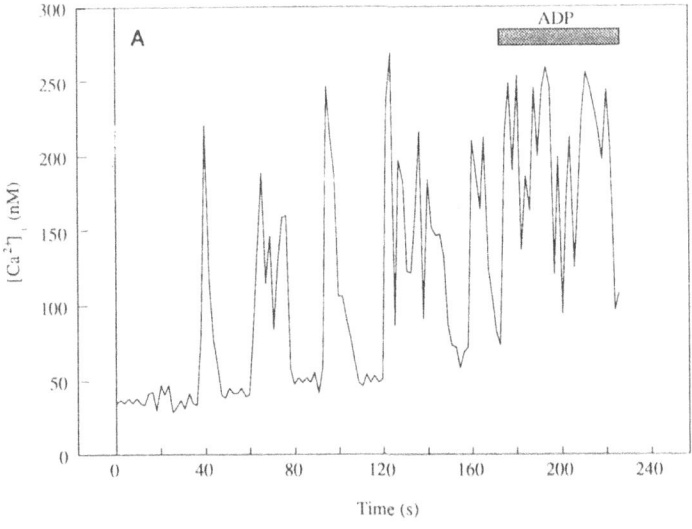

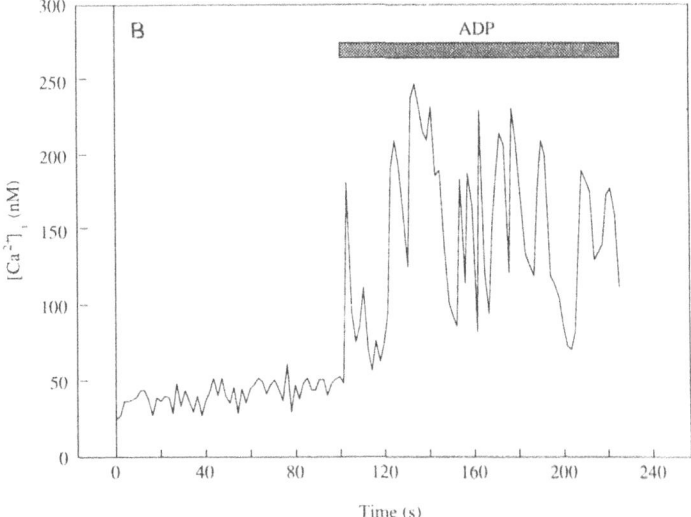

Figure 5   Oscillations in $[Ca^{2+}]_i$ in fibrinogen-bound human platelets. ADP (40 μM) was added as shown. Traces show responses of two platelets in the presence of 1 mM external Ca^{2+}. (a) shows a cell activated during attachment. (b) shows a cell which is only activated upon the addition of ADP.

(Heemskerk et al, 1992). This method secured the cells sufficiently for imaging, but many were activated during the attachment phase, and showed pronounced oscillations in $[Ca^{2+}]_i$ (Fig. 5A). Some platelets, however, appeared unactivated prior to agonist addition and responded to ADP with complex spikes in $[Ca^{2+}]_i$ (Fig. 5B). Activation during immobilization was prevented by the presence of prostacyclin, the fibrinogen receptor antagonist, Arg-Gly-Asp-Ser, and apyrase. We have subsequently shown that activation can be prevented by apyrase alone, indicating that it may result from the release of ADP from nearby platelets which are activated by contact with uncoated areas of glass (Heemskerk, Vis, Feijge, Hoyland, Mason & Sage, unpublished observations).

In platelets allowed to attach under inhibitory conditions, ADP (40 μM) evoked spikes in $[Ca^{2+}]_i$ with a frequency of 5/min in the presence of external Ca^{2+} (Heemskerk et al, 1992). Spikes with a slightly lower peak value and slightly lower frequency were also evoked by ADP in the absence of extracellular Ca^{2+}. No clear relationship between ADP concentration and spike frequency was observed.

Thrombin (0.1 U/ml), in the presence of external Ca^{2+}, evoked relatively long lasting elevations in $[Ca^{2+}]_i$ without oscillation. In other cell types, some agonists have been shown to evoke oscillations at low concentrations and sustained elevations in $[Ca^{2+}]_i$ at higher concentrations (e.g. Jacob, Merritt, Hallam & Rink, 1988). This seems to be the case with thrombin in platelets. At a concentration of 0.01U/ml, thrombin evokes spikes in $[Ca^{2+}]_i$ in both the presence and absence of external Ca^{2+} (Heemskerk, Vis, Feijge, Hoyland, Mason & Sage, unpublished observations).

Mechanism of Oscillation

In most cells, the basis of oscillation appears to be the periodic release of Ca^{2+} from intracellular stores. Since agonists evoke spiking in the absence of external Ca^{2+}, this also appears to be the case in human platelets. Interestingly, we observed spiking in the absence of external Ca^{2+} which continued for at least 5 minutes, whereas in many cells spiking ceases soon after external Ca^{2+} is removed (e,g. Jacob et al, 1988; Petersen, Gallacher, Wakui, Yule, Petersen & Toescu, 1991). This may be due to substantial intracellular Ca^{2+} stores in human platelets (Heemskerk, Feijge, Rietman & Hornstra, 1991), and the fact that stimulation by weak agonists results in little Ca^{2+} efflux, with most of the Ca^{2+} released from intracellular stores being resequestered (Rink & Sage, 1987). In a recent report of oscillations in $[Ca^{2+}]_i$ evoked by serotonin in rabbit platelets, external Ca^{2+} was shown to be essential (Nishio, Ikegami & Segawa, 1991). This may indicate that rabbit, like rat platelets, may have relatively small intracellular Ca^{2+} stores (Heemskerk et al, 1991).

The precise mechanism by which oscillations or spikes in $[Ca^{2+}]_i$ are generated is not understood. Several models have been proposed, which differ essentially in whether the Ins 1,4,5-P_3 concentration is assumed to oscillate or not (Berridge, 1990). In platelets, as in other cells, it is likely that oscillation is initiated by the formation of at least low concentrations of Ins 1,4,5-P_3, as reported to be stimulated by both low concentrations of thrombin (Billah & Lapetina, 1982) and ADP (Daniel et al., 1986). However, some workers report no detectable Ins 1,4,5-P_3 production in ADP-stimulated human platelets (Fisher et al., 1985). A possible explanation of this finding is that ADP-evoked Ins 1,4,5-P_3 production is small and transient (Rink & Sage, 1990). This suggests that sustained ADP-evoked Ca^{2+} spiking may occur in the absence of continued Ins 1,4,5-P_3 production, favouring a model of Ca^{2+}-induced Ca^{2+}-release. In tentative support of the presence of such a mechanism in human platelets, we find in single cell studies that the Ca^{2+}-ATPase inhibitor, thapsigargin, induces a sudden rise in $[Ca^{2+}]_i$, with a variable delay (Heemskerk, Vis, Feijge, Hoyland, Mason & Sage, unpublished observations). This suggests that a gradual rise in $[Ca^{2+}]_i$ in the presence of TG may be triggering a large and sudden store discharge by Ca^{2+}-induced Ca^{2+}-release.

Interestingly, a recent report describes oscillations in diacylglycerol (DAG) concentration in thrombin-stimulated suspensions of human platelets (Werner, Bielawska & Hannun, 1992). This is suggestive of periodic negative feedback by DAG on phospholipase C, leading to oscillations in DAG and so, presumably, Ins 1,4,5-P_3. This could in turn underlie thrombin-evoked oscillations in $[Ca^{2+}]_i$. This suggestion is all the more tempting since the thrombin-evoked oscillations in DAG concentration show a similar frequency to the oscillations in $[Ca^{2+}]_i$ evoked by this agonist. However, a population of platelets in suspension would be expected to show asynchronous responses, as seen in immobilized cells (Heemskerk et al., 1992). This accounts for why oscillations in $[Ca^{2+}]_i$ are not observed in platelet populations. It is therefore not clear what, if any, relationship exists between the reported oscillations in DAG and those in $[Ca^{2+}]_i$.

In platelets, as in other non-excitable cells, further single cell work is required if the mechanisms generating oscillations in [Ca^{2+}]i are to be understood.

Acknowledgements

S.O.S. is a Royal Society 1983 University Research Fellow. P.S. is an MRC Research Student.

REFERENCES

Agnew, W.S., 1987, Proteins that bridge the gap, *Nature* 334:299.

Alonso, M.T., Alvarez, J., Montero, M., Sanchez A. and Garcia-Sancho, J., 1991, Agonist-induced Ca^{2+} influx into human platelets is secondary to the emptying of intracellular Ca^{2+} stores, *Biochem. J.* 280:783.

Alvarez, J., Montero, M. and Garcia-Sancho, J., 1992, Cytochrome P450 may regulate plasma membrane Ca^{2+} permeability according to the filling state of the intracellular Ca^{2+} stores, *FASEB J.* 6:786.

Authi, K.S. & Crawford, N., 1985, Inositol 1,4,5-trisphosphate-induced release of sequestered Ca^{2+} from highly purified human platelet intracellular membranes, *Biochem. J.* 230:247.

Berridge, M.J., 1990, Calcium oscillations, *J. Biol. Chem.* 265:9583.

Billah, M.M. and Lapetina, E.G., 1982, Evidence for multiple metabolic pools of phosphatidylinositol in stimulated platelets, *J. Biol. Chem.* 257:11856.

Brass, L.F. and Joseph, S.K., 1985, A role for inositol trisphosphate in intracellular Ca^{2+} mobilisation and granule secretion in platelets, *J. Biol. Chem.* 260:15172.

Brüne, B. and Ullrich, V., 1991, Different calcium pools in human platelets and their role in thromboxane A$_2$ formation, *J. Biol. Chem.* 266:19232.

Daniel, J.L., Dangelmaier, C.A., Selak, M. and Smith, J.B., 1986, ADP stimulates IP$_3$ formation in human platelets, *FEBS Lett.* 206:299.

Fisher, G.J., Bakshian, S. and Baldassare, J.J., 1985, Activation of human platelets by ADP causes a rapid rise in cytosolic free Ca^{2+} without hydrolysis of phosphatidylinositol-4,5-bisphosphate, *Biochem. Biophys. Res. Commun.* 129:958.

Hallam, T.J., Poenie, M. and Tsien, R.Y., 1986, Homogeneity of ADP- and thrombin-stimulated rises in [Ca^{2+}]i in fura-2-loaded platelet populations revealed by fluorescence ratio image processing, *J. Physiol.* 377:123P.

Heemskerk, J.W.M., Feijge, M.A.H., Rietman, E. and Hornstra, G., 1991, Rat platelets are deficient in internal Ca^{2+} release and require influx of extracellular Ca^{2+} for activation, *FEBS Lett.* 284:223.

Heemskerk, J.W.M., Hoyland, J., Mason, W.T. and Sage, S.O., 1992, Spiking in cytosolic calcium concentration in single fibrinogen-bound fura-2-loaded human platelets, *Biochem. J.* 283:379.

Horn, R. and Marty, A., 1988, Muscarinic activation of ionic currents measured by a new whole-cell recording method, *J. Gen. Physiol.* 92:145.

Hoth, M. and Penner, R., 1992, Depletion of intracellular calcium stores activates a calcium current in mast cells, *Nature* 355:353.

Irvine, R.F., 1990, 'Quantal' Ca^{2+} release and the control of Ca2+ entry by inositol phosphates - a possible mechanism, *FEBS Lett.* 263:5.

Jackson, T.R., Patterson, S.I., Thastrup, O. and Hanley, M.R., 1988, A novel tumour promoter, thapsigargin, transiently increases cytoplasmic free Ca^{2+} without generation of inositol phosphates in NG115-401L neuronal cells, *Biochem. J.* 253:81.

Jacob, R., Merritt, J.E., Hallam, T.J. and Rink, T.J., 1988, Repetitive spikes in cytoplasmic calcium evoked by histamine in human endothelial cells, *Nature* 335:40.

Jones, G.D. and Gear, A.R.L., 1990, Rapid blood platelet activation: continuous- and quenched-flow versus stopped-flow approaches, *Biochem. J.* 265:305.

Kass, G.E.N., Duddy, S.K., Moore, G.A. and Orrenius, S., 1989, 2,5-di-(t-butyl)-1,4-benzohydroquinone elevates cytosolic Ca^{2+} concentration by mobilising the inositol 1,4,5-trisphosphate-sensitive Ca^{2+} pool, *J. Biol. Chem.* 264:15192.

Kuno, M., and Gardner, P., 1987, Ion channels activated by inositol 1,4,5- trisphosphate in plasma membrane of human T-lymphocytes, *Nature* 326:301.

Kwan, C.Y. and Putney, J.W., 1990, Uptake and intracellular sequestration of divalent cations in resting and methacholine-stimulated mouse lacrimal acinar cells, *J. Biol. Chem.* 265:678.

Lückhof, A. and Clapham, D.E., 1992, Inositol 1,3,4,5-tetrakisphosphate activates an endothelial Ca^{2+} permeable channel, *Nature* 355:356.

Mahaut-Smith, M.P., Sage, S.O. and Rink, T.J., 1990, Receptor-activated single channels in intact human platelets, *J. Biol. Chem.* 265:10479.

Mahaut-Smith, M.P., Sage, S.O. and Rink, T.J., 1992, Rapid ADP-evoked currents in human platelets recorded with the nystatin permeabilized patch technique, *J. Biol. Chem.* 267:3060.

Meldolesi, J., Clementi, E., Fasolato, C., Zacchetti, D. and Pozzan, T., 1991, Ca^{2+} influx following receptor activation, *Trends Pharm. Sci.* 12:289.

Merritt, J.E. and Rink, T.J., 1987, Regulation of cytosolic free calcium in fura-2-loaded rat parotid acinar cells, *J. Biol. Chem.* 262:17362.

Montero, M., Alvarez, J. and Garcia-Sancho, J., 1991, Agonist-induced Ca^{2+} influx in human neutrophils is secondary to the emptying of intracellular calcium stores, *Biochem. J.* 277:73.

Muallem, S., Khademazad, M. and Sachs, G., 1990, The route of Ca^{2+} entry during reloading of the intracellular Ca^{2+} pool in pancreatic acini, *J. Biol. Chem.* 265:2011.

Nishio, H., Ikegami, Y. and Segawa, T., 1991, Fluorescence digital image analysis of serotonin-induced calcium oscillations in single blood platelets, *Cell Calcium* 12:177.

Ozaki, Y., Yatomi, Y. and Kume, S., 1992, Evaluation of platelet calcium ion mobilization by the use of various divalent ions, *Cell Calcium* 13:19.

Petersen, O.H., Gallacher, D.V., Wakui, M., Yule, D.I., Petersen, C.C.H. and Toescu, S.C., 1991, Receptor-activated cytoplasmic Ca^{2+} oscillations in pancreatic acinar cells:Generation and spreading of Ca^{2+} signals, *Cell Calcium* 12:135.

Putney, J.W., 1986, A model for receptor-regulated Ca^{2+} entry, *Cell Calcium* 7:1.

Putney, J.W., 1990, Capacitative calcium entry revisited, *Cell Calcium* 11:611.

Rengasamy A. and Feinberg, H., 1988, Inositol 1,4,5-trisphosphate-induced calcium release from platelet plasma membrane vesicles, *Biochem. Biophys. Res. Commun.* 150:1021.

Rink, T.J., 1988, A real receptor-operated calcium channel?, *Nature* 334:649.

Rink T.J. and Sage, S.O., 1987, Stimulated calcium efflux from fura-2-loaded human platelets, *J. Physiol.* 393:513.

Rink, T.J. and Sage, S.O., 1990, Calcium signalling in human platelets, *Ann. Rev. Physiol.* 52:431.

Sage, S.O., 1992, Receptor-mediated calcium entry, *Current Biology 2* (in press).

Sage, S.O., Merritt, J.E., Hallam, T.J. and Rink, T.J., 1989, Receptor-mediated calcium entry in fura-2-loaded human platelets stimulated with ADP and thrombin. Dual wavelength studies with Mn^{2+}, *Biochem. J.* 258:923.

Sage, S.O., Reast, R. and Rink, T.J., 1990, ADP evokes biphasic calcium influx in fura-2-loaded human platelets. Evidence for calcium entry regulated by the intracellular calcium store, *Biochem. J.* 265:675

Sage, S.O. and Rink, T.J., 1987, The kinetics of changes in intracellular calcium concentration in fura-2-loaded human platelets, *J. Biol. Chem.* 262:16364.

Sage, S.O. and Rink, T.J., 1990, Reply to:'Rapid blood platelet activation: continuous- and quenched-flow versus stopped-flow approaches', *Biochem. J.* 265:306.

Sage, S.O., Rink, T.J. and Mahaut-Smith, M.P., 1991, Resting and ADP-evoked changes in cytosolic free sodium concentration in human platelets loaded with the indicator SBFI, *J. Physiol.* 441:559.

Sage, S.O., Sargeant, P., Merritt, J.E., Mahaut-Smith, M.P. and Rink, T.J., 1992, Agonist-evoked Ca^{2+} entry in human platelets, *Biochem. J.* (in press).

Sargeant, P., Clarkson, W.D., Sage, S.O. and Heemskerk, J.W.M., 1992, Calcium influx evoked by Ca^{2+} store depletion in human platelets is more susceptible to cytochrome P-450 inhibitors than receptor-mediated calcium entry, *Cell Calcium* (submitted).

Thastrup, O., Foder, B. and Scharff, O., 1987, The calcium mobilising and tumour promoting agent, thapsigargin, elevates the platelet cytoplasmic free calcium concentration to a higher steady state level. A possible mechanism of action for the tumour promotion, *Biochem. Biophys. Res. Commun.* 142:654.

Von Tscharner, V., Prod'hom, B., Baggiolini, M. and Reuter, H., 1986, Ion channels in human neutrophils activated by a rise in free cytosolic calcium concentration, *Nature* 324:369.

Vindlacheruvu, R.R., Rink, T.J. and Sage, S.O., 1991, Lack of evidence for a role for the lipoxygenase pathway in increases in cytosolic calcium evoked by ADP and arachidonic acid in human platelets, *FEBS Lett.* 292:196.

Vostal, J.G., Jackson, W.L. and Schulman, N.R., 1991, Cytosolic and stored calcium antagonistically control tyrosine phosphorylation of specific platelet proteins, *J. Biol. Chem.* 266:16911.

Werner, M.H., Bielawska, A.E. and Hannun, Y.A., 1992, Multiphasic generation of diacylglycerol in thrombin-activated human platelets, *Biochem. J.* 282:815.

Zschauer, A., van Breemen, C., Buhler, F.R. and Nelson, M.T., 1988, Calcium channels in thrombin-activated human platelet membranes, *Nature* 334:703.

Ca²⁺ HOMEOSTASIS AND INTRACELLULAR POOLS IN HUMAN PLATELETS

Kalwant S. Authi

Platelet Section
Thrombosis Research Institute
Manresa Road, Chelsea
London SW3 6LR. U.K.

1. INTRODUCTION

Ca²⁺ homeostasis

Ca^{2+} plays a key role in triggering many platelet functions and is perhaps the most important single intracellular mediator of cell function. Knowledge of its elevation and regulation is paramount to our appreciation of the mechanisms of platelet activation. Several of the important protein entities involved in the regulation of Ca^{2+} levels (i.e. Ca^{2+}ATPases and receptors for inositol phosphates will form the basis of this chapter. Activation of platelets involve either direct (e.g. with thrombin, PAF, vasopressin, thromboxane, ADP), or indirect (via the formation of secondary agonists e.g. as with collagen) elevation of the cytosolic levels of Ca^{2+}. The rise in cytosolic Ca^{2+} levels can vary from basal levels of $\approx 100nM$ to low micromolar levels depending upon the agonist (Rink & Hallam 1984). Indeed, direct elevation of Ca^{2+} levels without surface receptor occupancy by ionophores such as A23187 results in strong activation of platelets (White et al 1974) including expression of prothrombinase activities on the surface (Bevers et al 1983).

Ca²⁺ Elevation Mechanisms

When platelets are activated by agonists Ca^{2+} elevation in the cytosol occurs via the release of Ca^{2+} from intracellular stores and influx from the outside medium. The relationship between surface receptor occupancy, phosphoinositide metabolism and Ca^{2+} elevation is illustrated in Fig. 1 . Stimulation of surface receptors by agonists such as thrombin and thromboxane leads to the activation of G proteins which in turn stimulates the phospholipase C induced hydrolysis of phosphatidylinositol (4,5) bisphosphate (PIP_2) yielding inositol (1,4,5) trisphosphate ($In(1,4,5)P_3$) and diacylglycerol (DAG). This relationship is well established for platelet agonists such as thrombin, thromboxane A_2, PAF, and vasopressin where the surface receptors for these agonists have been cloned and are known to belong to the family of G protein linked receptors (see chapter by Brass et

Mechanisms of Platelet Activation and Control, Edited by
K.S. Authi *et al.*, Plenum Press, New York, 1993

al , this volume). Presently controversy still surrounds the agonist ADP which elevates cytosolic Ca^{2+} but is a poor effector for $In(1.4.5)P_3$ formation.

In(1,4,5)P_3 causes Ca^{2+} release from intracellular stores via its intracellular receptor and this role is well established (Berridge and Irvine 1989), however the biochemical characterisation of the intracellular pools mobilised - of which there are at least two, requires much clarification. When tested on permeabilised cells and isolated membranes In(1,4,5)P_3 only releases a fraction (40-70%) of the total non-mitochondrial stored Ca^{2+} that is available for rapid release indicating the existence of In(1,4,5)P_3 sensitive and insensitive pools (see later). The mechanisms associated with Ca^{2+} release from the In(1,4,5)P_3 insensitive stores are unknown and a number of possibilities may be important. It is possible that linkage between the intracellular pools occurs upon agonist stimulation allowing maximal release to occur via the In(1,4,5)P_3 receptor (IP$_3$R). This linkage may exhibit GTP dependence . Additionally Ca^{2+} release from the In(1,4,5)P_3 sensitive pool may initiate Ca^{2+} release from the In(1,4,5)P_3 insensitive pool via a Ca^{2+} induced Ca^{2+} release mechanism (CICR), analogous to the process occuring via the ryanodine receptor in skeletal muscle.

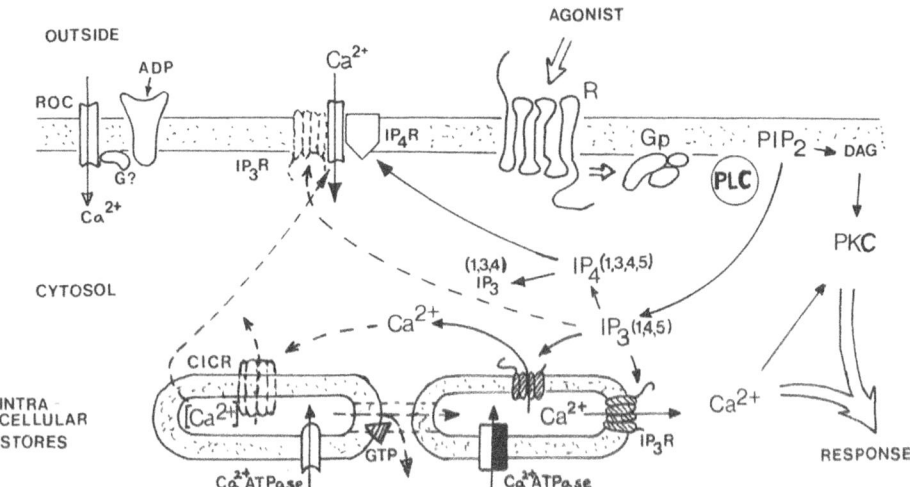

Figure 1.   Mechanisms of cytosolic Ca^{2+} elevation in human platelets. For complete explanation see text. Complete lines (__) represent established activities, dashed lines (---) represent possible mechanisms, some of which are controversial. Agonists such as thrombin, acting on G protein linked receptors lead to the production of In(1,4,5)P_3 resulting in Ca^{2+} release from intracellular stores via the IP$_3$R and Ca^{2+} influx via a number of mechanisms. The relationship between the IP$_3$R and IP$_4$R to the Ca^{2+} influx channel is yet to be established. Ca^{2+} release from the In(1,4,5)P_3 insensitive pool may occur via a possible CICR mechanism or a linkage of the two pools may occur. ADP may also induce Ca^{2+} influx via directly gating a ROC.

Sequestration of Ca^{2+} into intracellular pools occurs via ATP driven Ca^{2+} pumps ($Ca^{2+}Mg^{2+}$ATPases or Ca^{2+}ATPases). The properties of the Ca^{2+}ATPases have been well studied and both the plasma membrane (PM) type and the sarco- endo-plasmic reticulum type Ca^{2+}ATPases (SERCA) have been cloned. The properties of the platelet Ca^{2+}ATPases characterised so far and their relationship to intracellular Ca^{2+} pools will be discussed in detail in this chapter.

During platelet activation there is also Ca^{2+} influx from the extracellular medium.

The mechanisms associated with Ca^{2+} influx are poorly understood. It is probable that agonists which induce phosphoinositide turnover via G protein linked receptors induce Ca^{2+} entry via the formation of second messengers and that the entry pathway is somehow regulated by the level of Ca^{2+} in the intracellular stores. The store regulated Ca^{2+} entry pathway has been termed the capacitative Ca^{2+} model (Putney 1986,1990) which states that it is the depletion of the intracellular stores that somehow regulates the opening of a Ca^{2+} channel in the plasma membrane. It is also been suggested that second messengers such as $In(1,4,5)P_3$ and inositol 1,3,4,5-tetrakisphosphate ($In(1,3,4,5)P_4$) may directly interact with plasma membrane receptors involved in Ca^{2+} entry (Kuno and Gardner 1987, Irvine 1990). Additionally Ca^{2+} entry may arise as a consequence of surface receptors directly opening Ca^{2+} channels i.e. receptor operated Ca^{2+} channels (ROC). There is evidence that the platelet agonist ADP may activate Ca^{2+} via such a mechanism (see Sage et al, this volume). Whether these activities can occur through intermediate G proteins, or by directly inducing conformational changes is not known.

2. MAINTENANCE OF CYTOSOLIC Ca^{2+} CONCENTRATIONS

Ca^{2+} Removal across the Plasma Membranes

It is well accepted that platelets maintain cytosolic concentrations of Ca^{2+} at $\approx$ 80-120nM under resting conditions. As the external Ca^{2+} concentrations are at millimolar levels (and these may also be reached in the intracellular storage organelles such as the endoplasmic reticulum (E.R.)) the platelet maintains a $\approx$ 10,000:1 gradient across its plasma and intracellular membranes. The mechanisms utilised to maintain these gradients are not fully understood. There is general agreement that the intracellular E.R. type organelles possess ATP-dependent Ca^{2+} pumps active at low Ca^{2+} concentrations ($< 1\mu M$) that are able to translocate Ca^{2+} into the organelle (Statland et al 1969, Robblee et al, 1983; Kaser-Glanzmann et al, 1977; Hack et al, 1986). In the presence of Ca^{2+} mobilisers like $In(1,4,5)P_3$ these stores are rapidly released. The mitochondria also possess cation translocating activities but these are generally active at higher Ca^{2+} concentrations ($> 1\mu M$) (Brass, 1984).

The role played by the platelet PM in maintaining cytosolic Ca^{2+} levels low are still controversial. In many other cells the PM Ca^{2+}ATPases have been well established (for review see Carafoli, 1992). The PM Ca^{2+}ATPase is known to be stimulated by calmodulin and has a larger molecular size (124-136kDa) than its counterpart on intracellular membranes (size 100-115 kDa). Both belong to the P class of Ca^{2+}ATPases and are inhibited by vanadate (Carafoli, 1992). The PM Ca^{2+}ATPase can be phosphorylated by cAMP dependent protein kinase (PKA) directly leading to an increase in activity and also by protein kinase C (PKC) although this effect is unclear. Presently 4 genes are known to encode the PM Ca^{2+}ATPase - PMCA1, PMCA2, PMCA3 and PMCA4. To date however no definitive demonstration of a PM Ca^{2+}ATPase in platelets has been described. A number of initial studies using membranes of varying purity did suggest its presence (Enyedi et al, 1986; Enouf et al, 1987a, 1988, 1989) but others using highly purified plasma membrane fractions obtained by different techniques failed to confirm these observations (Steiner & Lusher, 1985; Hack et al, 1986; Fauvel et al, 1986). Enyedi et al (1986) indicated the presence of two distinct Ca^{2+} pumps in platelets on the basis of biochemical characteristics, tryptic fragmentation patterns and comparison with pumps present in skeletal muscle and red cells. Both migrated as 100-110 kDa proteins on their SDS-PAGE systems, neither were stimulated by calmodulin (a property of PM Ca^{2+} pump) but with one the formation of the steady state phosphoenzyme complex was stimulated by La^{3+} (a property of PM Ca^{2+}ATPases). Enouf et al (1987a, 1988, 1989) confirming these studies further suggested that one of these was present on the plasma and the other on intracellular membranes. However it is probable that "both" of these ATPases are intracellular in origin as the plasma membrane fractions used were impure (prepared using sucrose cushions) and cross contamination certain. This latter interpretation is likely as in their recent study use of an antibody that specifically recognises the 135 kDa PM Ca^{2+}ATPase failed to detect this protein in platelet membranes (Papp et al 1991).

High voltage free flow electrophoresis has allowed the preparation of platelet plasma (surface) and intracellular membranes fractions of high purity. The technique involves the use of neuraminidase treatment of intact platelets to reduce the surface negative charge, followed by homogenisation, sorbitol density gradient centrifugation and high voltage free flow electrophoresis (Menashi et al 1981, see also review by Crawford et al 1992). The membrane fractions obtained have well defined protein and lipid composition. The plasma membranes are rich in cytoskeletal proteins (particularly actin and myosin), all the platelet surface glycoproteins and in sphingomyelin (Menashi et al 1981, Lagarde et al 1982, Hack & Crawford 1984). The intracellular membranes are characteristically devoid of actin, myosin, surface glycoproteins and sphingomyelin but are rich in enzymes that release and metabolise arachidonic acid (Lagarde et al 1982, Carey et al 1982, Hack et al 1984, Authi et al 1985), as well as Ca^{2+}ATPase activities, ATP dependent Ca^{2+} sequestration and a binding site for In(1,4,5)P_3 (Menashi et al 1981, Authi and Crawford 1985, Hack et al 1986,1988a, Authi 1992). Plasma membranes prepared by free flow electrophoresis do not sequester Ca^{2+} or exhibit Ca^{2+}ATPase activities (Menashi et al 1984, Hack et al 1986), and these finding are in good aggreement with the absence of Ca^{2+}ATPase activities in plasma membranes prepared using percoll gradients and a wheat germ agglutinin affinity column (Fauvel et al 1986, Steiner & Luscher 1985). If these findings are correct then other mechanisms of Ca^{2+} control across the plasma membrane may be important. Brass (1984) identified two pools of Ca^{2+} in resting platelets using ^{45}Ca labelling experiments; a cytosolic pool which was rapidly exchangeable with the extracellular medium (t½=17 mins) and regulated by a Na^+/Ca^{2+} exchange mechanism; and an ER Ca^{2+} pool which was ATP-dependent and was slowly exchangeable (t½ = 300 mins) from which Ca^{2+} is released by In(1,4,5)P_3. An ATP driven Ca^{2+} extrusion pump was again not identified and PGI_2 which raises cAMP levels was not found to affect the sizes of either pool. Purified platelet surface membranes have been shown to contain a Na^+/Ca^{2+} exchange activity (Rengasamy et al, 1987) indicating that the exchanger may contribute to Ca^{2+} removal across the plasma membrane.

Recently the controversy of a PM Ca^{2+}ATPase in platelets has again been addressed in studies with Quin II overloaded platelets using low platelet counts, where a Ca^{2+} extrusion mechanism was identified in activated platelets and rates of cytosolic Ca^{2+} decrease measured after elevation to micromolar levels with ionomycin (Johansson & Haynes 1988, Johansson et al 1992). The authors indicate that in their system the extrusion system was more important than the Na^+/Ca^{2+} exchanger at Ca^{2+} levels less than 400nM but at higher levels the exchanger was more prominant. However they also indicated that the exchanger was poorly resolved in their system compared to the extrusion pump. Further, cAMP and cGMP elevation stimulated the extrusion system by stimulating the V_{max} (Johansson et al 1992, Johansson and Haynes 1992). Whether the cyclic nucleotides exert their effects via phosphorylation of a possible PM Ca^{2+}ATPase as was suggested remains to be established as in intact platelets phosphorylation of a 125-135 kDa protein (the size of a PM Ca^{2+}ATPase) has not been observed by either cAMP or cGMP elevating agents (Waldman et al 1986, 1987). Interestingly a 130 kDa protein has been shown to be phosphorylated by both A kinase and G kinase stimulation in isolated platelet membranes but the identity of this protein which may also be prominant in smooth muscle remains to be determined (Waldman et al 1986, 1987).

Another aspect of Ca^{2+} control has been described involving elevated Ca^{2+} removal from the cytosol via activation of PKC. Studies on Fura 2-labelled platelets and saponin permeabilised platelets indicate that PKC can stimulate removal of Ca^{2+} across the plasma membrane (Pollock et al 1987) and into intracellular stores (Yoshida & Nachmias, 1987). Whether PKC can influence Ca^{2+}ATPase activities in platelets is not known but in the red cell the Ca^{2+} ATPase has been shown to be phosphorylated by PKC leading to an increase in activity (Wang et al, 1991) and in smooth muscle cells PKC activation has been shown to stimulate both the calcium pump and Na^+/Ca^{2+} exchange activities (Furakawa et al, 1989; Vigne et al, 1988). In the probable absence of a PM Ca^{2+}ATPase it would be of interest to examine possible regulation of the Na^+/Ca^{2+} exchanger in platelets by protein phosphorylation.

Intracellular Membrane Ca^{2+} Stores and Pumps

Much more is known about the characteristics of the Ca^{2+} sequestration mechanisms of platelet intracellular Ca^{2+} stores which are also often referred to as the dense tubular system (DTS). Dean and Sullivan (1982) showed that the Ca^{2+} pump shares many similarities in the mechanism of Ca^{2+} translocation with the sarcoplasmic reticulum (S.R.) Ca^{2+} pump of skeletal muscle tissues. In that study antibodies to the purified SR pump were shown to cross react and recognise 2 polypeptides in the 100 kDa range in platelets (referred to here as 100 kDa and 95 kDa) with the lower molecular species initially thought to be a proteolytic product (Dean & Sullivan, 1982). The Ca^{2+} uptake processes are also similar as they are Ca^{2+}-dependent, stimulated by oxalate or phosphate which act as sinks in the lumen of the organelle, and Ca^{2+} translocation was associated with phosphorylation of the 100 kDa polypeptide(s) on aspartate residues. Fischer et al (1985) revealed that the platelet and SR pumps were structurally distinct based on tryptic digestion products and time course of loss of Ca^{2+} transport function during proteolysis. Further kinetic characterisation of the platelet pump (particularly La^{3+} stimulation of the lower 95kDa species) and tryptic analysis (particularly the formation of a 80kDa fragment derived from the 95 kDa species which was not seen in other intracellular pumps) suggested that there may be two distinct Ca^{2+}ATPase activities in platelets (Enyedi et al 1986, Enouf et al 1987a, 1988, 1989). The absence of any Ca^{2+}ATPase activities in highly purified plasma membrane fractions (Hack et al 1986) would suggest that "both" are present on intracellular membranes.

Molecular biology has revealed (so far) 3 genes that code for the intracellular Ca^{2+} ATPase of muscle and non-muscle tissues, SERCA 1, 2 & 3 genes (Brandl et al, 1986, Burk et al 1989). SERCA 1 (a + b forms - alternatively spliced) codes for the Ca^{2+}ATPase in the SR muscle tissue, and SERCA 2 (also a + b) of which the "a" product is expressed in heart and "slow" SR muscle, and "b" in non-muscle tissue. Recently the 100 kDa polypeptide of the platelet ATPase has been identified as a SERCA 2b product using molecular cloning from a HEL cell cDNA library, PCR amplification of mRNA from platelets and immune recognition using an antibody raised to a SERCA 2b specific epitope (Papp et al 1991, Enouf et al 1992). This SERCA 2b antiserum was raised to a peptide derived from 44 amino acids at the C terminal end of the predicted sequence specific for this isotype (Wuytack et al 1989). Interestingly the SERCA 2b antibody does not recognise the smaller of the two bands seen in many SDS-PAGE studies using platelet lysates or membranes. The isotype of the lower 95 kDa species remains to be determined and is likely not to be a proteolytic product of the SERCA 2b isoform.

A number of antibodies have been described which recognise the intracellular membrane Ca^{2+}ATPase (Dean & Sullivan, 1982; Fisher et al, 1985; Hack et al, 1988a, Wuytack et al 1989) with some of these recognising both 100 kDa and 95 kDa bands. In the study of Hack et al (1988a) 4 monoclonals were raised to the Ca^{2+}ATPase using purified intracellular membranes (by free flow electrophoresis), all recognising a single polypeptide with whole platelets. One, PL/IM 430, inhibited Ca^{2+} uptake into intracellular membranes by 60-70% but without inhibition of the Ca^{2+} ATPase, or an effect on the phosphoenzyme intermediate. The apparent descriminatory property of PL/IM 430 in inhibiting Ca^{2+} uptake without affecting Ca^{2+}ATPase activities and phosphoenzyme complex formation makes this a unique antibody with its site of recognition being at or close to the Ca^{2+} translocation site on the cytoplasmic surface. Further studies are required to epitope map the site and to detemine the isotype of the Ca^{2+}ATPase that is recognised. Interestingly PL/IM 430 does not totally inhibit Ca^{2+} uptake when measured using intracellular membranes or saponin permeabilised platelets with the maximal inhibition being 80% and 65% respectively (Hack et al 1988a, 1988b). In saponin permeabilised platelets PL/IM 430 does not inhibit Ca^{2+} uptake into the $In(1.4.5)P_3$ sensitive stores as Ca^{2+} release by $In(1,4,5)P_3$ is unaffected (Fig.2) and the EC_{50} for $In(1,4,5)P_3$ is unaffected (Hack et al 1988b). Indeed coincubation of $In(1,4,5)P_3$ with PL/IM 430 at the start of ATP dependent $^{45}Ca^{2+}$ uptake results in almost total inhibition of Ca^{2+} uptake. This would suggest that PL/IM 430 predominantly recognises and inhibits Ca^{2+} uptake into the $In(1,4,5)P_3$ insensitive Ca^{2+} store. The distribution of the SERCA 2b Ca^{2+}ATPase amongst $In(1,4,5)P_3$ sensitive and insensitive Ca^{2+} pools clearly needs to be determined and would reveal whether each pool is served by a distinct but related Ca^{2+} pump.

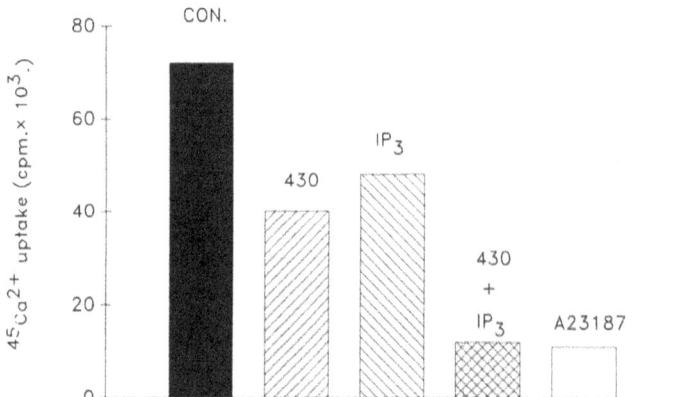

Figure 2.   Inhibition of ATP dependent $^{45}Ca^{2+}$ uptake into intracellular stores of saponin permeabilised platelets by a Ca^{2+}ATPase antibody (PL/IM 430) and $In(1,4,5)P_3$. Experimental conditions as in Hack et al (1988b). Agents are added at zero time with incubation carried out for 20min. $^{45}Ca^{2+}$ uptake in the presence of A23187 represents non-specific $^{45}Ca^{2+}$ binding. Con=control.

Recently the use of inhibitors of Ca^{2+}ATPase have also contributed significantly to the roles played by Ca^{2+}ATPases in cell regulation. Two agents in particular have been used extensively namely thapsigargin (Tg) and 2,5-di(t-butyl)-1,4-benzohydroquinone (tBuBHQ). Both agents cause an increase in cytosolic Ca^{2+} in a large number of different cell types. The mechanism of action of both agents does not involve the formation of $In(1,4,5)P_3$ or the direct opening of a calcium channel as the rate of calcium release is slower than in agonist stimulation or with calcium ionophores, but an inhibition of the Ca^{2+} ATPase (Kass et al 1989, Thastrup et al 1990). Our studies with the platelet membrane Ca^{2+} ATPase measurements indicates that both agents are effective inhibitors, with Tg being approximately 16 times more potent than tBuBHQ (Fig.3, and Authi et al 1993). Both agents also do not affect the basal Mg^{2+}ATPase activity and thus interfere only with the activity associated with Ca^{2+} translocation. Both compounds have been shown to inhibit the formation of the phosphoenzyme intermediate which is a pre-requisite for Ca^{2+} translocation (Papp et al 1991, Kijima et al 1991, Sagara et al 1992). It is now known that Tg is effective against all SERCA type Ca^{2+} ATPases (Sagara et al 1992) with no effect on the Ca^{2+}ATPase present on the plasma membrane. There is some evidence that tBuBHQ may show some selectivity towards inhibition of the phosphoenzyme complex formation of the 95kDa species of the Ca^{2+}ATPase present in platelets (Papp et al 1992). The use of these two inhibitors in intact cells to elucidate the contributions of intracellular pools to the activation process will be discussed later.

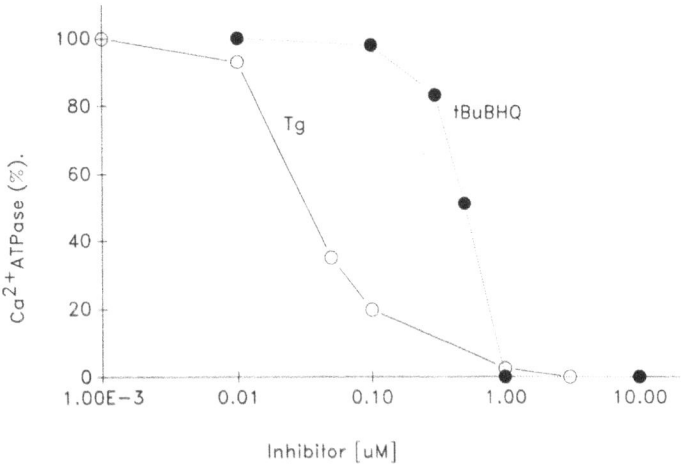

Figure 3.   Inhibition of platelet mixed membrane Ca^{2+}ATPase by Tg and tBuBHQ. The data is taken from Authi et al (1993).

Cyclic AMP Regulation of the Platelet Intracellular Ca^{2+} Pumps.

Intracellular Ca^{2+} pumps can be regulated by cAMP which acts via cAMP dependent protein kinase (PKA) leading to the phosphorylation of target proteins. While the PM Ca^{2+}ATPase are known to be phosphorylated directly by PKA , intracellular Ca^{2+} pumps are not but may be regulated by phosphorylation of a regulatory protein called phospholamban. Ca^{2+}ATPases of the SERCA 1 and 2 classes can bind phospholamban (Grover & Khan 1992) but the key factor is whether the tissue in question contains phospholamban or an analogous protein. In platelets the regulation of Ca^{2+}ATPase by cAMP is still under question. It was observations by Kaser-Glansmann et al (1977) on platelet microsomes that suggested cAMP stimulated Ca^{2+} uptake. This has been supported by a number of subsequent studies and a 22 kDa membrane protein which is phosphorylated by PKA has been implicated as having an analogous role to phospholamban in the heart (Enouf et al 1985, 1987b, Adunyah & Dean 1987, Hettasch & LeBreton 1987, Fisher & White 1987, Courvazier et al 1992). Using chlorotetracycline labelled platelets (monitoring Ca^{2+} levels in the DTS) Tao et al (1992) has again suggested that dibutryl-cAMP and forskolin stimulates the Ca^{2+} pump. Further Enouf et al (1987b) suggested that cAMP could even increase $In(1,4,5)P_3$ induced Ca^{2+} release by phosphorylation of the 22 kDa protein. However a number of studies using platelet microsomes have refuted these observations when not observing any correlation between phosphorylation and effects on Ca^{2+} sequestration (White et al 1989, O'Rourke et al 1989). They suggest that it was impurities in preparations of the PKA used (such as phosphate ions) that stimulated the Ca^{2+} sequestration observed and that even the protein kinase inhibitors used may have contained impurities inhibiting Ca^{2+} uptake . Additionally it is not certain whether platelets contain phospholamban as antibodies to cardiac phospholamban do not recognise relevant proteins in platelets (Adunyah et al 1988) and partially purified preparations of the 22 kDa protein do not affect Ca^{2+}ATPase activities (Fisher & White 1989). The 22 kDa protein has recently been identified as the small molecular weight G protein rap 1B and is a substrate for PKA (Ohmori et al 1989, White et al 1990, Siess et al 1990). Its function is not certain although it has been implicated in cytoskeletal interactions (White et al ,this volume) and in the regulation of phospholipase C (Lapetina & Ferrel, this volume). Using plasma membranes prepared with sucrose cushions (not pure), Courvazier et al (1992) suggest that rap 1B is located on plasma membranes making its relationship to regulation of intracellular Ca^{2+} pumps even more questionable. Localisation of rap 1B to the plasma membranes has also been shown using membranes prepared by free flow electrophoresis (White et al, this volume) again pointing to a role that is distinct from that of regulating intracellular Ca^{2+} pumps. Fig.4 summarises the mechanisms associated with maintaining Ca^{2+} levels low in human platelets.

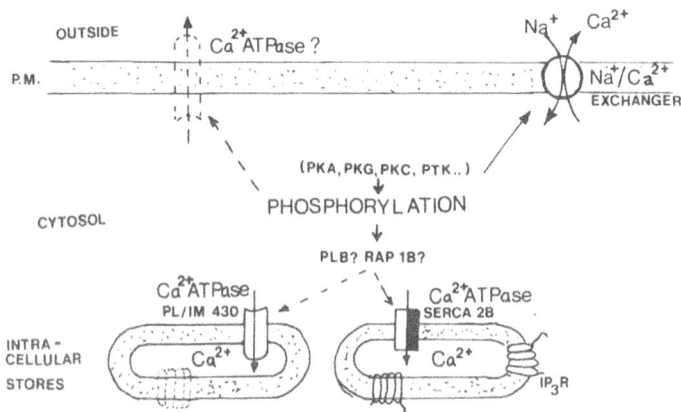

Figure 4. Mechanisms of cytosolic Ca^{2+} reduction in human platelets. Ca^{2+} is sequestered into intracellular stores by 2 Ca^{2+}ATPases, a SERCA 2B isoform and another recognised by PL/IM 430 antibody, the latter inhibiting Ca^{2+} sequestration into an $In(1,4,5)P_3$ insensitive store. Possible regulation of intracellular Ca^{2+}ATPase by a phospholamban (PLB) type mechanism is still controversial. Plasma membranes contain a Na^+/Ca^{2+} exchanger but its regulation by phosphorylation is to be established. The presence of a PM Ca^{2+}ATPase is controversial.

3. MECHANISMS OF Ca^{2+} ELEVATION IN PLATELETS

Inositol (1,4,5) trisphosphate

The role of $In(1,4,5)P_3$ in causing Ca^{2+} release from platelet intracellular stores is well documented. Its formation by agonists from PIP_2 and metabolism via phosphorylation to In(1,3,4,5)P$_4$ and dephosphorylation to In(1,3,4)P$_3$ and subsequently to inositol have also been well described (Daniel et al 1989, King et al 1990, for review see Daniel 1990). However its role in Ca^{2+} influx and the relationship of the intracellular pool of Ca^{2+} sensitive to In(1,4,5)P$_3$ to other pools is still poorly defined. In this section these latter points will be discussed in addition to the recently described properties of the IP$_3$R itself.

Binding studies have been used extensively to both characterise the IP$_3$R and to isolate the In(1,4,5)P$_3$ sensitive organelle. The binding site (or receptor) has been characterized in many different tissues as well as platelets with Kd varying from 0.1-80nM depending upon the tissue, method of preparation of membranes and the binding assay conditions (for review see Ferris & Snyder 1992). Binding studies using [^{3}H]In(1,4,5)P$_3$ on platelet membranes reveal similar information as other tissues with two sites identified at alkaline pH with Kd's varying between 0.1-23nM (O'Rourke & Feinstein, 1990; Hwang, 1991). Heparin is a potent but non-specific antagonist. Of probable physiological importance, Ca^{2+} and ATP have also been shown to inhibit In(1,4,5)P$_3$ binding (EC$_{50}$=40μM and 0.5mM, respectively). Additionally of importance is that In(1,3,4,5)P$_4$ has been shown to be a poor antagonist for the In(1,4,5)P$_3$ receptor (O'Rourke & Feinstein 1990). The binding of In(1,4,5)P$_3$ has been suggested to occur on arginine residues at a site distinct from Ca^{2+} release activity and the latter could be inhibited by the use of NN'-dicyclohexyl carbodi-imide (DCCD) which modifies carboxyl groups. The presence of distinct sites on the receptor protein has been confirmed with the use of monoclonal antibodies recognising different epitopes of the In(1,4,5)P$_3$ receptor particularly with 18A10, which recognises part of the C terminal end, and inhibits Ca^{2+} release by In(1,4,5)P$_3$ without affecting its binding (Miyawaki et al 1991, Nakade et al 1991). Our initial study using membranes prepared by free flow electrophoresis indicated that specific binding of [^{32}P] In(1,4,5)P$_3$ occured with intracellular membranes (Fig. 5), which showed a five fold enrichment of binding with respect to mixed membranes (Authi 1992). Scatchard analysis reveals a K$_d$=80nM with a high binding capacity (23pmol/mg protein). In some studies where the particulate fractions obtained are only partially purified and contain components of plasma and intracellular organnelles the In(1,4,5)P$_3$ binding has been observed to co-purify with plasma membranes (Guillemette et al 1988, Rossier et al 1989,1991) and even nuclei (Malviya et al 1990). The In(1,4,5)P$_3$ sensitive organelle has been suggested be a specialised part of or distinct from the endoplasmic reticulum (Rossier et al 1991), and has also been referred to as the calciosome (Volpe et al 1990). It has been suggested to also contain the Ca^{2+} binding protein calreticulin which allows high concentrations of Ca^{2+} to be present in the lumen without precipitation (Treves et al 1990, Milner et al 1991, Van Delden et al 1992). Our future studies will determine if calreticulin is also localised in intracellular membranes obtained with free flow electrophoresis.

The receptor for In(1,4,5)P$_3$ has been purified from a number of tissues (but so far not from platelets) (eg. Supattapone et al 1988a), cloned (Furuichi et al 1989, Mignery et al 1989) and has remarkable similarity to the ryanodine receptor which is the Ca^{2+} channel in skeletal muscle. Presently it can be assumed that the receptor in platelets will have similar properties to that described in other tissues but its fine regulation may differ according to the isotype that is expressed. Presently there are at least 3 and probably 4 genes that code for the IP$_3$R (Sudhof et al 1991, Ross et al 1992). Full length sequences for 2 genes (type I and type II) with 69% homology have been described (Sudhof et al 1991). The type I gene has two major splice sites which on the translated protein contributes to 15 and 40 amino acids which are present between the In(1,4,5)P$_3$ binding domain and 2 serine phosphorylation sites (Nakagawa et al 1991, Danoff et al 1991). Sequence information from probably two additional genes have also been published (Sudhof et al 1991, Ross et al 1992). The isotype(s) of the human platelet In(1,4,5)P$_3$ receptor(s) is yet to be described.

The receptor is a 250 kDa protein which as a tetramer (recognised as a four leaf

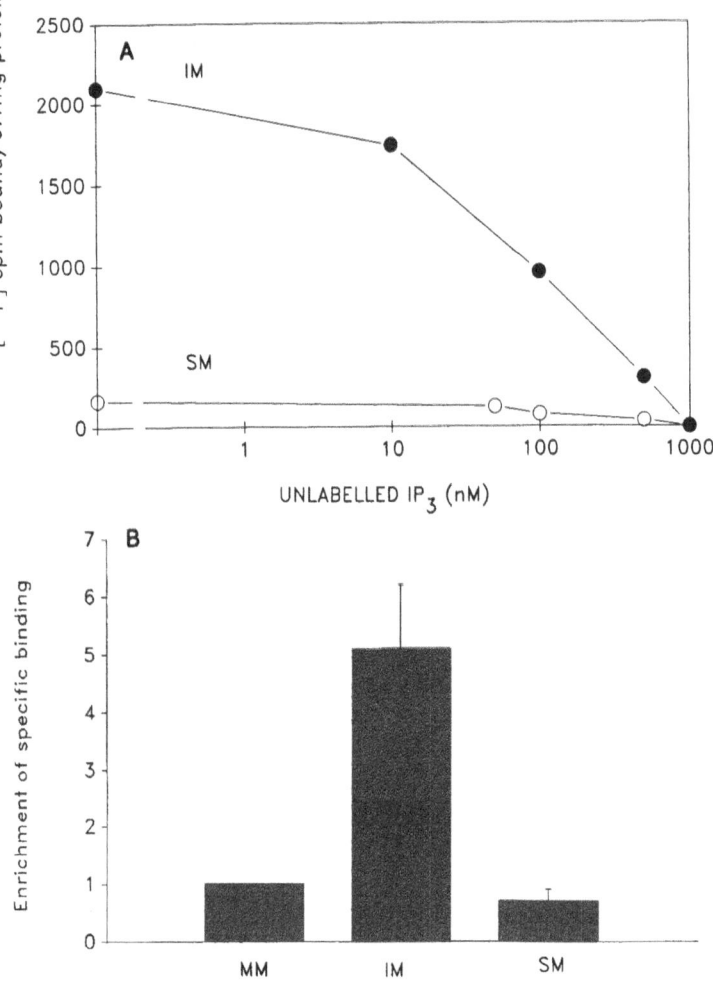

Figure 5. Binding and displacement of $[^{32}P]In(1,4,5)P_3$ to intracellular (IM) and surface (plasma, PM) membranes (A) prepared by free flow electrophoresis, and enrichment of binding with respect to mixed membranes. Data is taken from Authi (1992).

clover structure in electron micrographs) forms a Ca^{2+} channel. Studies using peptides derived from the cloned receptor and monoclonal antibodies have shown that sequences near the N terminal region bind $In(1,4,5)P_3$ with the channel portion of the molecule being near the C terminal end (Nakade et al 1991). Phosphorylation of the purified receptor by PKA uncouples $In(1,4,5)P_3$ binding from Ca^{2+} release and probably provides the major form of receptor regulation (Supattapone et al 1988b, Ferris et al 1991). Studies are still to be carried out in platelet systems as the mode of action of cAMP is still controversial. O'Rourke et al (1989) found no effect of the catalytic subunit of PKA on Ca^{2+} uptake or $In(1,4,5,)P_3$ induced Ca^{2+} release in platelet membranes. Tohmatsu et al (1989a) have reported inhibition of $In(1,4,5)P_3$ induced Ca^{2+} release by cAMP in saponin permeabilised platelets but this was not affected by an inhibitor of PKA. Recently Quinton and Dean (1992) have reported modest (30%) inhibition of $In(1,4,5)P_3$ induced Ca^{2+} release by PKA and phosphorylation of a partially purified 250 kDa protein binding $In(1,4,5)P_3$. However differentiation of this from actin binding protein which is also a 250kDa protein and phosphorylated by PKA needs to be carried out. This issue will probably become more clear once immunoprecipitation of the (phosphorylated or not) receptor is carrried out using In(1,4,5)P₃ receptor antibodies.

Ca²⁺ release from intracellular stores.

A number of initial studies reported Ca²⁺ release by In(1,4,5)P₃ from particulate preparation of human platelets that were either enriched in or highly purified intracellular membranes (O'Rourke et al 1985, Authi & Crawford 1985, Adunyah & Dean 1985). All of these studies yielded similar information. Ca²⁺ release by In(1,4,5)P₃ was rapid but transient, probably reflecting breakdown of In(1,4,5)P₃ to its metabolites, and confined to less than half of the total Ca²⁺ sequestered in the presence of ATP. Release of Ca²⁺ from saponin permeabilised platelets is equally rapid but not transient and the extent of release (not rate) is dependent upon the concentration of In(1,4,5)P₃ added (Hack et al 1988b). If the saponin levels are strictly controlled In(1,4,5)P₃ induced Ca²⁺ release can lead to aggregation and secretion of dense granule constitutents in human platelets which is dependent on the formation of thromboxane (Authi et al 1986,1987., Watson et al 1986). Thus In(1,4,5)P₃ induced Ca²⁺ release is tightly linked to phospholipase A₂ activation and thromboxane synthesis. Ca²⁺ release from saponin permeabilised platelets is again not total but a significantly larger proportion (up to 70%) than that seen in membrane preparations - but again indicating a significant In(1,4,5)P₃ insensitive Ca²⁺ pool. Factors important for Ca²⁺ release from the In(1,4,5)P₃ insensitive fraction are not certain. In saponin permeabilised platelets the Ca²⁺ATPase inhibitors Tg and tBuBHQ, at concentrations that totally inhibit the Ca²⁺ pump, release more Ca²⁺ than In(1,4,5)P₃ and from a pool that totally overlaps the In(1,4,5)P₃ sensitive pool (Authi et al 1993). In this system co-addition of In(1,4,5)P₃ with either inhibitor releases the same quantity of Ca²⁺ as either inhibitor alone. This makes Tg and tBuBHQ useful to deplete the In(1,4,5)P₃ sensitive pool in intact platelets as membrane penetrating In(1,4,5)P₃ analogues are not presently available. The effects of Tg and tBuBHQ would be useful to define the consequence of Ca²⁺ release from the In(1,4,5)P₃ sensitive pool in intact platelets. In indomethacin treated Fura 2 loaded platelets both agents elevate [Ca²⁺]ᵢ to approximately 300nM in the presence of extracellular Ca²⁺. In the presence of extracellular EGTA this is reduced reflecting only Ca²⁺ release from intracellular stores which from experiments carrried out in saponin permeabilised platelets would reflect Ca²⁺ release predominantly from an In(1,4,5)P₃ sensitive pool. Interestingly addition of thrombin to Tg or tBuBHQ treated platelets resulted in further Ca²⁺ elevation implying that Ca²⁺ from the In(1,4,5)P₃ insensitive pool was also released by thrombin (Authi et al 1993). Similar results are seen with U46619 although the levels of Ca²⁺ elevated are less than with thrombin. This implies that agonists either generate other messengers which release Ca²⁺ from the In(1,4,5)P₃ insensitive pool or if In(1,4,5)P₃ is the only Ca²⁺ mobilising message generated then linkage of intracellular pools takes place allowing further Ca²⁺ release via the In(1,4,5)P₃ receptor. Brune and Ullrich (1991,1992) have reported similar findings with Tg and tBuBHQ with agonists but suggest that Ca²⁺ pools sensitive to Ca²⁺ATPase inhibitors and agonists (In(1,4,5)P₃) are distinct and that cyclic nucleotides could accelerate uptake into the Ca²⁺ATPase inhibitors sensitive pool. However no studies were carried out in permeabilised cells or membrane preparations to confirm their suggestion that the Ca²⁺ATPase inhibitor and In(1,4,5)P₃ sensitive pools do not overlap.

When examining functional responses the results obtained with Tg and tBuBHQ differ in the extents of the responses obtained. In indomethacin treated platelets tBuBHQ induced shape change with no aggregation but Tg also induced a slow aggregation and secretory response even though similar levels of Ca²⁺ were achieved. This suggested that Tg had additional actions to Ca²⁺ elevation. (In the absence of indomethacin Tg induced a full aggregatory response with the amplification mediated by thromboxane formation

(Brune & Ullrich 1990, Authi et al 1993). In the presence of indomethacin both agents showed synergistic secretory responses with dioctanoyl glycerol (which activates PKC) again indicating that Ca^{2+} release by the two agents occured from a functionally relevant pool which synergised with protein kinases (Authi et al 1993). When [^{3}H] arachidonic acid release and protein phosphorylation was measured Tg again showed a greater ability to release [^{3}H]AA and a greater phosphorylation of pleckstrin than tBuBHQ (Authi et al 1993). Presently studies are underway to determine if Tg has any direct effect on protein kinases. The above studies would suggest that tBuBHQ would be a better tool to study the relevance of the $In(1,4,5)P_3$ sensitive pool in intact platelets than Tg.

A number of other agents have been shown to mobilise Ca^{2+} from platelet intracellular stores but their relevance is not certain. These include AA and neomycin with the latter agent probably affecting GTP dependent processes. In other tissues, GTP has been shown to cause Ca^{2+} release via a different mechanism to that of $In(1,4,5)P_3$ (Henne & Soling, 1986) and also from a different subpopulation of the endoplasmic reticulum (Henne et al, 1987). Under certain conditions it has also been shown to stimulate Ca^{2+} sequestration and has been suggested as a mechanism for refilling $In(1,4,5)P_3$ sensitive pools (Mullaney et al 1988). In platelets little information exists regarding the role of GTP in Ca^{2+} flukes. It is an ineffective Ca^{2+} mobilising agent in saponin permeabilised platelets (Authi et al, 1988). The aminoglycoside antibiotics neomycin and streptomycin, which at high doses (mM) inhibit PLC (Cockroft and Gomperts 1985), are also potent mobilisers of Ca^{2+} from platelet intracellular stores in the micromolar range ($EC_{50} = 15\mu M$, K.S. Authi unpublished observations) and induce AA release (Nakashima et al, 1987) and aggregation of saponin permeabilised platelets similar to $In(1,4,5)P_3$ (Polascik et al, 1987). How neomycin induces mobilisation of Ca^{2+} from intracellular stores and from which store, is not known. In saponin permeabilised platelets $^{45}Ca^{2+}$ release by neomycin is not affected by heparin under conditions where $In(1,4,5)P_3$ induced release is totally blocked (fig. 6), clearly indicating that the $In(1,4,5)P_3$ receptor is not involved. Neomycin has been shown to stimulate GTPase activities in platelet membranes (Hermann & Jakobs, 1988) and thus indicates support for an involvement for GTP in Ca^{2+} regulation. Further studies on the nature of the G proteins stimulated by neomycin in platelets may reveal the importance of this mechanism in Ca^{2+} regulation. A number of studies have demonstrated the ability of AA to mobilise Ca^{2+} in isolated membranes which is not dependent upon its conversion to eicosanoids (Authi & Crawford, 1985; Tohmatsu et al, 1989b; Fisher et al, 1990). The latter study indicated that AA releases Ca^{2+} via an ionophoric mechanism. However in intact platelets Ca^{2+} mobilisation induced by AA is dependent upon the formation of thromboxane A_2 and at high concentrations ($>50\mu M$) AA is even able to inhibit Ca^{2+} mobilization by other agonists via an increase in cAMP levels (Kowalska et al, 1988). Even higher concentrations (i.e. $100\mu M$) are lytic and results in not only elevation of Ca^{2+} (independent of thromboxane production) but also direct activation of protein phosphorylation (Kowalska et al, 1988; Nishikawa et al, 1988). Interestingly, collagen which is able to induce large increases in AA release is unable to induce significant Ca^{2+} elevation (Pollock et al, 1986). Thus the relevance of AA-induced Ca^{2+} mobilization is yet to be proved.

The Ca^{2+} release process in skeletal and cardiac muscle occurs via a CICR mechanism involving the ryanodine receptor. Pharmacological agents acting at CICR channels include caffeine and ryanodine. Studies using these agents indicate that CICR mechanisms are also present in non-muscle cells and may be responsible for the complex spatial and temporal organisation of the Ca^{2+} signal seen in many single cell studies using fluorescent indicators (for review see Berridge 1991). The initiation of Ca^{2+} oscillations is thought to require the formation of $In(1,4,5)P_3$ with the CICR mechanism required for the propagation of the oscillatory signal involving co-ordinated uptake and release cycles. Single cell studies with platelets are more difficult to carry out due to the small size of the

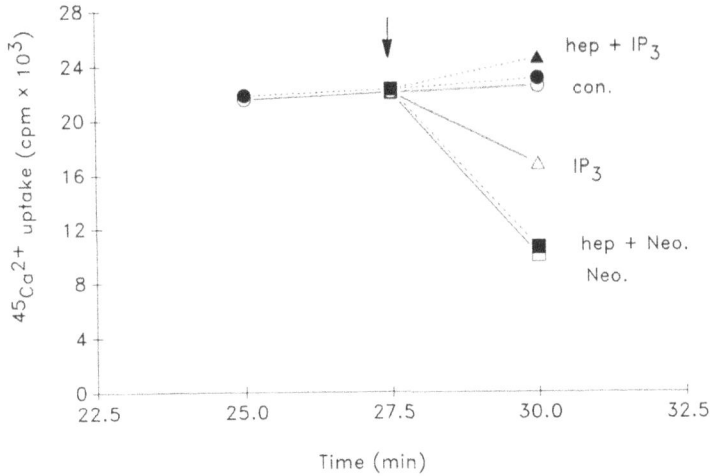

Figure 6. Inhibition of In(1,4,5)P$_3$ (IP$_3$) but not neomycin (Neo) induced ^{45}Ca^{2+} release from saponin permeabilised platelets. Incubation conditions as described in Hack et al [1988b]. Dashed lines represent incubations in the presence of 50μg/ml heparin. IP$_3$ (20μM) and neomycin (30μM) were added at 27.5min with incubations stopped at 30min by vacuum filtration.

cell limiting resolution, but a number of remarkable recent reports suggest similar phenomenon to be present in platelets (Tsunoda et al 1988, Heemskerk et al 1992). Recently cyclic ADP ribose (cADPr) , a metabolite of NAD$^+$, has been shown to be a potent agent at activating CICR channels in sea urchin egg homogenates (Galione et al 1991), and the enzyme converting NAD$^+$ to cADPr has been shown to be widespread in animal tissues (Rusinko & Lee 1989). Studies of the effects of caffeine, ryanodine and cADPr have yet to be fully described in platelets, but preliminery studies in our laboratory indicate that caffeine does elevate cytosolic Ca^{2+} levels in platelets (at 5-15mM) but does not cause any functional responses and its predominant effect is inhibitory on platelet function (Bokkala, Kakkar & Authi, manuscript in preparation). The relevance of the Ca^{2+} elevated and its origin is under investigation. Ryanodine is not active in Fura 2 labelled platelets and the effects of cADPr on platelet systems remains to be tested.

Mechanisms of Ca^{2+} entry ; Involvement of second messengers

This section will deal primarily with the involvement of second messengers - a more complete discussion of Ca^{2+} influx mechanisms is also present in the chapter by Sage et al, this volume. Influx of Ca^{2+} from the extracellular medium can occur via a number of mechanisms all of which require definitive validation. Ca^{2+} influx may be mediated by intracellular messengers such as In(1,4,5)P$_3$ or In(1,3,4,5)P$_4$ through their actions on intracellular stores where the state of the pools initiates Ca^{2+} entry, or directly via actions on the plasma membrane (Putney 1986, Kuno & Gardner 1987). Some agonists like ADP may be able to directly open a receptor operated Ca^{2+} channel (Sage & Rink 1985).

In 1986 Putney proposed the capacitance model for Ca^{2+} entry where the depletion of the intracellular stores by agonists (via In(1,4,5)P$_3$) somehow signalled the opening of the Ca^{2+} channel in the surface membrane so that the store may be replenished. Since then many studies have supported this model and it has been suggested that it is the concentration of Ca^{2+} in the intracellular store that initiates the opening of the plasma membrane Ca^{2+} channel (Putney, 1990). This mode of Ca^{2+} entry has received particular support from studies using Tg and tBuBHQ which by depleting intracellular stores in the absence of second messenger generation, induce Ca^{2+} entry. How depleted stores communicate with the plasma membrane Ca^{2+} channel is not understood but if such a link is important it is probably mediated by a protein. Cytochrome P450 has been suggested to link store depletion with Ca^{2+} entry in platelets by the finding that a cytochrome P450 inhibitor, econazole, inhibits Ca^{2+} and Mn^{2+} influx (Alonso et al 1991, Alvarez et al 1991). However this finding has not been fully supported by studies of Sargeant et al (1992) who found that although econazole inhibited Ca^{2+}ATPase mediated Mn^{2+} entry it was very weak with respect to agonist mediated Mn^{2+} entry, and using the stopped flow technique found that there was no effect on the initial phase of the Ca^{2+} signal. The IP_3R receptor has itself been suggested as the communication between the intracellular store and the plasma membrane Ca^{2+} channel (Irvine 1990). The IP_3R is certainly large enough but more importantly it has considerable sequence homology with the ryanodine receptor which does interact with the dihydropyridine receptor in skeletal muscle. Another protein suggested to be important in platelets is a 130kDa protein by Vostal et al (1992) who found that Ca^{2+} release from intracellular stores by agonists and Tg was accompanied by tyrosine phosphorylation of this 130kDa protein which was associated with Ca^{2+} influx. However the identity of this protein as well as a cause and effect relationship needs to be determined.

While the store depleted Ca^{2+} entry pathway clearly exists in platelets a number of studies have suggested that second messengers may directly gate Ca^{2+} channels in the plasma membrane. Kuno & Gardner (1987) first showed that $In(1,4,5)P_3$ opened ion channels in plasma membrane fractions in lymphocytes and in platelet plasma membrane fractions prepared using percoll gradients Rengasamy & Feinberg (1988) showed that $In(1,4,5)P_3$ could release Ca^{2+} loaded in the membranes by a Na^+/Ca^{2+} exchange mechanism, both studies pointing to the probable presence of IP_3R in plasma membranes. However in the latter study the dose response relationship showed no saturation questioning the specificity of the effect observed. Interestingly a number of earlier studies indicated that in some tissues $In(1,4,5)P_3$ binding activity co-eluted with plasma membranes enriched fractions although intracellular membrane contamination was evident (Guillemete et al 1988). More recently IP_3R's have been identified having carbohydrate residues such as sialic acid and N-acetyl glucosamine that are known to be associated (through post translational modification) with plasma membrane proteins (Khan et al 1992a, 1992b). In these studies a significant proportion (approximately 30%) of $In(1,4,5)P_3$ receptors in the thymus and olfactory cilia were plasma membrane associated and also had reduced specificity for $In(1,4,5)P_3$ with respect to other inositol phosphates, implying distinct biochemical properties from the intracellular membrane receptors. $In(1,4,5)P_3$ receptors on surface invaginated portions of the plasma membranes have also been localised by immunocytochemical staining using a monoclonal antibody (4CII, also to the $In(1,4,5)P_3$ receptor) in endothelial cells, smooth muscle and keratinocytes (Fujimoto et al 1992). There is therefore not only heterogeneity in the genes that code for IP_3R's but also post translational modification that confers localisation within subcellular compartments. The possibility of plasma membrane IP_3R in platelets is currently under study in our laboratory using purified membranes.

It should be stated that the levels of $In(1,4,5)P_3$ generated by thrombin in platelets are transient and will not explain the elevated Ca^{2+} levels that are sustained with Ca^{2+} influx. The major isomer of IP_3 formed and which persists is $In(1,3,4)P_3$ (Tarver et al 1987) which arises as a consequence of $In(1,4,5)P_3$ phosphorylation to $In(1,3,4,5)P_4$ and dephosphorylation to $In(1,3,4)P_3$. The $In(1,3,4)P_3$ has been shown to have a weak Ca^{2+} mobilising activity in Swiss 3T3 cells (Irvine et al 1986). It would be interesting to determine its specificity towards the plasma membrane IP_3R. Recently possible functions for $In(1,3,4,5)P_4$ particularly for Ca^{2+} entry, has attracted considerable attention as ATP is utilised in its formation from $In(1,4,5)P_3$ and the kinase in question is calmodulin regulated. $In(1,3,4,5)P_4$ in conjunction with $In(1,4,5)P_3$ has been shown to induce Ca^{2+} influx by measuring the activation of a Ca^{2+} dependent K^+ channel in the mouse lacrimal cell (Morris et al 1987). Recently in the endothelial cell $In(1,3,4,5)P_4$ has been shown to open channels that are permeable to Ca^{2+} and Mn^{2+} and this effect was not seen with $In(1,4,5)P_3$ (Luckhoff & Clapham 1992). While these two studies clearly indicate a role for $In(1,3,4,5)P_4$ in Ca^{2+} entry, the entry pathway is not solely dependent upon $In(1,3,4,5)P_4$ generation as non-phosphorylatable analogues of inositol trisphosphate (eg. $In(2,4,5)P_3$) have also been shown to induce Ca^{2+} entry in the mouse lacrimal cell (Bird et al 1991). Other suggested roles for $In(1,3,4,5)P_4$ include Ca^{2+} sequestration in the liver (Hill et al 1988) and Ca^{2+} release in bovine adrenal vesicles (Ely et al 1990, Yoo 1991). That an important function for $In(1,3,4,5)P_4$ is probable, is further supported by studies identifying specific binding sites in a number of tissues and with the purification of the $In(1,3,4,5)P_4$ binding protein(s). However there is presently no consensus as to which protein or proteins constitute the $In(1,3,4,5)P_4$ receptor. Theibert et al (1991,1992) purified two sets of $In(1,3,4,5)P_4$ binding proteins from rat brain and used photoaffinity labelling with ^{125}I-ASA-$In(1,3,4,5)P_4$ to detect proteins of molecular sizes of 182 kDa and a second set of proteins of 174 and 84kDa. Doni & Reiser (1991) using pig brain purified a 42kDa protein and with a similar photoaffinity label were unable to detect any higher molecular species (Reiser et al 1991). This may represent different $In(1,3,4,5)P_4$ binding proteins in different tissues but it should be noted that different conditions for $In(1,3,4,5)P_4$ binding and photoaffinity labelling (i.e. acid pH and presence of PO_4 - Doni & Reiser, neutral pH and no PO_4 - Theibert et al) were used in these studies. A recent study by Cullen & Irvine (1992) indicated that phosphate severely inhibits $In(1,3,4,5)P_4$ binding at neutral pH. So far there are no reported studies of $In(1,3,4,5)P_4$ binding to human platelet membranes. We have recently completed a study using membranes prepared by free flow electrophoresis. With mixed membranes two binding sites are observed with low and high affinity (Kd = 9.3 and 220nM) and these binding sites have been found to be selectively enriched in plasma membranes with similar binding constants (Kd = 28 and 314nM) (Cullen, Patel, Kakkar, Irvine & Authi, manuscript submitted). Importantly $In(1,4,5)P_3$ is a poor antagonist for $In(1,3,4,5)P_4$ binding clearly differentiating this from the IP_3R. This suggests that the $In(1,3,4,5)P_4$ receptor is located on plasma membranes and provides support for an involvement for $In(1,3,4,5)P_4$ in Ca^{2+} entry mechanisms in platelets. This demonstration of an $In(1,3,4,5)P_4$ binding site on highly purified plasma membranes means that we can entertain possible interactions of the $In(1,4,5)P_3$ and $In(1,3,4,5)P_4$ receptors as a mechanism for Ca^{2+} influx in a large number of cells. Studies are in progress to determine the functional relevance of the $In(1,3,4,5)P_4$ receptor and its possible regulation by protein phosphorylation.

Ca^{2+} influx via a receptor-operated Ca^{2+} channel

Ca^{2+} entry pathways stimulated by second messengers provide an explanation of the Ca^{2+} influx seen with most platelet agents such as thrombin, thromboxane, PAF etc. However, there is evidence from rapid time measurements using the stopped-flow technique that ADP is able to induce Ca^{2+} influx before release from intracellular stores (Sage & Rink, 1986; Sage et al, 1989, 1990). Indeed these studies indicated that Ca^{2+} influx by ADP was biphasic with the first phase linked to the opening of a ROC and the second probably linked to store depletion (Sage et al 1990). These findings are particularly interesting as ADP is a poor agonist at stimulating inositol phosphate formation and an action on receptor operated Ca^{2+} channel inducing Ca^{2+} influx would compensate for a poor second messenger response stimulating Ca^{2+} entry. However, it should be noted that in another approach using a continuous and quenched flow technique with not quite so rapid time measurements, Jones and Gear (1988) reported that ADP did give a similar Ca^{2+}

release pattern to thrombin with an initial release occurring from an intracellular store (see also Gear & Raha, this volume). While this controversy over the initial increase by ADP still remains to be resolved, the probability that ADP may directly gate a receptor-operated channel exists. In experiments using platelets attached to a patch clamp, Mahaut-Smith et al (1990, 1992) were able to record single channel activity induced by ADP but not by thrombin. Interestingly a divalent cation channel has been detected in thrombin stimulated platelet membranes fused to a planar lipid bilayer that was not seen in membranes prepared from control platelets (Zschauer et al, 1988). Amazingly this cation channel survived the fractionation procedure and it is of interest to determine its location as the published protocol only resulted in mixed membranes. This location is probably intracellular taking into account the failure of thrombin to induce any channel opening in the patch clamp studies of Mahaut-Smith et al (1990, 1992).

Concluding Statement

This short review has attempted to discuss the important protein entities that exert control on Ca^{2+} levels in platelets. While a number of the major determinants (eg Ca^{2+}ATPases, Na^+/Ca^{2+} exchanger, IP$_3$R) have been identified, we are still removed from identifying the Ca^{2+} channels present on the plasma membranes and the regulation of Ca^{2+} in stores insensitive to $In(1,4,5)P_3$. It is only now that we are beginning to appreciate the isotypes of the proteins expressed and the family of genes to which they belong. Additionally it is becoming clear that no one particular mechanism is sufficient to explain regulation of Ca^{2+} whether this be to keep the levels low in the resting state or the rise seen when the cell is stimulated. The next few years should see a proliferation of these complex mechanisms that regulate Ca^{2+} in the cytosol.

ACKNOWLEDGEMENTS

I would like to express sincere appreciation to all my colleagues (both past at the Royal College of Surgeons of England (particularly Professor N. Crawford, Dr. N. Hack, Dr. J.M. Wilkinson), and present, at the Thrombosis Research Institute (particularly Professor V. V. Kakkar, Dr. F. M. Munkonge, Dr. Shaila Bokkala and Mr. Y. Patel) for many detailed discussions of the work covered. The studies have been supported by grants from the British Heart Foundation and the Thrombosis Research Trust.

REFERENCES

Adunyah S.E. and Dean W.L., 1985, Inositol trisphosphate induced Ca^{2+} release from human platelet membranes, *Biochim. Biophy. Res. Commun.* 128:1274.

Adunyah S.E. and Dean W.L., 1987, Regulation of platelet membrane Ca^{2+} transport by cAMP- and calmodulin dependent phosphorylation, *Biochim. Biophys. Acta* 930:401.

Adunyah S.E., Jones L. R. and Dean W.L., 1988, Structural and functional comparison of a 22 kDa protein from internal human platelet membranes with cardiac phospholamban, *Biochim. Biophys. Acta* 941:63.

Alonso M.T., Alvarez J., Montero M., Sanchez A. and Garcia-Sancho J., 1991, Agonist induced Ca^{2+} influx into human platelets is secondary to emptying of intracellular Ca^{2+} stores, *Biochem. J.* 280: 783.

Alvarez J., Montero M., and Garcia- Sancho J. 1992, Cytochrome P-450 may link intracellular Ca^{2+} stores with plasma membrane Ca^{2+} influx, *Biochem. J.* 274:193.

Authi.K.S., 1992, Localisation of the IP$_3$ binding site on platelet membranes isolated by high voltage free flow electrophoresis, *FEBS Lett.* 298:173.

Authi. K. S. and Crawford. N., 1985, Inositol (1,4,5) trisphosphate induced release of sequestered calcium from highly purified human platelet intracellular membranes, *Biochem. J.* 230:247.

Authi. K.S., Lagarde. M. and Crawford. N., 1985, Diacylglycerol lipase activity in human platelet intracellular and surface membranes. Some kinetic properties and fatty acid specificity, *FEBS Letts.* 180:95.

Authi,K.S., Evenden.B.J., and Crawford. N., 1986, Metabolic and functional consequences of introducing Inositol (1,4,5) trisphosphate into saponin permeabilised platelets, *Biochem. J.* 233:707.

Authi.K.S., Hornby.E.J., Evenden,B.J., and Crawford.N., 1987). Inositol (1,4,5) trisphosphate induced rapid formation of thromboxane B_2 in saponin permeabilised human platelets. Mechanism of action, *FEBS Letts.* 213:95.

Authi.K.S., Rao.G.H.R., Evenden.B.J., and Crawford.N., 1988, The action of guanosine 5'-O(2thiodiphosphate) on thrombin induced activation and calcium mobilisation in intact and saponised human platelets, *Biochem. J.* 255:885.

Authi KS, Bokkala S., Patel Y, Kakkar VV & Munkonge F.M., 1993, Ca^{2+} Release from platelet intracellular stores by thapsigargin and 2,5 di-(t-butyl) 1,4- benzohydroquinone. Relationship to Ca^{2+} pools and relevance in platelet activation, *Biochem. J.* in press.

Berridge M.J., 1991, Cytoplasmic calcium oscillations. A two pool model, *Cell Calcium* 12:63.

Berridge M.J. and Irvine R.F., 1989, Inositol phosphates in cell signalling, *Nature* 341, 197.

Bevers E.M., Comfurius P. and Zwaal R.F.A., 1983, Changes in membrane phospholipid distribution during platelet activation, *Biochim. Biophys. Acta.* 736:57.

Bird G. S. J., Rossier M. F., Hughes A. R., Shears S. B., Armstrong D.L., and Putney Jr. J.W., 1991, Activation of Ca^{2+} entry into acinar cells by a non-phosphorylatable inositol trisphosphate, *Nature* 352:162.

Brandl C., Green N., Korczak B. and MacLennan D.H., 1986, Two Ca^{2+}ATPase genes; homologies and mechanistic implications of deduced amino acid sequences, *Cell* 44:597.

Brass L.F., 1984, Ca^{2+} homeostasis in unstimulated platelets, *J. Biol. Chem.* 259:12563

Brune B. and Ullrich V., 1990, Ca^{2+} mobilisation in human platelets by receptor agonist and Ca^{2+}ATPase inhibitors, *FEBS Letts,* 284:1.

Brune B. and Ullrich V., 1991, Different Ca^{2+} pools in human platelets and their role in thromboxane A_2 formation, *J. Biol. Chem.* 266:19232.

Brune B. and Ullrich V., 1992, Cyclic nucleotides and intracellular Ca^{2+} homeostasis in human platelets, *Eur. J. Biochem.* 207:607.

Burk S., Lytton J., MacLennan D.H. and Shull G., 1989, cDNA cloning, functional expression and mRNA distribution of a third organellar Ca^{2+}-pump, *J. Biol. Chem.* 264:18561.

Carafoli E., 1992, The plasma membrane calcium pump. Structure, function, regulation, *Biochim. Biophys. Acta. Bio-energetics*, 267:2115.

Carey F., Menashi S. and Crawford N., 1982, Localisation of cyclo-oxygenase and thromboxane synthase in human platelet intracellular membranes, *Biochem. J.* 204:847.

Cockcroft S. and Gomperts B.D., 1985, Role of guanine nucleotide binding protein in the activation of polyphosphoinositide phosphodiesterase, *Nature* 314:534.

Courvazier E., Enouf J., Papp B., Gunzburg J. de., Tavitian A. and Levy-Toledano S., 1992, Evidence for a role of rap 1 protein in the regulation of human platelet Ca^{2+} fluxes, *Biochem. J.* 281:325.

Crawford, N., Authi, K.S.,.and Hack, N., 1992, Isolation and characterisation of platelet membranes prepared by free flow electrophoresis, *Meth. Enzymol.* 215:5.

Cullen P.J. and Irvine R.F., 1992, Inositol 1,3,4,5-tetrakisphosphate binding sites in neuronal and non-neuronal tissues, *Biochem. J.* 288:149.

Daniel J. L., 1990, Inositol phosphate metabolism and platelet activation, *Platelets* 1:117.

Daniel J. L., Dangelmair C.A. and Smith J. B., 1989, Calcium modulates the generation of inositol 1,3,4-trisphosphate in human platelets by the activation of inositol 1,4,5-triphosphate 3-kinase, *Biochem. J.* 253:789.

Danoff S.K., Ferris C.D., Donata C. Fisher C. A., Munemitsu S., Ullrich A., Snyder S. H. and Ross C.A., 1991, Inositol 1,4,5-triphosphate receptors; Distinct neuronal and non-neurnal forms by alternative splicing differ in phosphorylation, *Proc. Natl. Aca. Sci. USA.* 88:2951.

Dean W.L. and Sullivan D.M., 1982, Structural and Functional properties of a Ca^{2+}ATPase from human platelets, *J. Biol. Chem.* 257:14390.

Doni F. and Reiser G., 1991, Purification of a high affinity inositol 1,3,4,5-tetrakisphosphate receptor from brain, *Biochem. J.* 275:453.

Ely J. A., Hunyadi L., Baukal A. J. and Catt K. J., 1990, Inositol 1,3,4,5-tetrakisphosphate stimulates Ca^{2+} release from bovine adrenal microsomes by a mechanism independent of inositol 1,4,5-trisphosphate receptor, *Biochem. J.* 268:333.

Enouf J., Bredoux R., Boucheix C., Mirshahi M., Soria C., and Levy-Toledano S., 1985, Possible involvement of two proteins (phosphoprotein and CD9 (p24)) in regulation of platelet calcium fluxes, *FEBS Lett.* 183:398.

Enouf J., Bredoux R., Bourdeau N., and Levy-Toledano S., 1987a, Two different Ca^{2+} transport systems are associated with plasma and intracellular human platelet membranes, *J. Biol. Chem.* 262:9293.

Enouf J., Giraud F., Bredoux R., Bourdeau N., and Levy-Toledano S., 1987b, Possible role of a cAMP dependent phosphorylation in the calcium release mediated by inositol 1,4,5-trisphosphate in human platelet membrane vesicles, *Biochim. Biophys. Acta* 928:76.

Enouf J., Lompre A.M., Bredoux R., Bourdeau N., De la Bastie D. and Levy-Toledano S., 1988, Different sensitivity to trypsin of the human platelet plasma and intracellular membrane Ca^{2+} pumps, *J. Biol.* 263:13922.

Enouf J., Bredoux R., Bourdeau N., Sarkadi B. and Levy-Toledano S., 1989, Further characterisation of the plasma membrane and intracellular membrane associated Ca^{2+} transport systems, *Biochem. J.* 263:547.

Enouf J., Bredoux R., Papp B., Djaffer I., Lompre A.M., Keiffer N., Gayet O., Clemetson K., Wuytack F. and Rosa J-P., 1992, Human platelets express the SECRCA 2b isoform of the Ca^{2+} transport ATPase, *Biochem. J.* 286:135.

Enyedi A., Sarkadi B., Foldes-Papp Z., Monostory S. and Gardos G., 1986, Demonstration of two distinct Ca^{2+} pumps in human platelet membrane vesicles, *J. Biol. Chem.* 261:9558.

Fauvel J., Chap H., Rogues V., Levy-Toledano S. and Douste-Blazy L., 1986, Biochemical characterisation of plasma membranes and intracellular membranes isolated from human platelets using percoll gradients, *Biochim. Biophys. Acta* 856:155.

Ferris C.D. and Snyder S.H., 1992, Inositol 1,4,5,-trisphosphate associated calcium channels, *Annu. Rev. Physiol.* 54:469.

Ferris C.D., Cameron A.M., Bredt D.S., Huganir R.L. and Snyder S.H., 1991, Inositol 1,4,5-trisphosphate receptor is phosphorylated by cyclic AMP-dependent protein kinase at serine 1755 and 1589, *Biochim. Biophys. Res. Commun.* 175:192.

Fisher T.H. and White II. G.C., 1987, Partial purification and characterisation of thrombolamban, a 22kDa cAMP dependent protein kinase substrate in platelets, *Biochim. Biophys. Res. Commun.* 149:700.

Fisher T.H. and White II. G.C., 1989, cAMP -dependent protein kinase substrates in platelets. Evidence that thrombolamban, a 22kDa substrate and the Ca^{2+}ATPase are not associated proteins, *Biochem. Biophys. Res. Commun.* 159:644.

Fisher T.H., Campbell K. P. and White II G.C., 1985, Evidence that platelet and skeletal sarcoplasmic reticulum Ca^{2+}ATPase are structurally distinct, *J. Biol. Chem.* 260:8996.

Fischer T.H., Griffin A.M., Barton D.W. and White II G. C., 1990, Kinetic evidence that arachidonate induced Ca^{2+} efflux from platelet microsomes involves a carrier type ionophoric mechanism, *Biochim. Biophys. Acta* 1022:215.

Fujimoto T., Nakade S., Miyawake A., Mikoshiba K., and Ogawa K., 1992, Localisation of inositol 1,4,5-trisphosphate receptor-like protein in plasmalemmal caveolae, *J. Cell Biol.* 119:1507.

Furakawa K-I., Tawada Y., Shigekawa M., 1989, Protein kinase C activation stimulates plasma membrane Ca^{2+} pump in cultured vascular smooth muscle cells, *J. Biol. Chem.* 264:4844.

Furuichi T., Yoshikawa S., Miyawaki A., Wada K., Maeda N. and Mikoshiba K., 1989, Primary structure and functional expression of the inositol 1,4,5-trisphosphate binding protein P400, *Nature* 342;32.

Galione A., Lee H.C., and Busa W.B., 1991, Ca^{2+} induced Ca^{2+} release in sea urchin egg homogenates; modulation by cADP ribose, *Science* 253:1143.

Grover A.K. and Khan I., 1992, Calcium pump isoforms, diversity, selectivity and plasticity, *Cell Calcium* 13:9.

Guillemette G., Balla T., Baukal A.J. and Catt K.J., 1988, Characterisation of inositol 1,4,5-trisphosphate receptors and Ca^{2+} mobilisation in a hepatic plasma membrane fraction, *J. Biol. Chem.* 263:4541.

Hack N. and Crawford N., 1984, Two dimensional polyacrylamide gel electrophoresis of the proteins and glycoproteins of purified human platelet surface and intracellular membranes, *Biochem. J.* 222:235.

Hack N., Carey F. and Crawford N., 1984, The inhibition of platelet cyclooxygenase by aspirin is associated with the acetylation of a 72kDa polypeptide in the intracellular membranes, *Biochem. J.* 223:105.

Hack N. Croset M. and Crawford N., 1986, Studies on the bivalent cation activated ATPase activities of highly purified human platelet surface and intracellular membranes, *Biochem. J.* 233:661.

Hack N. Wilkinson J.M. and Crawford N., 1988a, A monoclonal antibody (PL/IM 430) to platelet intracellular membranes which inhibits the uptake of Ca^{2+} without affecting the Ca^{2+}Mg^{2+}ATPase, *Biochem. J.* 250:355.

Hack.N. Authi.K.S., and Crawford. N., 1988b, Introduction of antibody (PL/IM 430) to a 100 kDa protein into permeabilised platelets inhibits intracellular sequestration of calcium, *Bioscience Rep.* 8:379.

Heemskerk J.W.M., Hoyland J., Mason W.T. and Sage S.O., 1992, Spiking in cytosolic calcium concentration in single fibrinogen bound Fura 2 loaded human platelets, *Biochem. J.* 283:379.

Henne V. and Soling H-D., 1986, Guanosine 5'-triphosphate releases calcium from rat liver and guinea pig parotid gland endoplasmic reticulum independently of inositol 1,4,5-trisphosphate, *FEBS Lett.* 202: 267.

Henne V. Piiper A. and Soling H-D., 1987, Inositol 1,4,5-trisphosphate and 5'GTP induce release from different intracellular calcium pools, *FEBS Lett.* 218:153.

Herrmann E. and Jakobs K.H., 1988, Stimulation and inhibition of human platelet membrane high affinity GTPase by neomycin, *FEBS Lett.* 229:49.

Hettasch J.M. and Lebreton G.C., 1987, Modulation of Ca^{2+} fluxes in isolated platelet vesicles; effects of cAMP dependent protein kinase and protein kinase inhibitor on Ca^{2+} sequestration and release, *Biochim. Biophys. Acta* 931:41.

Hill T.D., Dean N.M. and Boynton A.L., 1988, Inositol 1,3,4,5-tetrakisphosphate induces calcium sequestration in rat liver cells, *Science* 242:1176.

Hwang S-B., 1991, Specific binding of tritium labelled inositol 1,4,5-trisphosphate to human platelet membranes; ionic and GTP regulation, *Biochim. Biophys. Acta* 1064:351.

Irvine R.F., 1990, "Quantal" calcium release and the control of Ca^{2+} entry by inositol phosphates - a possible mechanism, *FEBS Lett.* 263:5.

Irvine R.F., Letcher A.J., Lander D.J., and Berridge M.J., 1986, Specificity of inositol phosphate stimulated Ca^{2+} mobilisation from Swiss 3T3 cells, *Biochem. J.* 240:301

Johansson J.S. and Haynes D.H., 1988, Deliberate quin 2 overload as a method for in situ characterization of active calcium extrusion systems and cytoplasmic calcium binding. Application to the human platelet, *J. Membr. Biol.* 104:147

Johansson J.S. and Haynes D. H., 1992, Cyclic GMP increases the rate of the Ca^{2+} extrusion pump in intact platelets but has no direct effect on the dense tubular Ca^{2+} accumulation system, *Biophim. Biophys. Acta* 1105:40.

Johansson J.S., Nied L.E. and Haynes D.H., 1992, Cyclic AMP stimulates the Ca^{2+}ATPase mediated Ca^{2+} extrusion from human platelets, *Biochim. Biophys. Acta.* 1105:19.

Jones G.D. and Gear A.L.R., 1988, Subsecond Ca^{2+} dynamics in ADP and thrombin stimulated platelets; a continuous flow approach using Indo-1, *Blood* 71:1539

Kaser-Glansmann R., Jakabova M., George J.N. and Luscher E.F., 1977, Stimulation of Ca^{2+} uptake in platelet membrane vesicles by adenosine 3'5'-cyclic monophosphate and protein kinase, *Biochim. Biophys. Acta* 466:429.

Kass G.E.N., Duddy S.K., Moore G.A. and Orrenhius S., 1989: 2,5-di(tert-butyl)-1,4,-benzohydroquinone rapidly elevates cytosolic Ca^{2+} concentration by mobilising the inositol 1,4,5-trisphosphate sensitive Ca^{2+} pool, *J. Biol. Chem.* 264:15192.

Khan A.A., Steiner J.P., Klein M.A., Schneider M.F. and Snyder S.H., 1992a, IP_3 receptor; localisation to plasma membrane of T cells and co-capping with the T cell receptor, *Science* 257:815.

Khan, A.A., Steiner, J.P. and Snyder S.H., 1992b, Plasma membrane inositol 1,4,5-trisphosphate receptor of lymphocytes; selective enrichment of sialic acid and unique binding specificity, *Proc. Natl. Acad. Sci. USA.* 89:2849.

Kijima Y., Ogunbunmi E. and Fleischer S., 1991, Drug action of thapsigargin on the Ca^{2+} pump protein of sarcoplasmic reticulum, *J. Biol. Chem.* 266:12912.

King W.G., Downes C.P., Prestwich D.G., and Rittenhouse S.E., 1990, Ca^{2+} stimulated and protein kinase C inhibitable accumulation of inositol 1,3,4,6-tetrakisphosphate in human platelets, *Biochem. J.* 270:125.

Kowalska M.A., Rao A.K., and Disa J., 1988, High concentrations of exogenous arachidonate inhibit calcium mobilisation in platelets by stimulation of adenyl cyclase, *Biochem. J.* 253:255.

Kuno M. and Gardner P., 1987, Ion channels activated by inositol 1,4,5-trisphosphate in plasma membrane of human T-lymphocytes, *Nature* 355:356.

Lagarde M., Guichardant M, Menashi S. and Crawford N., 1982, The phospholipid and fatty acid composition of human platelet surface and intracellular membranes isolated by high voltage free flow electrophoresis, *J. Biol. Chem.* 256:3100.

Luckhoff P. and Clapham D.E., 1992, Inositol 1,3,4,5-tetrakisphosphate activates an endothelial Ca^{2+} permeable channel, *Nature* 326:301.

Mahaut-Smith M.P., Sage S.O. and Rink T.J., 1990, Receptor activated single channels in intact human platelets, *J. Biol. Chem.* 265:10479.

Mahaut-Smith M.P., Sage S.O. and Rink T.J., 1992, Rapid ADP evoked currents in human platelets recorded with the nystatin permeabilised patch technique., *J. Biol. Chem.* 267:3060.

Malviya A.N., Rogue P. and Vincendon G., 1990, Stereospecific inositol 1,4,5-trisphosphate binding to isolated rat liver nuclei: Evidence for inositol trisphosphate receptor mediated calcium release from the nucleus, *Proc. Natl. Aca. Sci. USA.* 87:9270.

Menashi S., Wientroub H., and Crawford N., 1981, Characterisation of human platelet surface and intracellular membranes by free flow electrophoresis, *J. Biol. Chem.* 256:4095.

Menashi. S., Authi.K.S., Carey.F.,and Crawford. N., 1984, Characterisation of the calcium sequestering process associated with human platelet intracellular membranes isolated by free flow electrophoresis, *Biochem. J.* 222:413.

Mignery G.A., Sudhof T.C., Takei K. and De Camilli P., 1989, Putative receptor for inositol 1,4,5-trisphosphate similar to ryanodine receptor, *Nature* 342:192.

Milner R.E., Baksh S., Shemanko C., Carpenter M.R., Smillie L., Vance J.E., Opas M. and Michalak M., 1991, Calreticulin, not calsequestrin is the major calcium binding protein of smooth muscle sarcoplasmic reticulum, *J. Biol. Chem.* 266:7155.

Miyawaki A., Furuichi T., Ruou Y., Yoshikawa S., Nakagawa T., Saitoh T. and Mikoshiba K., 1991, Structure function relationships of the mouse inositol 1,4,5-trisphosphate receptor, *Proc. Natl. Acad. Sci. USA.* 88:4911.

Morris A.P., Gallacher D.V., Irvine R.F., and Peterson O.H., 1987, Synergism of inositol trisphosphate and tetrakisphosphate in activating Ca^{2+} dependent K^+ channels, *Nature* 330:653.

Mullaney J.M., Yu M., Ghosh T.K. and Gill D.L., 1988, Ca^{2+} entry into the inositol 1,4,5-trisphosphate releasable calcium pool is mediated by a GTP regulatory mechanism, *Proc. Natl. Acad. Sci. USA.* 85:2499.

Nagakawa T., Okano H., Furuichi T., Aruga J. and Mikoshiba K., 1991, The subtypes of the mouse inositol 1,4,5-trisphosphate receptor are expressed in a tissue specific manner, *Proc. Natl. Acad. Sci. USA.* 88:6244.

Nakade S., Maeda N., and Mikoshiba K., 1991, Involvement of the C-terminus of the inositol 1,4,5-trisphosphate receptor in Ca^{2+} release analysed using region specific monoclonal antibodies, *Biochem. J.* 277:125.

Nakashima S., Tohmatsu T., Shirato L., Takenaka A., and Nozawa Y., 1987, Neomycin is a potent agent for arachidonic acid release in human platelets, *Biochim. Biophys. Res. Commun.* 146:820.

Nishikawa M., Hidaka H. and Shirakawa S., 1988, Possible involvement of direct stimulation of protein kinase C by unsaturated fatty acids in platelet activation, *Biochem. Pharmac.* 37:3079.

Ohmori T., Kikuchi A., Yamamoto K., Kim S. and Takai V., 1989, Small molecular weight GTP binding proteins in human platelet membranes, *J. Biol. Chem.* 264:1877.

O'Rourke F.A. and Feinstein M.B., 1990, The inositol 1,4,5-trisphosphate receptor binding sites of platelet membranes, *Biochem. J.* 267:297.

O'Rourke F.A., Halenda S.P., Zavioco G.B. and Feinstein M.B., 1985, Inositol 1,4,5-trisphosphate release Ca^{2+} from a Ca^{2+} transporting membrane vesicle fraction derived from human platelets, *J. Biol. Chem.* 260:956.

O'Rourke F.A., Zavioco G.B. and Feinstein M.B., 1989, Release of Ca^{2+} by inositol 1,4,5-trisphosphate in platelet membrane vesicles is not dependent on cAMP dependent protein kinase, *Biochem. J.* 257:715.

Papp B., Enyedi A., Kovacs T., Sarkadi B., Wuytack F., Thastrup O., Gardos G., Bredoux R., Levy-Toledano S. and Enouf J., 1991, Demonstration of two forms of the Ca^{2+} pumps by Thapsigargin inhibition and radioimmunoblotting in platelet membrane vesicles, *J. Biol. Chem.* 266:14593.

Papp B. Enyedi A., Paszty K., Kovacs T., Sarkadi B., Gardos G., Magnier C., Wuytack F. and Enouf J., 1992, Simultaneous presence of two distinct endoplasmic reticulum type calcium isoforms in human cells, *Biochem. J.* 288:297.

Polascik T., Gofrey P.P. and Watson S.P., 1987, Neomycin cannot be used as a selective inhibitor of inositol phospholipid hydrolysis in intact or semi-permeabilised human platelets, *Biochem. J.* 243:815.

Pollock W.K., Rink T.J., and Irvine R.F. (1986) Liberation of [³H] arachidonic acid and changes in cytosolic free Ca^{2+} in Fura 2 loaded human platelets stimulated by ionomycin and collagen, *Biochem. J.* 235:869.

Pollock W.K., Sage S.O. and Rink T.J., 1987, Stimulation of Ca^{2+} efflux from Fura 2 loaded platelets activated by thrombin or phorbol myristate acetate, *FEBS Lett.* 210:132.

Putney J.W. Jr., 1986, A model for receptor regulated calcium entry, *Cell Calcium* 7:1.

Putney J.W. Jr., 1990, Capacitative calcium entry revisited, *Cell Calcium* 11:611.

Quinton T.M. and Dean W.L., 1992, Cyclic AMP dependent phosphorylation of the inositol 1,4,5-trisphosphate receptor inhibits Ca^{2+} release from platelet membranes, *Biochim. Biophys. Res. Commun.* 184:893.

Reiser G., Schafer R., Donie F., Hulser E., Nehls-Sahabandu M., and Mayr G.W., 1991, A high affinity inositol 1,3,4,5-tetrakisphosphate receptor protein from brain is specifically labelled by a newly synthesised photoaffinity analogue, N-(4-azidosalicyl)aminoethanol(1)-1-phospho-D-myo-inositol 3,4,5-trisphosphate, *Biochem. J.* 280:533.

Rengasamy A., Soura S. and Feinberg H., 1987, Platelet Ca^{2+} homeostasis; Na/Ca^{2+} exchange in plasma membrane vesicles, *Thrombos. Haemost.* 57:337.

Rengasamy A. and Feinberg H., 1988, Inositol 1,4,5-trisphosphate induced Ca^{2+} release from platelet plasma membrane vesicles, *Biochim. Biophys. Res. Commun.* 150:1021.

Rink T. J. and Hallam T.J., 1984, What turns platelets on, *TIBS* 9:215.

Robblee L.S., Shepro D. and Belamarich F.A., 1983, Calcium uptake and associated adenosine triphosphatase activity of isolated platelet membranes, *J. Gen. Physiol.* 1:462.

Ross C.A., Danoff S.K., Schell M.J., Snyder S.H. and Ullrich A., 1992, Three additional inositol 1,4,5-trisphosphate receptors; Molecular cloning and differential localisation in brain and peripheral tissues, *Proc. Natl. Acad. Sci. USA.* 89:4265.

Rossier M.F., Capponi A.M., and Vallotton M.B., 1989, The inositol 1,4,5-trisphosphate binding site in adrenal cortical cells is distinct from the endoplasmic reticulum, *J. Biol. Chem.* 264:14078.

Rossier M.F., Bird G.S.J. and Putney J.W.Jr., 1991, Subcellular distribution of the calcium storing inositol 1,4,5-trisphosphate sensitive organnelle in rat liver, *Biochem. J.* 274:643.

Rusinko N. and Lee H-C., 1989, Widespread occurence in animal tissues of an enzyme catalysing the conversion of NAD^+ into a cyclic metabolite with intracellular Ca^{2+} mobilising activity, *J. Biol. Chem.* 264:11725.

Sagara Y., Wade J.B. and Inesi G., 1992, A conformational mechanism for the formation of a dead-end complex by the sarcoplasmic reticulum ATPase with thapsigargin, *J. Biol. Chem.* 267:1286.

Sage S.O. and Rink T.J., 1986, Kinetic differences between thrombin induced and ADP-induced Ca^{2+} influx and release from internal stores in Fura 2 loaded human platelets, *Biochem. Biophys. Res. Commun.* 136:1124.

Sage S.O., Merrit J.E., Hallam T.J and Rink T.J., 1989, Receptor mediated Ca^{2+} entry in Fura 2 loaded human platelets stimulated with ADP and thrombin. Dual wavelength studies with Mn^{2+}, *Biochem. J.* 258:923.

Sage S.O., Reast R. and Rink T.J., 1990, ADP evokes biphasic Ca^{2+} influx in Fura 2 loaded human platelets. Evidence for Ca^{2+} entry regulated by the intracellular Ca^{2+} store, *Biochem. J.* 265:675.

Sargeant P., Clarkson W.D., Sage S.O. and Heemskerk J.W.M., 1992, Calcium influx evoked by Ca^{2+} store depletion in human platelets is more susceptable to cytochrome P-450 inhibitors than receptor regulated Ca^{2+} entry, *Cell Calcium* 13:553.

Siess W., Winegar D.A. and Lapetina E.G., 1990, Rap 1b is phosphorylated by protein kinase A in intact human platelets, *Biochem. Biophys. Res. Commun.* 170:944.

Statland B.E., Heagen B.M. and White J.G., 1969, Uptake of calcium by platelet relaxing factor, *Nature* 223:521.

Steiner B. and Lusher E.F., 1985, Evidence that the platelet plasma membrane does not contain a $(Ca^{2+} + Mg2+)$-dependent ATPase, *Biochim. Biophys. Acta.* 818:299.

Sudhof T.C., Newton C.L., Archer III B.I., Ushkaryov Y.A. and Mignery G.A., 1991, Structure of a novel IP_3 receptor, *EMBO J.* 10:3199.

Supattapone S., Worley P.F., Baraban J.M. and Snyder S.H., 1988a, Solubilisation, purification and characterisation of an inositol trisphosphate receptor, *J. Biol. Chem.* 263:1530.

Supattapone S., Danoff S.K., Theibert A., Joseph S., Steiner J. and Snyder S.H., 1988b, Cyclic AMP-dependent phosphorylation of a brain inositol trisphosphate receptor decreases its release of calcium, *Proc. Natl. Acad. Sci. USA.* 85:8747.

Tao J., Johansson J.S. and Haynes D.H., 1992, Stimulation of dense tubuler Ca^{2+} uptake in human platelets by cAMP, *Biochim. Biophys. Acta.* 1105:29.

Tarver A.P., King W.G. and Rittenhouse S.E., 1990, Inositol 1,4,5-trisphosphate and inositol 1,2 cyclic 4,5-trisphosphate are minor components of total mass of inositol trisphosphate in thrombin stimulated platelets, *J. Biol. Chem.* 262:17268

Thastrup O., Cullen P., Drobak B.K., Hanley M.R. and Dawson A.P., 1990, Thapsigargin, a tumour promoter discharges intracelullar Ca^{2+} stores by specific inhibition of the endoplasmic reticulum Ca^{2+}ATPase, *Proc. Natl. Acad. Sci. USA.* 87:2466.

Theibert A.B., Estevez V.A., Ferris C.D., Danoff S.K., Barrow R.K., Prestwich G.D. and Snyder S.H., 1991, Inositol 1,3,4,5-tetrakisphosphate and inositol hexakisphosphate receptor proteins. Isolation and characterisation from rat brain, *Proc. Natl. Acad. Sci. USA.* 88:3165.

Theibert A.B., Estevez V.A., Mourey R.J., Maracek J.F. Barrow R.K., Prestwich G.D. and Snyder S.H., 1992, Photoaffinity labelling and characterisation of inositol 1,3,4,5-tetrakisphosphate and inositol hexakisphosphate binding proteins, *J. Biol. Chem.* 267:9071.

Tohmatsu T., Nishida A., Nagao S., Nakashima S. and Nozawa Y., 1989a, Inhibitory action of cAMP on inositol 1,4,5-trisphosphate induced Ca^{2+} release in saponin permeabilised platelets, *Biochim. Biophys. Acta.* 1013:190.

Tohmatsu T., Nakashima S. and Nozawa Y., 1989b, Evidence for Ca^{2+} mobilising action of arachidonic acid in human platelets, *Biochim. Biophys. Acta.* 1012:97

Treves S., DeMattei., Lanfredi M., Villa A., Green N.M., MacLennan D.H., Meldolesi J. and Pozzan T., 1990, Calreticulin is a candidate for a calsequestrin like function in Ca^{2+} storage compartments (calcisomes) of liver and brain, *Biochem. J.* 271:473.

Tsunoda Y., Matsuno K. and Tashiro Y., 1988, Spatial distribution and temporal change of cytoplasmic free calcium in human platelets, *Biochem. Biophys. Res. Commun.* 156:1152.

Van Delden C., Favre C., Spat A., Cerny E., Krause K-H. and Lew D.P., 1992, Purification of an inositol 1,4,5-trisphosphate-binding calreticulin-containing intracellular compartment of HL-60 cells, *Biochem. J.* 281:651.

Vigne P., Breittmayer J-P., Duval D., Freline C. and Luzdunski M., 1988, The Na^+/Ca^{2+} antiporter in aortic smooth muscle cells, *J. Biol. Chem.* 263:8078.

Volpe P., Krause K-H., Hashimoto S., Zorzato F., Pozzan T., Meldolesi J. and Lew D.P., 1990, 'Calciosome', a cytoplasmic organnelle, the inositol 1,4,5-trisphosphate sensitive Ca^{2+} store of non-muscle? *Proc. Natl. Aca. Sci. USA.* 85:1091.

Vostal J.G., Jackson W.L. and Shulman N.R., 1992, Cytosolic and stored calcium antagonistically control tyrosine phosphorylation of specific platelet proteins, *J. Biol. Chem.* 266:16911.

Waldman R., Bauer S., Gobel C., Hofman F., Jacobs K.H. and Walter U., 1986, Demonstration of cGMP dependent protein kinase and cGMP dependent phosphorylation in cell free extracts of platelets, *Eur. J. Biochem.* 158:203.

Waldman R., Nieberding M. and Walter U., 1987, Vasodilator stimulated protein phosphorylation in platelets is mediated by cAMP- and cGMP-dependent protein kinases., *Eur. J. Biochem.* 167:441.

Wang K.W.W., Wright L.C., Machan C.L., Allen B.G., Conigrave A.G. and Rougfogalis B.D., 1991, Protein kinase C phosphorylates the carboxyl terminus of the plasma membrane Ca^{2+}ATPase from human erythrocytes, *J. Biol. Chem.* 266:9078.

Watson S.P. Ruggeiro M., Abrahams S.L. and Lapetina E.G., 1986, Inositol trisphosphate induces aggregation and release of 5-hydroxytryptamine from saponin permeabilised human platelets, *J. Biol. Chem.* 261:5368.

White J. G., Rao G.H.R. and Gerrard J.M., 1974, Effects of ionophore A23187 on blood platelets. I. Influence on aggregation and secretion, *Am. J. Pathol.* 77:135.

White II. G.C., Barton D.W., White T.E. and Fisher T.H., 1989, Cyclic AMP-dependent protein kinase does not increase calcium transport in platelet microsomes, *Thrombos. Res.* 56:575.

White T.E., Lacal J.C., Reep B., Fisher T.H., Lapetina E.G. and White II. G.C., 1990, Thrombolamban, the 22 kDa platelet substrate of cyclic AMP dependent protein kinase, is immunologically homologous with the ras family of GTP-binding proteins, *Proc. Natl. Aca. Sci. USA.* 87:758.

Wuytack F., Eggerman J.A., Raeymackers L., Pleggers L. and Casteels R., 1989, Antibodies against the non-muscle isoform of the endoplasmic reticulum Ca^{2+} transport ATPase, *Biochem. J.* 264:765.

Yoo S.H., 1991, Inositol 1,3,4,5-tetrakisphosphate induced Ca^{2+} sequestration into bovine adrenal medullary secretory vesicles, *Biochem. J.* 278:381.

Yoshida K. and Nachmias V.T., 1987, Phorbol ester stimulates calcium sequestration in saponised human platelets, *J. Biol. Chem.* 262:16048.

Zschauer A., van Breemen C., Buhler F.R. and Nelson M.T., 1988, Calcium channels in thrombin activated human platelet membrane, *Nature* 334:703.

THE USE OF INHIBITORS OF PROTEIN KINASES AND PROTEIN PHOSPHATASES TO INVESTIGATE THE ROLE OF PROTEIN PHOSPHORYLATION IN PLATELET ACTIVATION

Steve P Watson, Robert A Blake, Trevor Lane and Trevor R Walker

Department of Pharmacology
Mansfield Road
Oxford
OX1 3QT

INTRODUCTION

A diverse range of agents have been shown to activate human platelets through cell surface proteins, including thrombin, collagen, ADP, thromboxane A2, vasopressin, adrenaline, platelet activating factor and various monoclonal antibodies. In the majority of cases, activation is associated with stimulated formation of the second messengers inositol 1,4,5 trisphosphate (IP3) and 1,2-diacylglycerol (DG), produced by hydrolysis of the minor membrane phospholipid, phosphatidylinositol 4,5-bisphosphate; receptors for ADP and adrenaline are notable exceptions. IP3 releases Ca^{2+} from intracellular stores and DG activates protein kinase C. These two messengers interact synergistically in a way that is poorly understood to induce platelet shape change, aggregation and secretion.

The aim of this article is to describe work performed in the authors' laboratory over the last few years which has been designed to address the role of protein phosphorylation in the sequaela of events that lead to the above functional responses. The predominant approach in these studies has been to use a range of inhibitors of protein kinases and protein phosphatases in combination with receptor agonists and post-receptor stimuli e.g. phorbol esters and Ca^{2+} ionophores. It should be remembered throughout this chapter that nearly all of the inhibitors used cannot be described as specific, and that in all likelihood they are exerting additional effects to those considered in the text. Moreover, protein kinase C and Ca^{2+}-activated kinases are known to regulate the activity of additional, unidentified kinases which may also participate in aggregation and secretion (Ferrell & Martin, 1989). Since the identity of these kinases is largely unknown, the ability of the inhibitors described in this article to alter their activity cannot be readily tested. This recruitment of additional kinases is likely to be complicated further by the probable regulation of protein phosphatases through phosphorylation, which, in turn, may regulate further kinases. Nevertheless, by using a pharmacopoeia of agents, it is possible to make preliminary conclusions on the relative importance of IP3 and DG in triggering platelet activation. The specific questions to be addressed in this article concern the role of protein kinase C, tyrosine kinases and phosphorylation independent events in platelet activation. A more complete understanding of the mechanism of secretion and aggregation, however, will require a full molecular description of the two processes, including identification of the proteins that participate in vesicular fusion/granule secretion, and the control of activation of glycoprotein IIb/IIIa exposure, the receptor for fibrinogen which cross-links platelets leading to aggregation.

ROLE OF PROTEIN KINASE C

More than a decade has elapsed since Nishizuka and colleagues performed their pioneering studies on platelets in which they proposed that platelet activation is mediated through a synergism between protein kinases C and mobilisation of Ca^{2+} (Nishizuka, 1983). The essential foundation for this hypothesis was the observation that sub-maximal concentrations of a phorbol ester e.g. phorbol dibutyrate, and a Ca^{2+} ionophore, e.g. A23187, interact synergistically to stimulate secretion and aggregation. A maximally effective concentration of phorbol dibutyrate induces less than 5% secretion of [3H]5-hydroxytryptamine in 60s, but synergises with a sub-maximal concentration of A23187 (which on its own also produces less than 5% secretion within 60s), to produce a response which approaches that seen with a maximally-effective concentration of thrombin (Walker and Watson, 1993). On the other hand, there is a less marked synergism between the phorbol ester and a relatively high concentration of A23187 ($1\mu M$), in part, because at this concentration A23187 induces a relatively large response on its own (Table 1).

Studies of this kind, however, do not mimic the temporal and spatial pattern of activation of protein kinase C and mobilisation of Ca^{2+} that occur following platelet activation by a physiological agonist, and therefore do not establish the relative importance of protein kinase C and Ca^{2+} in triggering platelet activation. Moreover, there is evidence that, under certain conditions, protein kinase C is able to inhibit platelet activation. For

Table 1. Effect of Ro 31-8220 on secretion and phosphorylation.

	5-HT (% tissue level)	47kDa (% basal)	20kDa (% basal)
Thrombin	65.1±1.3	336±15	232±16
Thrombin + Ro 31-8220	4.3±1.2	114±8	159±15
A23187	33.5±4.9	212±20	176±6
A23187 + Ro 31-8220	13.6±2.9	76±6	132±9
PDBu	3.3±1.0	345±39	160±10
PDBu + Ro 31-8220	0.7±0.6	90±6	95±11
A23187 + PDBu	66.5±2.0	n.d.	n.d.
A23187 + PDBu + Ro 31-8220	11.7±4.6	n.d.	n.d.

Results are expressed as % + s. e. mean of tissue levels of [3H]5-hydroxytryptamine [5HT] or % + s. e. mean of basal phosphorylation from at least three separate experiments. Intact platelets were stimulated for 1 min following pre-incubation for 1 min with Ro 31-8220 ($3\mu M$). Concentrations were thrombin (1 Unit/ml), A23187 ($1\mu M$) and PDBu (300nM). n.d. - not determined.

example, the now established ability of protein kinase C to inhibit agonist-induced formation of IP3 was first demonstrated in the human platelet (Watson and Lapetina, 1984). Further, protein kinase C has also been reported to promote the extrusion of Ca^{2+} in platelets (Pollock et al, 1987; Watson and Hambleton, 1989). In order to establish the role of protein kinase C in platelet activation, therefore, it is necessary to study either the effect of inhibitors of DG metabolism, so as to increase agonist stimulated formation of DG, or to study the effect of inhibitors of protein kinase C on platelet functional responses. DG undergoes two routes of metabolism in platelets. It is converted to phosphatidic acid by the action of DG kinase or it is metabolised by diacylglycerol lipase to monoacylglycerol and stearic acid. At least three forms of DG-kinase (DGK I, II and III) have been identified in platelets (Yada et al, 1990). The DG-kinase inhibitor R59022, described by de Chaffoy de Courcelles et al (1985), inhibits DGK II and III but has little effect on DGK I (Yada et al, 1990). We have observed a small potentiation of platelet aggregation and secretion to thrombin in the presence of R59022, and the related DG-kinase inhibitor R59949 (de Chaffoy de Courcelles et al, 1989), at concentrations which cause approximately 50% inhibition of DG-kinase activity as judged by the conversion of the membrane-permeable diacylglycerol, dioctanoylglycerol, to dioctanoylphosphatidic acid, the product of DG kinase (Nunn et al, 1987; Rodriguez-Linares et al, 1991, respectively). In the study with R59022 we also observed a marked increase in the level of DG in thrombin-stimulated platelets (a similar measurement was not performed with R59949). These results are consistent with a role of the DG/protein kinase C system in triggering aggregation and secretion. However, we were prevented from using higher concentrations of R59022 or R59949 in these studies because of non-specific effects. Moreover, the incomplete inhibition of the conversion of dioctanoylglycerol to dioctanoylglycerol phosphatidic acid by R59022 indicates that DGK-I is also involved in the metabolism of DG and that this approach cannot produce a complete inhibition of this route of metabolism.

STUDIES WITH PROTEIN KINASE C INHIBITORS

We have also carried out a comparative series of experiments with two compounds claimed to be inhibitors of protein kinase C, staurosporine (Tamaoki et al, 1986) and Ro 31-8220 (Davis et al, 1989). Staurosporine (Figure 1) is a microbial alkaloid which is now recognised as a relatively non-selective inhibitor of both serine-threonine and tyrosine kinases (Watson et al, 1988; Fallon, 1990; Kocher & Clementson, 1991); it inhibits kinase activity through competition with ATP thereby accounting for its non-selective action. An analogue of staurosporine, Ro 31-8220 (Figure 1), designed by Davis et al, (1989), shows a far greater selectivity to protein kinase C relative to a number of other protein kinases. However, despite the fact that, to date, there is no published evidence to suggest that Ro 31-8220 is not selective to protein kinase C, the large number of kinases in existence and its mechanism of inhibition through competition with ATP make it unlikely that Ro 31-8220 will be entirely selective.

Maximally effective concentrations of both staurosporine (1-10 μM; Watson et al, 1988) and Ro 31-8220 (10 μM; Figure 2) were found to inhibit completely thrombin induced phosphorylation of the major protein kinase C substrate in human platelets, a protein of 47kDa (pleckstrin) with unknown function (Imaoka et al, 1983; Tyers et al, 1989). In contrast, staurosporine inhibited completely phosphorylation of myosin light chain (Watson et al, 1988), a substrate for both protein kinase C and myosin light chain kinase (MLCK) (Naka et al, 1983), while Ro 31-8220 caused only a partial inhibition (Figure 2). Analyses of tryptic digests demonstrate that Ro 31-8220 inhibits the protein kinase C component of phosphorylation of myosin light chains but not that induced by MLCK (Walker and Watson, 1993). We have also shown that staurosporine inhibits

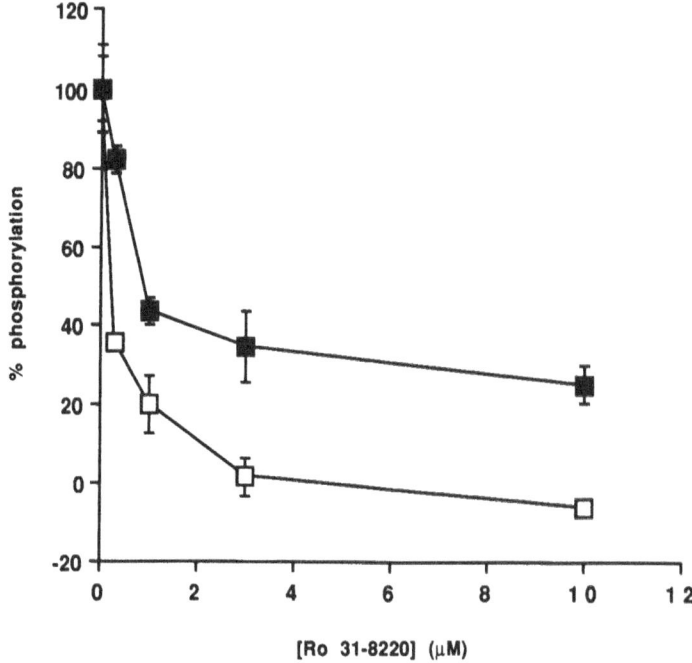

Figure 1   Structures of staurosporine and Ro 31-8220

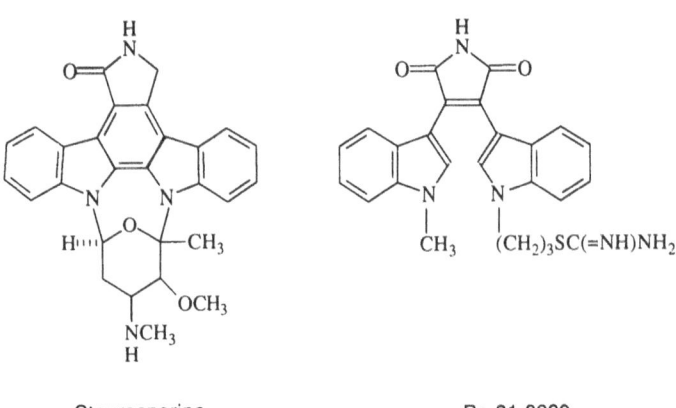

Staurosporine   Ro 31-8220

Figure 2   Effect of Ro 31-8220 on platelet protein phosphorylation. Platelets were prelabelled with
[^{32}P$_i$] and incubated with Ro 31-8220 for 1 minute at varying concentrations as indicated
prior to stimulation with thrombin (1 Unit/ml). Phosphorylation of pleckstrin and myosin
light chain was localised by autoradiography following polyacrylamide gel electrophoresis
(11%), excised from the gel and counted in a liquid scintillation counter. Results are
expressed as mean + s.e.m. (n=3-5) of % increase in phosphorylation induced by
thrombin.

tyrosine phosphorylation of a large number of proteins in platelets, while this is unaffected
by Ro 31-8220 (Blake et al, 1993). These data demonstrate the increased selectivity of Ro
31-8220 for protein kinase C relative to staurosporine, although we cannot rule out the
possibility that Ro 31-8220 is also able to inhibit other unidentified kinases in platelets over
a similar concentration range.

We have used maximally effective concentrations of staurosporine and Ro 31-8220 to investigate the role of protein phosphorylation and protein kinase C, respectively, in functional responses induced by thrombin and the non-receptor stimuli, phorbol dibutyrate and A23187. Ro 31-8220 causes a marked inhibition of thrombin-induced secretion of [^{3}H]5-hydroxytryptamine from approximately 65% of total tissue levels released in 60s to approximately 5%; however, secretion is not inhibited completely (Table 1). Similar results are seen with staurosporine (not shown). The degree of inhibition of secretion by maximally effective concentrations of staurosporine or Ro 31-8220 is not significantly different, and no further inhibition is observed when the two inhibitors are given together (not shown). These results indicate a common mechanism of inhibition of secretion by staurosporine and Ro 31-8220, which is most likely mediated through the inhibition of protein kinase C. The staurosporine/Ro 31-8220 - insensitive component of secretion may be mediated through the mobilisation of Ca^{2+} and, since staurosporine is a non-selective inhibitor of kinases, appears to be phosphorylation independent. It therefore appears that on their own protein kinase C and Ca^{2+} elicit a relatively small, but significant secretory response and that it is their synergistic interaction which is the major stimulus in triggering secretion by agonists such as thrombin.

Further evidence in support of this mechanism comes from studies with phorbol dibutyrate and A23187. The remarkable synergy observed between a maximal concentration of phorbol dibutyrate and a sub-maximal concentration of A23187 (Walker & Watson, 1993) is similar to that proposed above for secretion induced by thrombin. The relatively large secretory response to higher concentrations of A23187 (Table 1) may therefore be explained by an activation of protein kinase C. Consistent with this, high concentrations of A23187 induce a significant phosphorylation of the major protein kinase C substrate in platelets, the 47kDa protein (Siess & Lapetina, 1988) which can be inhibited completely by Ro 31-8220 (Table 1). Moreover, secretion of [^{3}H]5-hydroxytyptamine induced by A23187 and phorbol dibutyrate, is inhibited from more than 60% of tissue levels to below 20% in the presence of 10μM Ro 31-8220 or 10μM staurosporine (Table 1). The larger Ro 31-8220/staurosporine-insensitive component of secretion induced by A23187, relative to thrombin, most likely reflects differing spatial and temporal patterns of Ca^{2+} mobilisation between the stimuli.

A similar mechanism of activation appears to occur in regard to aggregation induced by thrombin or by the combination of A23187 and phorbol dibutyrate. On its own a maximal concentration of phorbol dibutyrate induces a very slow rate of aggregation relative to that seen with thrombin or with high concentrations of A23187. The rate of aggregation induced by A23187 (not shown) or thrombin (Figure 3) is slightly reduced by maximally effective concentrations of Ro 31-8220 or staurosporine (Watson et al, 1988; Walker & Watson, 1993) providing further evidence for a role of protein kinase C in this response. Moreover, the inhibitory actions of Ro 31-8220 and staurosporine are not additive suggesting that they are mediated through a similar mechanism. Bearing in mind the arguments described above in relation to secretion, it seems likely that the staurosporine/Ro 31-8220 resistant component of aggregation is mediated through a Ca^{2+}-mediated, phosphorylation independent pathway (Watson and Hambleton, 1989). Under normal physiological conditions, this component synergises with protein kinase C.

In contrast to the results for secretion and aggregation, staurosporine and Ro 31-8220 have very different effects on thrombin-induced shape change. Whilst staurosporine inhibits completely the shape change response induced by thrombin, Ro 31-8220 has no apparent effect or appears to increase the degree of shape change possibly through the inhibition of aggregation (Figure 3). The most likely explanation for these results is that shape change is a phosphorylation-driven event, presumably mediated by MLCK (Daniels et al, 1984), which does not involve protein kinase C. This is consistent

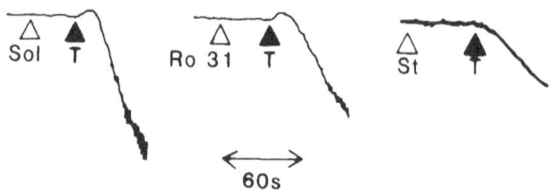

Figure 3   Ro 31-8220 and staurosporine inhibit of aggregation induced by thrombin; however, only staurosporine blocks shape change. Shown is an example trace that is representative of more than ten separate experiments. Thr (Thrombin, 1U/ml); St (staurosporine, 10μM); Ro (Ro 31-8220, 10μM).

with the inability of protein kinase C-activation to elicit shape change (Rink et al, 1983).

STUDIES WITH A PSEUDOSUBSTRATE INHIBITOR OF PROTEIN KINASE C

As mentioned above, the selectivity of Ro 31-8220 is not fully established, and it is possible that it also inhibits additional, uncharacterised protein kinases which play a role in secretion. One approach to this problem is to use a further inhibitor of protein kinase C of a different structural class and with a different mechanism of inhibition. Therefore we have investigated the action of the pseudosubstrate peptide inhibitor of protein kinase C designed by House and Kemp (1987). This pseudosubstrate peptide inhibits protein kinase C at its catalytic site through competition with cellular substrates. In order to deliver the inhibitor to platelets, it is necessary to permeabilize the plasma membrane which can be achieved by incubation with the detergent saponin in a buffer designed to mimic the intracellular composition of the cell. Under these conditions, preliminary studies demonstrate that, at a concentration of 30μM, the pseudosubstrate peptide inhibits completely thrombin-induced [^{32}P]orthophosphate incorporation into the 47kDa substrate for protein kinase C, the [^{32}P]orthophosphate being donated by [^{32}P]ATP (see also King et al, 1991).

The use of a Ca^{2+}-buffering system made up of defined Ca^{2+}:EGTA ratios (Fabiato & Fabiato, 1979), prevents thrombin-induced changes in "intracellular" Ca^{2+}. Therefore, under these conditions the secretory response to thrombin shown in Figure 4 is most likely mediated by generation of DG. The ability of the pseudosubstrate inhibitor of protein kinase C to block completely thrombin-induced secretion is consistent with a physiological role of the DG/protein kinase C pathway in triggering this response and therefore is in agreement with the studies described above with Ro 31-8220.

In support of this, phorbol dibutyrate also stimulated secretion in permeabilized platelets and this was also inhibited completely by the pseudosubstrate inhibitor. The ability of phorbol dibutyrate to stimulate secretion was enhanced at higher concentrations of Ca^{2+} again demonstrating a synergistic interaction. The ability of the pseudosubstrate peptide to block secretion by 2μM Ca^{2+} alone may reflect the ability of Ca^{2+} to prime the activation of protein kinase C (Siess & Lapetina, 1988) and is consistent with the marked synergism between Ca^{2+} and phorbol ester described previously in electropermeabilized platelets (Coorsen et al,1990). The studies with the pseudosubstrate inhibitor therefore are in accordance with those for Ro 31-8220 i.e. that protein kinase C plays an important role in triggering secretion through a synergism with Ca^{2+} and that it is the synergism which is the major stimulus in triggering this response.

The arguments developed in this section for the roles of Ca^{2+}, protein kinase C and protein phosphorylation in platelet responses are summarised in Figure 5.

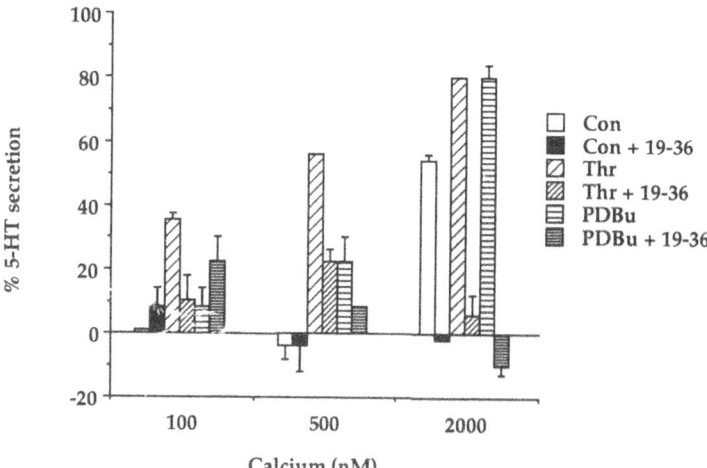

Figure 4   The pseudosubstrate inhibitor of protein kinase C inhibits secretion-induced by thrombin in permeabilized platelets. Twice-washed platelets (2×10^8/ml) were suspended in buffer at $37°C$ (composition in mM: KCl 120, MgCl$_2$ 4, NaCl 25, NaH$_2$PO$_4$ 1, EGTA 1, [γ-^{32}P]ATP 0.625, HEPES 15, indomethacin 0.01 - after King et al, 1991). The buffer also contained Ca^{2+} to give the predicted concentrations shown in the figure (Fabiato and Fabiato, 1979). The pseudosubstrate inhibitor (30μM), $\alpha\alpha$ 19-36, was given 4 min before challenge with 1 Unit/ml thrombin (Thr) or 300nM phorbol dibutyrate (PDBu); Con=control. The incubation was terminated after 5 minutes. Results are expressed as mean $\pm$ range for an experiment performed in duplicate which is representative of two other similar studies. $\alpha\alpha$ 19-36 had no effect on secretion induced by thrombin in intact platelets (not shown).

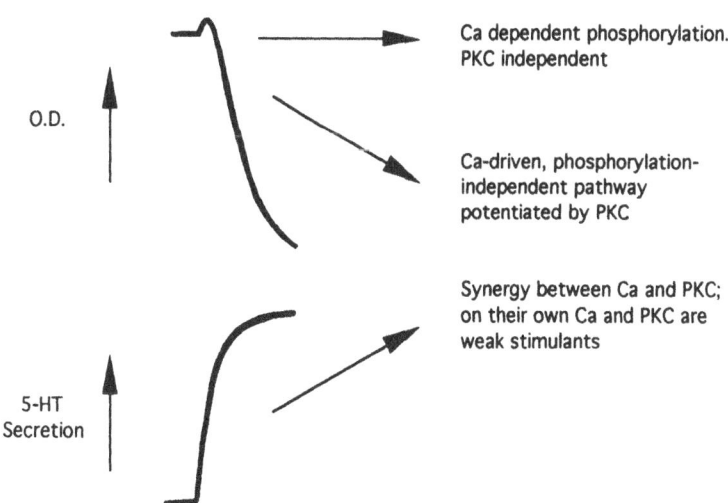

Figure 5   Proposed scheme for the role of protein kinase C, protein phosphorylation and Ca^{2+} mobilisation in triggering platelet responses.

TYROSINE PHOSPHORYLATION IN PLATELET ACTIVATION

The above section focussed on the relative contributions of IP3/Ca^{2+} and DG/protein kinase C in triggering platelet responses mediated by phospholipase C-coupled receptors. However, as discussed above, other kinases may also play a role in activation downstream from the two messengers. For example, platelet activation mediated by these receptors is also associated with marked tyrosine phosphorylation of a number of platelet proteins (Nakamura and Yamamura, 1989; Ferrell and Martin, 1988) although the functional significance and mechanism of this effect is unclear (for review see Shattil and Brugge, 1991). Platelets contain very high levels of certain tyrosine kinases, e.g. pp 60c-src (0.3% of platelet protein), and it has been reported that tyrosine kinase inhibitors are able to block thrombin-induced platelet activation (Salari et al, 1990; Levitzki, 1991). Their data are consistent therefore, with a possible role of tyrosine phosphorylation in platelet activation by phospholipase C coupled receptors.

In order to address the role of tyrosine phosphorylation in further detail, we have used vanadate to inhibit tyrosine phosphatases. Tyrosine protein phosphatase activity in platelets is among the highest that has been found in all tissues (Lerea et al, 1989) and the significance of this is compounded further by the remarkably high turnover rate of this class of enzyme which is of the order of one to three orders of magnitude greater than the activity of tyrosine kinases. A number of studies report that vanadate ions either in the presence of hydrogen peroxide or in permeabilized platelets induce a rapid activation of platelets in association with increased tyrosine phosphorylation (Lerea et al, 1989, Inazu et al, 1990). The present studies were therefore undertaken to investigate the pathway through which vanadate induces platelet activation and to ascertain whether tyrosine phosphorylation has a role downstream from IP3 and DG in stimulating functional responses.

In untreated platelets the level of tyrosine phosphorylation of proteins is relatively low. However, following challenge with vanadate in saponin permeabilized platelet (Figure 6) or with vanadate/hydrogen peroxide in intact platelets (Inazu et al, 1990; Blake et al, 1993) there is a marked and rapid enhancement of tyrosine phosphorylation in a range of proteins with peak increases in tyrosine phosphorylation seen within 5 min.

In contrast, if platelets are treated with okadaic acid, an inhibitor of the serine/threonine protein phosphates, PP1 and PP2A, there is a very slow increase in phosphorylation which continues to increase up to 60 minutes (Figure 7). This comparison illustrates the remarkably high levels and turnover rates of tyrosine kinase activity in platelets. The increase in tyrosine phosphorylation by vanadate/hydrogen peroxide is accompanied by stimulation of shape change, aggregation and release responses, similar to those seen with phospholipase C-coupled receptors (Inazu et al, 1990; Blake et al, 1993).

The similarity in the pattern of shape change, aggregation and secretion responses to vanadate/hydrogen peroxide with that seen to thrombin suggests that these stimuli may induce activation through a common mechanism. It was therefore of interest to investigate whether vanadate/hydrogen peroxide is able to induce activation of phospholipase C and associated events. In platelets pre-labelled with [^{32}P] orthophosphate, vanadate/hydrogen peroxide stimulated a marked increase in phosphorylation of the 20kDa and 40kDa proteins consistent with the activation of the phosphoinositide pathway (Blake et al, 1993). Moreover, vanadate/hydrogen peroxide also induce a marked increase in [^{3}H]inositol phosphates in the presence of the cyclo-oxygenase inhibitor, indomethacin, confirming direct activation of phospholipase C (Blake et al, 1993).

In addition to its ability to block protein tyrosine phosphatases, however, vanadate has also been reported to activate G proteins and it was therefore conceivable that activation of the phosphoinositide pathway is independent of the increase in tyrosine phosphorylation. In order to test this possiblity, we used the non-selective inhibitor of protein kinases,

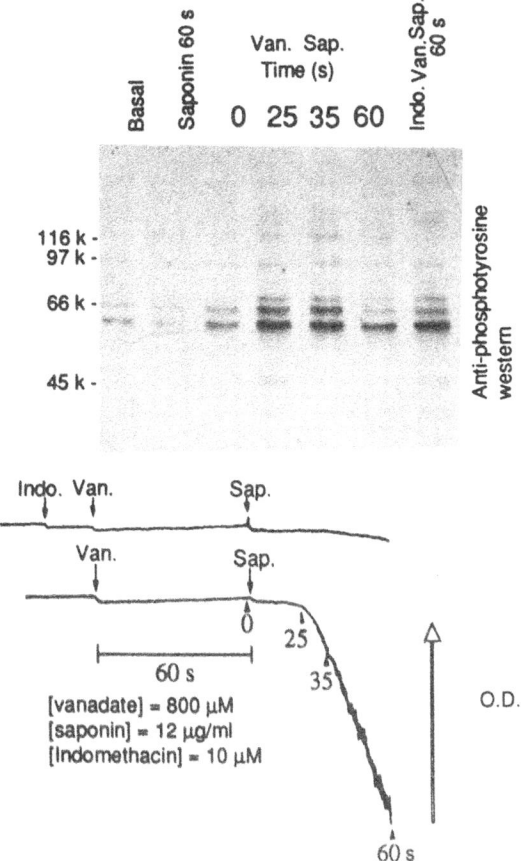

Figure 6   Changes in protein phosphotyrosine during vanadate stimulation of saponised platelets. [^{32}P$_i$]labelled platelets were pretreated with indomethacin (indo) and vanadate (van) before permeabilisation-induced by saponin. Incubations were stopped at the times shown and samples analysed by SDS-PAGE electrophoresis (11% gels) followed by blotting and exposure to antiphosphotyrosine antibody (PY20). Changes in protein phosphotyrosine during vanadate stimulation of saponised platelets in the presence and absence of indomethacin.

staurosporine. Staurosporine (10 μM) markedly inhibits tyrosine phosphorylation and formation of [^{3}H]inositol phosphates induced by vanadate/hydrogen peroxide (Blake et al, 1993) but has been reported previously to have no effect on the formation of inositol phosphates by thrombin (Watson et al, 1988). These data suggest that the mechanism of activation of phospholipase C by vanadate/hydrogen peroxide is through tyrosine phosphorylation and not by direct activation of the G protein coupled to phospholipase C. It is therefore tempting to speculate that vanadate/hydrogen peroxide stimulates increased phosphoinositide metabolism by increased tyrosine phosphorylation of phospholipase Cγ as has been reported to occur for a number of tyrosine kinase-linked receptors in other cell types (Rhee, 1991). However, as described in the Chapters of this book by Lapetina et al and by Nozawa et al, the predominant species of phospholipase C in platelets, phospholipase Cγ1, does not appear to undergo tyrosine phosphorylation on platelet activation by thrombin and other G protein-coupled receptors. The degree of tyrosine phosphorylation induced by vanadate/hydrogen peroxide is far greater than that induced by agonists such as thrombin, however, and it is therefore possible that the greater stimulation

time (min) | 1 | 3 | 10 | 20 |

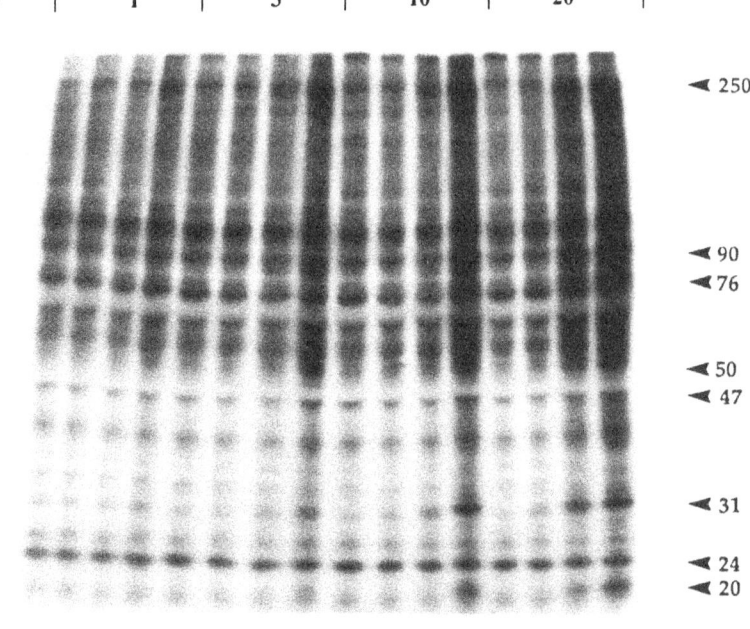

◀ 250

◀ 90
◀ 76

◀ 50
◀ 47

◀ 31

◀ 24
◀ 20

O.A. (μM) 0 0.1 0.3 1   0 0.1 0.3 1   0 0.1 0.3 1   0 0.1 0.3 1

Figure 7   Effect of okadaic acid on protein phosphorylation. Platelets were prelabelled with [^{32}P$_i$] and incubated with okadaic acid (1 μM) for varying times. Separation of proteins was achieved by gradient polycrylamide gel electrophoresis (7.5%-15%), and autoradiography of the resulting gel indicates the extent of phosphorylation. Incubation time in the presence of okadaic acid (OA) is indicated at the top of the figure and the concentration indicated at the bottom. Relative molecular weights (in kDa) of distinct proteins are shown to the right of the figure. Reproduced from Walker & Watson, (1992). Br. J. Pharmacol. 105 627-631 with kind permission.

of tyrosine kinase activity by vanadate/hydrogen peroxide induces a significant phosphorylation of phospholipase Cγ1 on tyrosine residues. Alternatively the species of phospholipase Cγ which mediates the effects of vanadate/hydrogen peroxide in platelets may be the γ$_2$ species which is present in much lower levels (Banno et al, 1989). A more likely alternative explanation, however, is that vandate/hydrogen peroxide elicits an increase in tyrosine phosphorylation of an unidentified protein which is then able to associate with phospholipase C γ by interaction between the newly phosphorylated tyrosine residue and the SH2 domain on phospholipase C γ1 leading to activation (Koche et al, 1991). A more speculative possibility is that vanadate/hydrogen peroxide stimulates an increase in tyrosine phosphorylation of a protein possibly associated with the cytoskeleton, on tyrosine residues which ordinarily exerts an inhibitory action on phospholipases C. (We have recently shown that vanadate/hydrogen peroxide induces tyrosine phosphorylation of PLCγ [Blake et al, 1993]). The ability of staurosporine to inhibit platelet activation by vanadate/hydrogen peroxide is consistent with the mechanism of activation illustrated in Figure 8. Thus, an increase in tyrosine phosphorylation by vanadate/hydrogen peroxide leads to activation of phospholipases C and generation of the two messengers IP3 and DG.

These studies, however, do not rule out the possibility that an increase in tyrosine phosphorylation by vanadate/hydrogen peroxide is able to induce platelet activation independent of phosphoinositide metabolism because staurosporine causes an almost complete inhibition of tyrosine phosphorylation. Studies with Ro 31-8220 are of potential

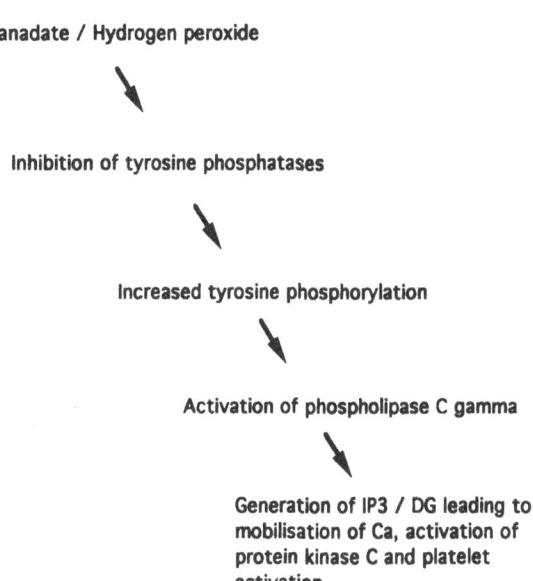

Vanadate / Hydrogen peroxide

Inhibition of tyrosine phosphatases

Increased tyrosine phosphorylation

Activation of phospholipase C gamma

Generation of IP3 / DG leading to
mobilisation of Ca, activation of
protein kinase C and platelet
activation

Figure 8   Proposed mechanism of platelet activation by vanadate / hydrogen peroxide.

significance in this regard because the increased tyrosine phosphorylation induced by vanadate/hydrogen peroxide in the presence of Ro 31-8220 is apparently unaltered (Blake et al, 1993). Thus the ability of Ro 31-8220 to reduce secretion and aggregation induced by vanadate/hydrogen peroxide to a similar degree to the inhibition of these responses to thrombin provides indirect evidence that tyrosine phosphorylation does not have a significant role in these events (Blake et al, 1993). Both thrombin and vanadate/hydrogen peroxide induce a similar degree of aggregation and secretion in the presence of 10 μM Ro 31-8220 despite the much greater increase in tyrosine phosphorylation induced by vanadate/hydrogen peroxide relative to thrombin.

There are a number of caveats that have to be made in regard to this conclusion. Although Ro 31-8220 has no apparent effect on the degree of tyrosine phosphorylation induced by thrombin, we cannot rule out the possibility that it inhibits tyrosine phosphorylation of a minor protein which has an essential role in platelet activation. Similarly, thrombin may induce maximal phosphorylation of certain key substrates on tyrosine residues which have a role in platelet activation: the much larger, more general phosphorylation induced by vanadate/hydrogen peroxide may therefore occur on proteins that do not play a significant role in activation.

Taken together, therefore, there data provide preliminary evidence that tyrosine phosphorylation is able to induce platelet activation through the stimulation of phospholipase C. On the other hand, no evidence has been obtained to suggest that tyrosine phosphorylation contributes directly to the secretion and aggregation responses induced by thrombin. Future work will need to establish the mechanism through which an increase in tyrosine phosphorylation is able to induce phospholipase C activation, and to study the importance of this pathway in the activation of platelets by other agonists. For example, a number of 'weak' platelet agonists e.g. ADP, induce secretion through an aggregation-dependent pathway. It is now established that the binding of fibrinogen to the integrin receptor, glycoproteins IIb-IIIa, leads to tyrosine phosphorylation of a number of substrates and it is conceivable that this is involved in the aggregation-dependent secretory response. Further, as illustrated in Figure 9, collagen-induced platelet activation is

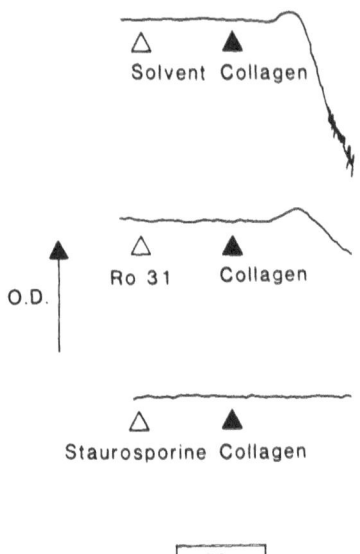

Figure 9   Collagen (10mg/ml) induced a activation of platelets in inhibited completely by staurosporine (10mM) but only partially by Ro 31-8220.

inhibited completely by staurosporine but only reduced by Ro 31-8220. This is consistent with the activation of a tyrosine kinase by collagen having an essential role in platelet stimulation. Since it is known that activation of platelets by collagen is mediated through the stimulation of phosphoinositide metabolism (Watson et al, 1985), it is possible that collagen-induces platelet activation through an initial activation of a tyrosine kinase leading to an increase in phospholipase C activity analogous to that induced by vanadate/hydrogen peroxide.

ACKNOWLEDGEMENTS

T.R.W. and R.A.B. are British Heart Foundation Scholars. S.P.W. is a Royal Society Research Fellow. We thank Roche for the kind gift of Ro 31-8220.

REFERENCES

Banno, Y., Yu, A., Nakashima, T., Honma, Y., Takenawa, T. and Nozawa, Y. 1990, Purification and characterisation of a cytosolic phosphoinositide-phospholipase C (γ 2-type) from human platelets, *Biochem. Biophys. Res. Commun.* 167:396.

Blake, R. A., Walker, T. R. and Watson, S. P. 1993, Activation of human platelets by peroxovanadate is associated with tyrosine phosphorylation of PLCγ and formation of inositol phosphates, *Biochem. J.* 290: 471.

Cohen, P. 1990, Okadaic acid: a new probe for the study of cellular regulation, *Trends Biochem Sci.* 15: 98.

Coorsen, J. R., Davidson, M. M. L. and Haslam, R. J. 1990, Factors affecting dense and α-granule secretion from electropermeabilized human platelets: Ca^2-independent actions of phorbol ester and GTPγS, *Cell Regulation* 1:1027.

Daniel, J. L., Molish, I. R., Rigmaiden, M. and Stewart, G. 1984, Evidence for a role of myosin phosphorylation in the initiation of the platelet shape change response, *J. Biol. Chem.* 259: 9826.

Davis, P. D., Hill, C. H., Keech, E., Lawton, G., Nixon, J. S., Sedgwick, A. D., Wadsworth, J., Westmacott, D. and Wilkinson, S. E. 1989, Potent selective inhibitors of protein kinase C, *FEBS Lett.* 156:1250.

de Chaffoy de Courcelles, D., Roevens, P., and Belle, H. V. 1985, R59022 a diacylglycerol kinase inhibitor: Its effect on diacylglycerol and thrombin-induced C kinase activity in human platelets, *J. Biol. Chem.* 260:15762.

de Chaffoy de Courcelles, D., Roevens, P., Belle, H. V., Somers, Y. and de Clerck, F. 1989, The role of endogenously formed diacylglycerol in the propagation of termination of platelet activation, *J. Biol. Chem.* 264:3274.

Fabiato, A. and Fabiato, F. 1979, Ca^{2+}/EGTA buffer systems, *J. Physiol.* (Paris) 75:463.

Fallon, R. J. 1990, Staurosporine inhibits a tyrosine protein kinase in human hepatoma cell membranes, *Biochem. Biophys. Res. Commun.* 170:1191.

Ferrell, J. E. and Martin, G. S. 1988, Platelet tyrosine-specific protein phosphorylation is regulated by thrombin, *Mol. Cell. Biol.* 8:3603.

Ferrell Jr., J. E. and Martin, G. S. 1989, Thrombin stimulates the activities of multiple previously unidentified protein kinases in platelets, *J. Biol. Chem.* 264:20723.

House, C., and Kemp, B. E. 1987, Protein kinase C contains a pseudosubstrate prototype in its regulatory domain, *Science* 238:1726.

Imaoka, T., Lynham, J. A. and Haslam, R. J. 1983, Purification and characterization of the 47,000-dalton protein phosphorylated during degranulation of human platelets, *J. Biol. Chem.* 258:11404.

Inazu, T., Taniguchi, T., Yanagi, S. and Yamamura, H. 1990, Vanadate and hydrogen peroxide cause platelet activation, *Biochem. Biophys. Res. Commun.* 170:259.

Kikkawa, U., Kishimoto, A. and Nishizuka, Y. 1989, The protein kinase C family: heterogeneity and its implications, *Annu. Rev. Biochem.* 58:31.

King, W., Lucera, G. L., Sorinsky, A., Zhang, J. and Rittenhouse, S. E. 1991, Protein kinase C regulates the stimulated accumulation of 3-phosphorylated phosphoinositides in platelets, *Biochem. J.* 278:475.

Koch, C. A., Anderson, D., Moran, M. F., Ellis, C. and Pawson, T. 1991, SH2 and SH3 domains: elements that control interactions of cytoplasmic signalling proteins, *Science* 252:668.

Kocher, M., and Clementson, K. J. 1991, Staurosporine both activates and inhibits serine/threonine kinases in human platelets, *Biochem. J.* 275:301.

Lerea, K. M., Tonks, N. K., Krebs, E. G., Fischer, E. H. and Giomset, J. A. 1989, Vanadate and molybdate increase tyrosine phosphorylation in a 50-kilodalton protein and stimulate secretion in electropermeabilised platelets, *Biochemistry* 28:9286.

Levitzki, A. 1991, Tyrphostins as molecular tools and potential antiproliferative drugs, *Trends Pharmacol. Sci.* 12:171.

Naka, M., Nishikawa, M., Adelstein, R. S. and Hidaka, H. 1983, Phorbol ester-induced activation of human platelets is associated with protein kinase C phosphorylation of myosin light chains, *Nature* 306:490.

Nakamura, S.-I., and Yamamura, H. 1989, Thrombin and collagen induce rapid phosphorylation of a common set of cellular proteins on tyrosine in human platelets, *J. Biol. Chem.* 264:7089.

Nishizuka, Y. 1983, Calcium phospholipid turnover and transmembrane signalling, *Phil. Trans. R. Soc. Lond.* B 302:101.

Nunn, D. L. and Watson, S. P. 1987, A diacylglycerol kinase inhibitor, R59022, potentiates secretion by and aggregation of thrombin-stimulated human platelets, *Biochem. J.* 243:809.

Pollock, W. K., Sage, S. O. and Rink, T. J. 1987, Stimulation of Ca^{2+} efflux from fura-2 loaded platelets activated by thrombin or phorbol myristate acetate, *FEBS Letts.* 210:132.

Rhee, S. G. 1991, Inositol phospholipid-specific phospholipase C: interaction of the γ 1 isoform with tyrosine kinase, *Trends Biochem. Sci.* 16:297.

Rink, T. J., Sanchez, A., and Hallam, T. J. 1983, Diacylglycerol and phorbol ester stimulate secretion without raising cytoplasmic free calcium in human platelets, *Nature* 305:317.

Rodgriguez-Liñares, B., Walker, T. and Watson, S. P. 1991, The diacylglycerol kinase inhibitor, R59949, potentiates secretion but not phosphorylation of a 40kDalton protein in human platelets, *Biochem. Pharmacol.* 41:835.

Salari, H., Duronio, V., Howard, S.L., Demos, M., Jones, K., Reany, A., Hudson, A. T. & Pelech, S. L. 1990, Erbstatin blocks platelet activating factor-induced protein-tyrosine phosphorylation, polyphosphoinositide hydrolysis, protein kinase C activation, serotonin secretion and aggregation of rabbit platelets, *FEBS Lett.* 263:104.

Shattil, S. J. and Brugge, J. S. 1991, Protein tyrosine phosphorylation and the adhesive functions of platelets, *Current Op. Cell Biol.* 3:869.

Siess, W. and Lapetina, E. G. 1988, Ca^{2+} mobilization primes protein kinase C in human platelets. Ca^{2+} and phorbol esters stimulate platelet aggregation and secretion synergistically through protein kinase C, *Biochem. J.* 255:309.

Tamaoki, T., Nomoto, H., Takahashi, I., Kato, Y., Morimoto, M. and Tomita, F. 1986, Staurosporine, a potent inhibitor of phospholipid/Ca^{2+} dependent protein kinase, *Biochem. Biophys. Res. Commun.* 135:397.

Tyers, M., Rachubinski, R. A., Stewart, M. I., Varriochio, A. M., Shorr, R. G. L., Haslam, R. J. and Harley, C. B. 1988, Molecular cloning and expression of the major protein kinase C substrate of platelets, *Nature* 333:470.

Watson, S. P. and Hambleton, S. 1989, Phosphorylation-dependent and -independent pathways of platelet aggregation, *Biochem. J.* 258:479.

Watson, S. P. and Lapetina, E. G. 1985, 1,2-Diacylglycerol and phorbol ester inhibit agonist-induced formation of inositol phosphates in human platelets: possible implications for negative feedback regulation of inositol phospholipid hydrolysis, *Proc. Natl. Acad. Sci.* 82:2623.

Watson, S. P., Reep, B., McConnell, R. T. and Lapetina, E. G. 1985, Collagen stimulates inositol trisphosphate formation in indomethacin-treated human platelets, *Biochem. J.* 226:831.

Watson, S. P., McNally, J., Shipman, L. and Godfrey, P. P. 1988, The action of the protein kinase C inhibitor, staurosporine, on human platelets; evidence against a negative feedback role for protein kinase C in platelet activation, *Biochem. J.* 249:345.

Walker, T. and Watson, S. P. 1993, The synergism between Ca^{2+} and protein kinase C is the major factor in determining the level of 5-HT secretion from human platelets, *Biochem. J.* 289:277.

Yada, Y., Ozeki, T., Kanoh, H. and Nozawa, Y. 1990, Purification and characterisation of cytosolic diacylglycerol kinases of human platelets, *J. Biol. Chem.* 265:19237.

SERINE/THREONINE KINASES IN SIGNAL TRANSDUCTION IN RESPONSE TO

THROMBIN IN HUMAN PLATELETS

USE OF 17—HYDROXYWORTMANNIN TO DISCRIMINATE SIGNALS

Kenneth J. Clemetson, Markus Kocher and Vinzenz von Tscharner

Theodor Kocher Institute
University of Berne
Freiestrasse 1
CH-3012 Berne, Switzerland

SUMMARY

Although the importance of protein kinases in platelet activation, particularly protein kinase C (PKC), is well established there remain many problems regarding the various phosphorylation cascades, the role of phosphatases and the importance of other serine/threonine and tyrosine kinases. A particular problem is the mechanism of activation of the fibrinogen receptor, GPIIb/IIIa, a critical step in aggregation. Although GPIIIa is phosphorylated (on threonine) neither the stoichiometry nor the minor changes on activation seem adequate to explain the response. Relatively unspecific inhibitors of PKC such as staurosporine prevent PO_4 incorporation into most kinase substrates but only inhibit platelet aggregation partially. However, staurosporine does induce activation and then inhibits several renaturable serine/threonine kinases, probably via phosphatases. Staurosporine did not, however, inhibit the platelet Ca^{2+} signal in response to thrombin but rather enhanced it. 17-Hydroxywortmannin (HWT), a fungal metabolite, has been shown to inhibit respiratory burst in neutrophils and causes haemorrhages. It was recently reported to be a myosin light chain kinase (MLCK) inhibitor and to inhibit PKC only at much higher concentrations. In platelets, HWT inhibits aggregation and partially inhibits phosphorylation of myosin light chain and P47 in thrombin-activated platelets. It also allows the discrimination of an early and a late phase in the cytoplasmic Ca^{2+} signal since at lower concentrations it only inhibits the late phase. The late phase of ATP release was also inhibited in a dose-dependent manner. The activation of most of the renaturable serine/threonine kinases was also inhibited by HWT. These results support earlier conclusions that the early phase of the Ca^{2+} signal is phospholipase C dependent but indicate that other mechanisms must be responsible for the late phase. The relative specificity of HWT for MLCK might indicate that this has an unexpected major role in controlling these late phase reactions including activation of GPIIb/IIIa or its clustering. However, staurosporine completely inhibits phosphorylation of myosin light chain by its kinase (as well as other kinases) and has the opposite effect on Ca^{2+} signals. Clearly, the interactions and feed-back mechanisms between these kinases are very complex but the results suggest that phosphatases acting together with their complementary kinases should also be considered as important platelet activation regulators. P47, long considered a major PKC substrate, may also be phosphorylated by MLCK.

INTRODUCTION

There is considerable evidence that protein kinases have an important role in platelet activation, based upon the effects of (more or less) specific inhibitors and on changes in protein phosphorylation that appear to be linked to the various activation events. The kinases fall into two main categories, serine/threonine specific and tyrosine specific. The best known representative of the first group is protein kinase C (PKC) and its activation via diacylglycerol or phorbol esters has also been thoroughly investigated (Nishizuka, Y., 1984). Another major platelet serine/ threonine kinase is myosin light chain kinase known in muscle to be an important regulator of contraction (Lebowitz and Cooke, 1978). However, there are clearly a much larger number of serine/threonine kinases as can be seen from the wide range of substrates in platelets and also from the large number of serine/threonine kinases recently described in other cells. A major problem remains the definition of substrates for these various kinases and the determination of the order in which they act in various phosphorylation cascades. Many substrates are also capable of being phosphorylated by different kinases, perhaps at different sites and this is one way in which a differentiated signal can be obtained. In addition, specific phosphatases are involved in modulating and controlling these reactions and in providing a certain degree of reversibility. At present we are still in the very early phase of sorting out these various steps and it is important to identify more kinases/phosphatases and their substrates and to have more specific inhibitors available to be able to dissect the different pathways. One major problem in platelets remains the mechanism of activation of the fibrinogen receptor, GPIIb/IIIa, which is necessary for aggregation to occur. GPIIIa is known to contain a phosphorylation site on threonine but neither the stoichiometry of this nor the changes observed in activated platelets appear adequate to account for the changes in fibrinogen binding (Hillery et al., 1991). The role of this phosphorylation is unclear. A major substrate for PKC in platelets is P47 (pleckstrin) which has been well characterised (Imaoka et al., 1983). However, its function still remains obscure. We have examined, in platelets activated by thrombin, a recently described group of renaturable serine/threonine kinases (Ferrell and Martin, 1989) and their response to two kinase inhibitors; staurosporine which has been used for a number of years as a PKC inhibitor but which is now known to be relatively unspecific, inhibiting not only other serine/threonine kinases but also tyrosine kinases; and 17-hydroxywortmannin (HWT), a derivative of the fungal metabolite wortmannin (Baggiolini et al., 1987), which is a potent inhibitor of the respiratory burst and exocytosis in neutrophils, monocytes and macrophages (Dewald et al. 1988) and also produces haemorrhaging in experimental animals (Abbas and Mirocha, 1988) implying that it also inhibits platelets.

METHODS AND MATERIALS

Isolation of Human Platelets

Buffy coats were obtained from the Central Laboratory, Swiss Red Cross Transfusion Service, Berne (FDA four-bag system and were washed as previously described (Kocher and Clemetson, 1991a). For studies on renaturable kinases washed platelets were finally suspended in 137 mM NaCl, 2mM KCl, 1 mM $MgCl_2$, 1 mM $CaCl_2$, 0.4 mM NaH_2PO_4, 5.6 mM glucose, 5 mM Hepes containing 1 unit apyrase/ml to a final concentration of 10^9 platelets/ml. Experiments were performed in 1.5 ml Eppendorf tubes gently shaken in a water bath at 37°C. Staurosporine (Sigma, St. Louis, USA) and 17-hydroxywortmannin (kindly supplied by Dr. T.G. Payne, Preclinical Research, Sandoz Ltd., Basle, Switzerland) were added from stock solutions in dimethyl sulphoxide to yield a final dimethyl sulphoxide concentration of 0.1% which did not affect phosphorylation patterns.

Polyacrylamide Gel Electrophoresis

Samples were taken at fixed times and solubilized immediately in SDS/PAGE sample buffer, vortexed and boiled for 2 min. The samples were separated by electrophoresis in 7.5% polyacrylamide gels in a mini-gel system with 30 μg of platelet protein loaded per lane.

Renaturation of Kinases

Proteins were transferred to PVDF membranes by semi-dry blotting. The blots were treated in 7 M guanidine hydrochloride, 50 mM Tris, 50 mM dithiothreitol, 2 mM EDTA, pH 8.3 for 1 hr at room temperature. Renaturation was performed as described by Ferrell and Martin (1989) in 100 mM NaCl, 50 mM Tris, 2 mM dithiothreitol, 2 mM EDTA, 0.1% Nonidet P-40, 1% BSA, pH 7.5 for 14-16 hr with gentle rocking. After blocking the membranes with 3% BSA, 30 mM Tris, pH 7.5 for 1 hr at room temperature the kinases were detected by treating the blots in 10 mM $MgCl_2$, 2 mM $MnCl_2$, 30 mM Tris, pH 7.5 containing 50 μl of $[\gamma\text{-}^{32}P]ATP$ after washing and treatment with 1M KOH for 10 min to reduce background the blots were neutralized, rinsed with water, dried and autoradiography was performed.

Two-Dimensional Gel Electrophoresis, Western Blotting and Detection of Proteins containing Phosphothreonine

Platelet membranes were prepared from washed platelets by sonication and differential centrifugation. Two-dimensional gel electrophoresis was performed as previously described (Bienz and Clemetson, 1989) on platelets and platelet fractions. Western blotting was carried out using a semi-dry technique from two dimensional gels onto PVDF membranes. Proteins containing phosphothreonine were detected using rabbit polyclonal antibodies prepared against phosphothreonine coupled to keyhole limpet haemocyanin followed by goat anti-rabbit second antibodies coupled to, alkaline phosphatase with 5-bromo-4-chloro-3-indolyl phosphate and nitro blue tetrazolium as substrate.

Labelling of Platelet Proteins with $^{32}PO_4$

After washing as described above, platelets were suspended in 20 mM Hepes, 140 mM NaCl and 5.5 mM glucose, pH 7.5 at 10^9 platelets/ml and were then labelled with 1 mCi/ml $Na_3^{32}PO_4$ by incubating for 1 h at 37°C. The labelled platelets were washed with 20 mM Hepes, 140 mM NaCl and 5.5 mM glucose, pH 7.5 and centrifuged at 2,000 g for 10 min. They were resuspended in 20 mM Hepes buffer to 10^9/ml and divided into aliquots depending on the experiment.

Measurement of Cytosolic Free Calcium using Fura-2

A platelet suspension (4.4 x 10^8 cells/ml, enough for four experiments at a time) was incubated for 20 min at 37°C with 1 μM Fura-2-AM (0.23 nmol of Fura-2-AM/10^8 cells) dispensed from a 1mM stock solution in DMSO. Loadings were carried out at appropriate time intervals to allow an uninterrupted supply which could then be used within a delay of 10 min. Fura-2 loaded platelets (10^8 cells/ml) were washed in test buffer and then stimulated in magnetically stirred polystyrol cuvettes (1ml) at 37°C by an agonist injected with a microsyringe. The binding of Fura-2 to calcium was followed fluorimetrically. The fluorescence of fully saturated Fura-2 (Fmax) was obtained in the presence of extracellular calcium by adding 1 μM (final concentration) of the calcium ionophore, ionomycin, to the cell suspension. The fluorescence is fully quenched (Fo) with 1 mM $MnCl_2$

RESULTS AND DISCUSSION

The inhibition in platelets of a wide range of phosphorylations by staurosporine was confirmed (Fig. 1). In controls on resting platelets, staurosporine appeared to lower the phosphorylation of some substrates below resting levels, implying that phosphatases were still active. As previously reported (Watson et al., 1988), staurosporine only partly

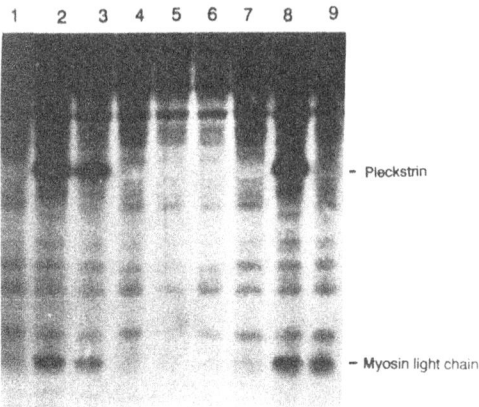

Figure 1   Effect of HWT and staurosporine on phosphorylation of platelet proteins in response to thrombin and PMA. Washed platelets were incubated with $^{32}PO_4$, then either alone or with varying concentrations of HWT or staurosporine. Apart from the control the platelets (5·10^8/ml) were then stimulated with either 0.5 U/ml thrombin or PMA (30 nM). After 3 min the platelets (5·10^8) were solubilized in 1% SDS by boiling for 2 min and directly analyzed by SDS-PAGE on a 13% polyacrylamide gel. The gel was dried and an autoradiogram prepared. The positions of the major proteins phosphorylated during activation, pleckstrin (P47) and myosin light chain (P20), are indicated. Lane 1) Unstimulated platelets; 2) Stimulated with PMA; 3) Stimulated with thrombin 4) Staurosporine (10^{-6} M); 5) Staurosporine plus PMA; 6) Staurosporine plus thrombin; 7) HWT (10^{-5} M); 8) HWT plus PMA; 9) HWT plus thrombin. These results are typical of 3 experiments with different platelet donors.

inhibited thrombin induced platelet aggregation, while completely inhibiting shape change and release. The effect on the renaturable serine/threonine kinases was examined. Aliquots of washed human platelets were treated with different amounts of the inhibitor and/or activated with thrombin. After gel electrophoresis and Western blotting the blots were blocked with BSA and renatured. The renatured blots were overlaid with [γ-^{32}P]ATP and, after incubation and washing, exposed to X-ray film. Contrary to what had been expected, it was found that several of these kinases were activated by staurosporine, even in resting platelets that had not been treated with thrombin (Fig. 2). Direct treatment of the renatured blots with staurosporine at the same time as [γ-^{32}P]ATP resulted in a strong inhibition of kinase activity. The kinase activity can be detected in this assay due to autophosphorylation and phosphorylation of BSA on the blot. The direct inhibition is due to non-covalent interaction with the catalytic subunit of the enzyme, probably by blocking the ATP-binding site, and activity is restored after renaturation of the enzyme, which removes the inhibitor.

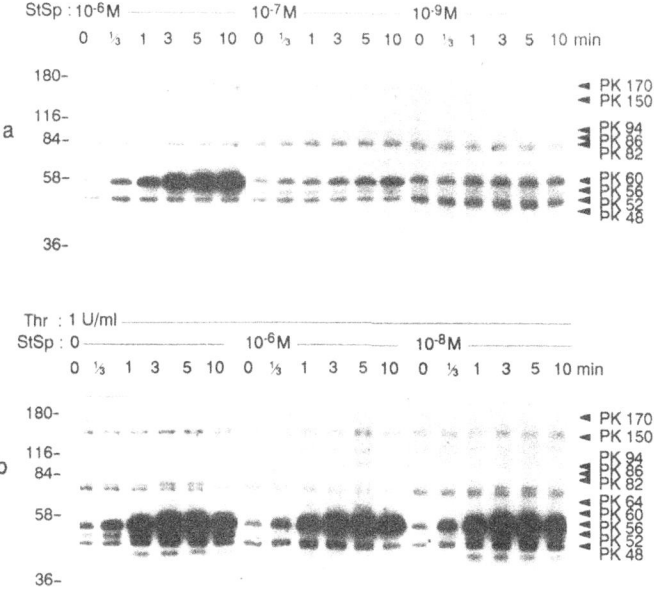

Figure 2   Effect of staurosporine treatment and thrombin stimulation of intact platelets on the activities *in vitro* of the renaturable protein kinases from human platelets. Autoradiogram of PVDF membranes with renatured kinases labelled in vitro with [γ-^{32}P]ATP. (a) Effect of staurosporine. Samples taken at times shown after addition of concentrations shown. (b) Effect of thrombin (1 U/ml) and staurosporine. Samples taken at times shown after addition of concentrations shown. The 0 min samples were taken just before addition of the agonists. Molecular mass is indicated in kDa. PK indicates protein kinase with molecular mass in kDa according to the nomenclature of Ferrell and Martin (1989)

The direct interaction is not responsible for the activation process. Since the activation proceeds while most, if not all, of the platelet kinases are inhibited it seems most likely that the activation occurs via covalent modification and dephosphorylations, produced by the imbalance between phosphatase and kinase activity. It is therefore of considerable interest that the activation of these kinases in thrombin-activated platelets shows considerable similarity to that obtained with staurosporine treatment and may indicate that the thrombin activation is also via phosphatases. Activation of these kinases by thrombin goes in parallel with fibrinogen occupation of GPIIb/IIIa indicating a regulatory step that staurosporine is possibly able to by-pass. This may help to explain why staurosporine does not completely inhibit aggregation and might also implicate phosphatase activity in the exposure of the fibrinogen receptor. The Ca^{2+} signal induced by thrombin in staurosporine-treated platelets is also not inhibited, as might have been predicted, but is rather enhanced (Fig 3)

As mentioned above GPIIIa is phosphorylated on threonine in resting platelets at about 2-4% (Fig 4) and shows only minor changes on activation. The kinase that is responsible for this phosphorylation has not yet been identified but is probably not PKC. Recent studies have shown that the cytoplasmic domains of GPIIb/IIIa play a critical role in maintaining the resting state of the extracellular domains. Whether the phosphothreonine form of GPIIIa has other properties is still unknown. There are, however, grounds for thinking that minor amounts of at least one different population might be expected. After years of controversy it now seems certain that fibrinogen is acquired in the platelet α-granules from the plasma via GPIIb/IIIa (Harrison et al., 1989) and there is a vesicular uptake mechanism that operates in megakaryocytes and probably also in platelets.

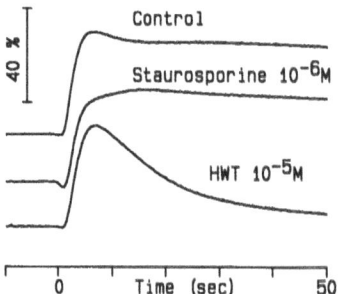

Figure 3   Effect of staurosporine and HWT on the thrombin-induced rise in $[Ca^{2+}]_i$ in platelets measured as % Fura-2 saturation. Platelets (10^8/ml), loaded with fura-2/AM in the presence of 1 mM $CaCl_2$ and resuspended in medium containing 1 mM $CaCl_2$ were preincubated at 37°C a) alone, or b) with 10^{-6} M staurosporine or c) with 10^{-6} M staurosporine and 10^{-5} M HWT or d) with 10^{-5} M HWT. The platelets were then stimulated with 0.5 U/ml thrombin.

Antibodies to GPIIb/IIIa were also observed to be transported in vesicles in platelets (Morgenstern et al., 1992). Fibrinogen is absent from a-granules in Glanzmann's thrombasthenia platelets where GPIIb/IIIa are missing. Thus, it appears likely that there is a population of GPIIb/IIIa molecules capable of binding fibrinogen, even in resting platelets, and that when bound it is directed into vesicles and towards the α-granules. The properties of the subpopulation of GPIIb/IIIa molecules capable of doing this and its distinguishing characteristics are still unknown but the variant phosphorylated on threonine provides an interesting candidate. Characterization of the kinase(s) and phosphatases acting on GPIIIa will be an essential step in understanding this process.

Unlike staurosporine, HWT inhibits platelet aggregation but has weaker effects on shape-change and release. Phosphorylation in thrombin-activated platelets was only partly inhibited and was restricted to pleckstrin (P47) (80-90% inhibition at 10^{-5}M) and myosin light chain (Fig.1) (about 50% inhibition at 10^{-5}M). Known substrates to cAMP and cGMP dependent kinases were not affected (not shown). When its effects on the renaturable serine/threonine kinases were examined it was found to inhibit their activation by thrombin almost completely (Fig. 5). HWT also affected the Ca^{2+} signal in thrombin-activated platelets in a dose-dependent way. While at lower concentrations it had little effect on the early phase of the Ca^{2+} signal (maximum at 6 sec), the late phase of the signal (maximum at 20 sec) was strongly inhibited (Fig. 3). Thus, HWT is capable of distinguishing between

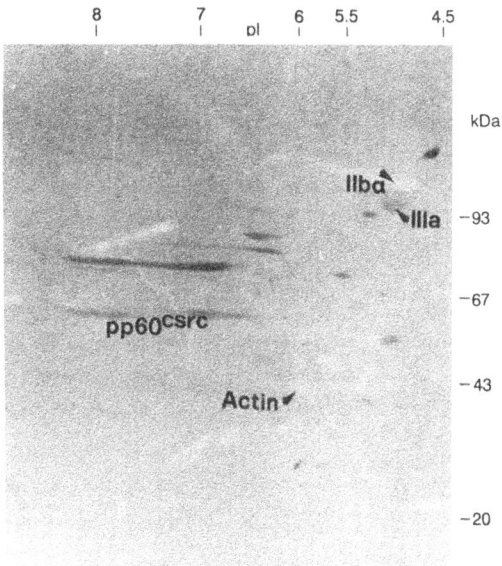

Figure 4   Western blot of a two-dimensional gel electrophoresis separation of platelet membranes stained for phosphothreonine. Although weakly stained compared to some other components such as pp60^{c-src}, GPIIIa gives a similar positive signal in both resting and activated platelets, while other glycoproteins such as GPIIb are negative.

different phases of thrombin induced signal pathways. HWT also inhibited the late phase of thrombin-induced ATP release from platelets, implying that this phase of both ATP secretion and the Ca^{2+} signal is controlled by the same mechanism. Thus, staurosporine and HWT, while both affecting kinase activities have very different effects on platelet activity. Recently, it has been suggested that HWT is one of the most specific inhibitors of myosin light chain kinase (MLCK) known, compared to its effect on PKC (Nakanishi et al., 1992). If this is considered to be its main effect in platelets can this explain the observations? Clearly an inhibitory effect on MLCK with little effect on PKC can explain the 50% inhibition of phosphorylation of myosin light chain observed since it is also phosphorylated by PKC. The major inhibition of phosphorylation of pleckstrin is, however, unexpected since this is a major PKC substrate. Phosphorylation of pleckstrin in platelets induced by phorbol esters was not affected by HWT (Fig. 1), confirming that HWT does not affect PKC directly. Thus, either pleckstrin is also a major substrate for MLCK or MLCK can regulate the activation of PKC in thrombin-activated platelets. In fact, based on the close parallel between phosphorylation of MLC and pleckstrin it had indeed been suggested earlier that pleckstrin might also be a substrate for MLCK (Harris, 1981). HWT has been reported to interact at or near the ATP binding site of MLCK (Nakanishi et al. 1992) and its specificity for MLCK has been established versus several other well characterised protein kinases. These studies cannot however exclude the possibility that other, still uncharacterised kinases, with ATP binding sites very similar to MLCK might also be inhibited by HWT, perhaps even at lower concentrations. That other MLCK specific inhibitors have at least some of the same consequences, while strengthening the case for a direct effect, can still not eliminate this completely. In fact some other "specific" MLCK inhibitors have also been reported to inhibit platelet aggregation in response to thrombin while H—7, a somewhat more specific PKC inhibitor than staurosporine, had little effect (Itoh et al., 1992). H-7 was earlier reported to potentiate serotonin release in platelets in

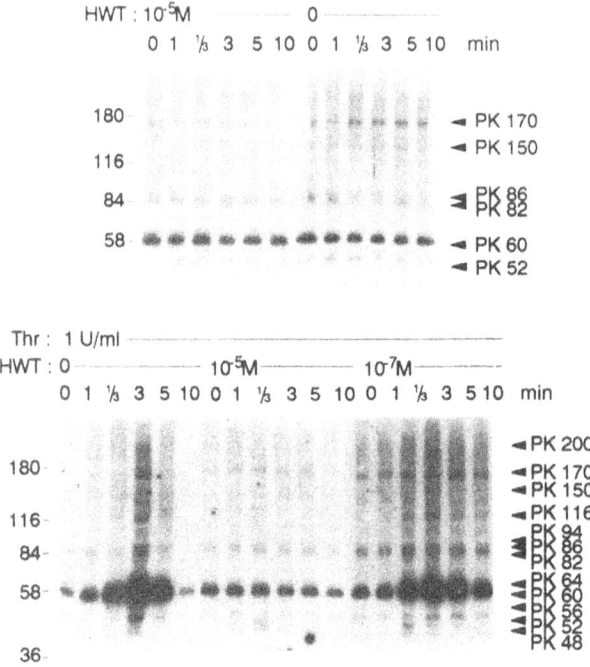

Figure 5   Effect of HWT treatment and thrombin stimulation in intact platelets on the *in vitro* activities of the renaturable kinases from human platelets. Autoradiogram of PVDF membranes with renatured kinases labelled in vitro with [γ-³²P]ATP. Platelets were preincubated for 10 min with HWT. Upper picture shows resting platelets preincubated with and without HWT. Lower picture shows platelets activated with 1 U/ml thrombin. The 0 time samples were taken just before adding thrombin.

response to thrombin. The importance of phosphorylation of MLC in platelet activation has been questioned since it shows little phosphorylation in platelets activated by oleyl acetyl glycerol (Kaibuchi et al., 1983), however, it cannot be excluded that phosphorylation on the PKC specific sites in the absence of phosphorylation on the MLCK specific sites may cause a different activity from that when neither are phosphorylated. How could the inhibition of MLCK affect aggregation under conditions where release and shape change are less affected? Aggregation is dependent not only on the expression of the GPIIb/IIIa fibrinogen receptor but also on its subsequent association with the cytoskeleton and cluster formation. The cytoskeleton involved is the heavy fraction, containing myosin, so that this might provide the possible link. The activation of the renaturable serine/threonine kinases goes in parallel with these aggregation induced changes (Ferrell and Martin, 1989) which might thus explain why their activation is also inhibited by HWT. There remains the problem of the Ca^{2+} signal and the inhibition of the late phase by HWT. The long lasting signal induced by thrombin and trypsin is fairly unusual, most Ca^{2+} signals induced by agonists in cells are fairly rapid. Addition of hirudin at any point after thrombin results in an immediate cessation of signal generation and the signal then sinks more rapidly. This has been interpreted to mean that a continual occupation of the thrombin receptor is necessary (Hoffmann and Markwardt, 1984). However, the recent cloning and characterisation of the thrombin receptor (Vu et al., 1991) requires a rethinking of the interpretation of this observation. Cleavage of the N-terminal domain by thrombin and binding of the new tethered terminal peptide to another domain of the receptor induces the signal which eventually is seen as a transient rise in Ca^{2+}. The persistance of the signal can be ascribed to a relatively slow on-rate followed by a rapid cleavage so that only a few molecules per second per platelet need to be split in order to maintain the Ca^{2+} level. The addition of hirudin, by inactivating the thrombin, prevents this continual slow cleavage and the decay curve is then the natural decay of the signal due to the desensitisation (inactivation?) of the cleaved receptors. Part of the late phase Ca^{2+} signal comes from the opening of plasma membrane channels but this is not sufficient to account for most of the late phase signal. Even so GPIIb/IIIa has been linked to the control of membrane Ca^{2+} channels (Rybak and Renzulli, 1989) and an inhibition of activation of GPIIb/IIIa by HWT might account for part of the decrease in the signal. Otherwise it would seem that the first few cleavages are capable of generating Ca^{2+} signal but, in the presence of HWT, later cleavages are not, implying an uncoupling between the receptor and the signal transduction leading to Ca^{2+} release. Ca^{2+} release from intracellular stores is thought to be dependent on IP_3 generation by phospholipase C action on phosphatidylinositol but is also controlled by cAMP levels. Since the thrombin receptor is linked to both the Gi and Gp proteins the induction of a Ca^{2+} signal can be related to both, and its inhibition to pathways induced by either or both. Since signals derived from the activation of phospholipase C are restricted to the early part while thrombin receptor continues to be cleaved later and phospholipase C shows persistant activation (Lazarowski and Lapetina, 1990) it must be assumed that the signalling pathway from phospholipase C is switched off in the late phase or that its substrate(s) have been depleted. The maintainance of a high Ca^{2+} signal, which also occurs in the absence of extracellular Ca^{2+}, could then be ascribed simply to continued deactivation of adenylate cyclase by Gi preventing the pumping of Ca^{2+} back into stores by a pump activated by cAMP-dependent kinase. The results obtained with HWT distinguishing between an early and a late Ca^{2+} signal in thrombin activation of platelets should allow a finer analysis of the signal transduction process occuring via this receptor. It remains to be demonstrated whether the effects of HWT are due to inhibition of MLCK alone or involve inhibition of other kinases. In the case of HWT there is no clear evidence for an increased activity of corresponding phosphatases. An effect on the activation of Gi or the pathways linked to it is an obvious candidate for further investigation. It should also be noted that HWT and staurosporine were found to have opposite effects on the activation of phospholipase D in neutrophils (Reinhold et al., 1990) and it remains a possibility that this enzyme may also be important in the signal pathway influenced by HWT in platelets.

Acknowledgements

The expert technical assistance of Barbara Hügli is gratefully acknowledged. We thank the Central Laboratory, Swiss Red Cross Transfusion Service, Berne for supplying buffy coats. This work was supported by the Swiss National Science Foundation Grants 31.25633.88 and 31.25700.88.

REFERENCES

Abbas, H.K., and Mirocha, C.J., 1988, Isolation and purification of a haemorrhagic factor (wortmannin) from Fusarium oxysporum (N17B), *Appl. Environ. Microbiol.* 54:1268.

Baggiolini, M., Dewald, B., Schnyder, J., Ruch, W., Cooper, P.H., and Payne, T.G., 1987, Inhibition of the phagocytosis-induced respiratory burst by the fungal metabolite wortmannin and some analogues, *Exp. Cell Res.* 169:408.

Bienz, D. and Clemetson, K.J., 1989, Human platelet glycoprotein Ia, *J. Biol. Chem.* 264:507.

Dewald, B., Thelen, M., and Baggiolini, M., 1988, Two transduction sequences are necessary for neutrophil activation by receptor agonists, *J. Biol. Chem.* 263:16179.

Ferrell, J.E. Jr and Martin, G.S., 1989, Thrombin stimulates the activities of multiple previously unidentified protein kinases in platelets, *J. Biol. Chem.* 264:20723.

Harris, H. Regulation of motile activity in blood platelets, *in:* Platelets in Biology and Pathology, Vol. 2, J.L. Gordon, ed., pp473-495, Elsevier/North-Holland, Amsterdam (1981).

Harrison, P., Wilbourn, B., Debili, N., Vainchencker, W., Breton-Gorius, J., Lawrie, S., Masse, J.M. Savidge, G.F. and Cramer, E.M., 1989, Uptake of plasma fibrinogen into the alpha-granules of human megakaryocytes and platelets, *J. Clin. Invest.* 84:1320.

Hillery, C.A., Smyth, S.S. and Parise, L.V., 1991, Phosphorylation of human platelet glycoprotein IIIa (GPIIIa), *J. Biol. Chem.* 266:14663.

Hoffmann, A. and Markwardt, F., 1984, Inhibition of the thrombin-platelet reaction by hirudin, *Haemostasis* 14:164.

Imaoka, T., Lynham, J.A. and Haslam, R.J., 1983, Purification and characterization of the 47,000 dalton protein phosphorylated during degranulation of human platelets, *J. Biol. Chem.* 258:11404.

Itoh, K., Hara, T. and Nobuhiko, S., 1992, Diphosphorylation of platelet myosin by myosin light chain kinase, *Biochim. Biophys. Acta* 1133:286.

Kaibuchi, K., Takai, T., Sawamura, M., Hoshijima, M., Fijikura, T., and Nishizuka, Y., 1983, Synergistic functions of protein phosphorylation and calcium mobilization in platelet activation, *J. Biol. Chem.* 258:6701.

Kocher, M. and Clemetson, K.J., 1991a, Staurosporine both activates and inhibits serine/threonine kinases in human platelets, *Biochem. J.* 275:301.

Kocher, M. and Clemetson, K.J., 1991b, Effects of 17-hydroxywortmannin on serine/threonine kinases in human blood platelets, *FEBS Lett.* 291:363.

Lazarowski, E.R. and Lapetina, E.G., 1990, Persistant activation of platelet membrane phospholipase C by proteolytic action of trypsin and thrombin, *Arch. Biochem. Biophys.* 276:265.

Lebowitz, E.A. and Cooke, R., 1978, Contractile properties of actinomyosin from human blood platelets, *J. Biol. Chem.* 253:5443.

Morgenstern, E., Ruf, A. and Patscheke, H., 1992, Transport of anti-glycoprotein IIb/IIIa antibodies into the alpha-granules of unstimulated blood platelets, *Thromb. Haemostas.* 67:121.

Naka, M., Nishikawa, M., Adelstein, R.S. and Hidaka,H., 1983, Phorbol ester-induced activation of human platelets is associated with protein kinase C phosphorylation of myosin light chains, *Nature* 306:490.

Nakanishi, S., Kakita, S., Takahashi, I., Kawahara, K., Tsukuda, E., Sano, T., Yamada, K., Yoshida, M., Kase, H. Matsuda, Y., Hashimoto, Y. and Nonomura, Y., 1992, Wortmannin, a microbial product inhibitor of myosin light chain kinase, *J. Biol. Chem.* 267:2157.

Nishizuka, Y., 1984, Turnover of inositol phospholipids and signal transduction, *Science* 225:1365.

Reinhold, S.L., Prescott, S.M., Zimmerman, G.A., and McIntyre, T.M., 1990, Activation of human neutrophil phospholipase D by three separable mechanisms, *FASEB J.* 4:208.

Rybak, M.E. and Renzulli, L.A., 1989, Ligand inhibition of the platelet glycoprotein IIb-IIIa complex function as a calcium channel in liposomes, *J. Biol. Chem.* 264:14617.

Vu, T.-K.H., Hung, D.T., Wheaton, V.I. and Coughlin, S.R., 1991, Molecular cloning of a functional thrombin receptor reveals a novel proteolytic mechanism of receptor activation, *Cell* 64:1057.

Watson, S.P., McNally, J., Shipman, L.J. and Godfrey, P.P., 1988, The action of the protein kinase C inhibitor, staurosporine, on human platelets, *Biochem. J.* 249:345.

TYROSINE PHOSPHORYLATION IN PLATELETS:

ITS REGULATION AND POSSIBLE ROLES IN PLATELET FUNCTIONS

Maurice B. Feinstein, Kevin Pumiglia and Lit-Fui Lau

Department of Pharmacology, School of Medicine
University of Connecticut Health Center
Farmington, CT, 06030, USA

INTRODUCTION

The specific phosphorylation of tyrosine residues in proteins discovered in 1979 [Eckhart et al, 1979] is catalyzed by a unique class of protein tyrosine kinases (PTKs). The *src* protein of Rous Sarcoma virus was the first protein tyrosine kinase (PTK) to be discovered, and it was also the first oncogene found to be a protein kinase [Collett et al, 1980; Hunter and Cooper, 1985]. Much of the great current interest in tyrosine phosphorylation stems from the discovery that the transforming activity of the virus is due to that enzyme, whose progenitor is the product of a normal cellular gene, the protooncogene *c-src*. A link between oncogenes and the control of normal cell growth was made when the EGF receptor was found to be a PTK whose kinase activity was necessary for its biological functions. PTKs were thereby implicated in signal transduction in normal cells in response to growth factors, as well as in neoplastic transformation. Various mutations (*e.g.* deletions of regulatory sequences or formation of fusion proteins) that result in the formation of constitutively active PTKs (*e.g. v-src, erbB, bcr/abl*), or overexpression of a normal protooncogene, can result in abnormal cell proliferation and neoplastic transformation.

The cellular PTKs are of 2 general classes: the transmembrane receptors with cytoplasmic tyrosine kinase domains, such as PDGF, CSF, IGF, FGR and insulin receptors [Ullrich and Schlessinger 1990], and the non-receptor class of PTKs. The latter do not span the membrane, but are associated with the inner surface of the plasma membrane. Myristoylation of the N-terminal glycine of *c-src* is necessary, but not sufficient for membrane association of the non-receptor PTKs and their transforming activity (Glover et al, 1988; Goddard et al, 1989 Feder and Bishop, 1991). Some of these PTKs are oncogenes from the transforming retroviruses; *i.e. abl, yes, fgr, fes*. The closely related *src* family consists of the *src, yes, fgr, lck, fyn, lyn,* and *hck* proteins of approximately 60 KDa. There are also larger PTKs such as *c-abl* (150 KDa) and *c-fes* (92 KDa).

The tyrosine kinase activity of the growth factor receptors is necessary for the transduction of some of their intracellular signals. For example, tyrosine phosphorylation has been demonstrated to be necessary for the activation of phospholipase Cγ-1 (PLCγ-1) by EGF (Mustelin et al, 1990). That enzyme generates the key intracellular second messengers diacylglycerol and Ins(1,4,5)P$_3$ from the membrane substrate PtdIns(4,5)P$_2$. The 3-kinase that phosphorylates PtdIns(4)P to generate PtdIns(3,4)P$_2$ can also be tyrosine phosphorylated and associate with growth factor receptors (Escobedo et al, 1991).

In general, tyrosine phosphorylation modulates the activity of some enzymes, and/or their molecular associations with other proteins. A key regulatory region of *src* is Tyr-527, which is inhibitory when phosphorylated, and Tyr-416 is activating when phosphorylated. Mutations at Tyr-527 produce activated kinases; *e.g.* *v-src* lacks Tyr-527 and is phosphorylated at Tyr-416. The N-terminal regulatory regions termed SH3 (*src* homology region 3) play a negative regulatory role and have sequence homology to some cytoskeletal and membrane proteins. The SH3 regions together with the positive regulatory SH2 regions are involved in the recognition and association with other tyrosine phosphorylated proteins which also contain such *src*-homology regions. In this way, complexes of tyrosine phosphorylated PTK-receptors with PLCγ1, GAP and PtdIns(4)P 3-kinase are believed to be assembled and lead to the transduction of coordinated signals (Anderson et al, 1990; Koch et al, 1991; Fantl et al, 1992).

This review is concerned with the role of tyrosine phosphorylation in platelet biochemistry and cellular functions. We shall discuss several facets of this topic; *i.e.* the platelet tyrosine kinases and phosphatases, the correlation of changes in tyrosine phosphorylation of proteins with specific responses, and the evaluation of the effects of inhibitors of tyrosine kinases and phosphatases.

TYROSINE PHOSPHORYLATION IN INTACT PLATELETS AND PLATELET TYROSINE KINASES AND PHOSPHATASES

Tyrosine phosphorylation of platelet proteins was first discovered by Tuy et al (1983) by measuring protein phosphorylation that was susceptible to alkaline hydrolysis. Subsequently the development of specific antiphosphotyrosine antibodies led to a great expansion of interest in the measurement of tyrosine phosphorylated proteins. It should be noted though, that the various antibodies that are available through investigators, or commercially, vary considerably in their efficacy. As a result the number of tyrosine phosphorylated proteins detected in platelets varies from 2 (Dhar and Shukla, 1990) to more than 20 (Ferrell and Martin, 1988, 1989; Pumiglia et al, 1992). Similarly, antibodies raised to the *src* family of non-receptor PTKs have proven to be valuable for determining the cellular and subcellular localization of these enzymes, as well as their associations with other proteins. Some of these antibodies possess specificity for a single PTK, such as monoclonal antibody mAb327 (Golden et al, 1986) which, unlike some commercial anti-*src* antibodies, is highly specific for pp60$^{c\text{-}src}$.

Although platelets are anucleate fragments of megakaryocytes they suprisingly contain very high levels of basal protein tyrosine phosphorylation and tyrosine kinase activity, due presumably mainly to pp60$^{c\text{-}src}$. Golden et al (1986) and Golden and Brugge (1989) determined that pp60$^{c\text{-}src}$ amounted to 0.2-0.4% of total platelet protein using mAb327. This *src* protein was found to be the major protein kinase activity in platelet membranes, and its level was much higher than in brain or lymphocytes, and was comparable to that in certain transformed cells. Golden and Brugge (1989) also reported finding a high *src* protein content in megakaryocytes, but no data was supplied. The platelet enzyme was purified by Feder and Bishop [1990]. Platelets also contain additional PTKs in much lower quantitity than *c-src*: *i.e.* pp60^{fyn}, pp62^{yes} (Horak et al 1990; Zhao 1991), pp54/pp58^{lyn}, pp62^{hck} (Huang et al, 1991). Recently the PTK pp72^{syk} was cloned and shown to comprise 0.1-0.2% of porcine platelet protein,and to be activated in intact platelets by wheat germ agglutinin [Ohta et al, 1992].

Because platelets are the product of a terminal cellular differentiation the tyrosine phosphorylated proteins and the PTKs may be merely remnants of the megakaryocytes (Okano et al, 1991), and their functions entirely related to the proliferation and differentiation of those cells in response to growth regulatory factors. On the other hand, it has been suggested that pp60^{c-src} and other the other PTKs might be involved in processes other than growth control, and that their presence is not necessarily correlated with a high state of cell proliferation.

Ever since the discovery of protein tyrosine phosphorylation in platelets the role of this process in platelet functions has been a subject of some intense study. Although tyrosine phosphorylation has been shown to rapidly increase when platelets are stimulated by typical platelet agonists, there is no definitive proof of a role for this biochemical pathway in platelet signal transduction processes. Furthermore, the mechanisms that account for the enhanced tyrosine phosphorylation are similarly obscure, as no changes in the activity of the *src* kinase or other PTKs were demonstrated as a result of platelet stimulation, with the exception of the recently described stimulation of pp72^{syk} by WGA (Ohta et al, 1992). The protein substrates for PTKs in intact platelets, except for the PTKs themselves, are largely unidentified, despite the fact that a number of substrates are known in other cell types; *e.g.* PLCγ-1, PtdIns(4)P 3-kinase, and some cytoskeletal proteins such as talin, fodrin, tubulin, MAP2, *tau*, and vinculin (Akiyama et al, 1986). GpIIb-IIIa can be tyrosine phosphorylated in vitro, but this has not been demonstrated to be a regulatory function in intact platelets (Findik et al, 1990; Elmore et al, 1990). Of great interest is the recent report that the protein pp125^{fak}, a PTK found at focal adhesions, was identified as a tyrosine phosphorylated substrate in platelets that may have associations with GpIIb-IIIa (Schaller et al, 1992).

Several studies have dealt with the subcellular localization of pp60^{c-src} in platelets. Dhar and Shukla [1991] reported that PAF increased the tyrosine phosphorylation of pp60^{c-src} in rabbit platelets, and caused the translocation of the enzyme from the cytosol to the membrane fraction. The extent of the translocation was not quantitated, but based on their figure 4 it appeared to be substantial. Using mAb327 we determined that pp60^{c-src} was virtually entirely (> 90%) present in the particulate fraction of sonicated platelets, and no translocation from cytosol to membrane was observable as a result of stimulation. Rendu et al (1989) used mAb327 to identify pp60^{c-src} in various fractions of sonicated platelets isolated by differential centrifugation on a metrizamide gradient. Western blots of the fractions showed about 2-fold more *src*/mg protein in the dense body fraction than in the plasma membrane fraction, from which they concluded that most of the *src* protein was in the dense bodies and might play a role in the secretion of their contents. However, their data actually shows that the plasma membrane fraction contained 200-fold more protein than the dense body fraction, from which it can be calculated that 99% of the *src* protein was actually located in the plasma membrane fraction. In addition, Sorisky et al (1992) found no difference in the pp60^{c-src} content of platelets that lacked dense granules. Cytochemistry revealed *src* protein to be confined to the plasma membrane and the surface connected canalicular system (Ferrell et al, 1990). The *src* protein may associate through its myristoylated N-terminal with a 50 kDa membrane protein (Feder and Bishop, 1990).

The balance between tyrosine kinase and phosphatase activities must be an important determinant of the steady-state level of tyrosine phosphorylation. The importance of the PTPases is suggested by the powerful effects that result from the use of the PTPase inhibitor pervanadate (*see below*). The tyrosine phosphatases found in cells are of two types, transmembrane and intracellular. The former include the leukocyte common antigen

(LCA, CD45) which is involved in signal transduction in T-cells by dephosphorylating a negative regulatory site on the PTK pp56^{lck} (Klausner and Samelson, 1991). Gu et al [1991] found high PTPase activity in platelets, and cloned many PTPases, including LCA, from a human megakaryoblastic cell line MEG-01. A novel PTPase designated MEG had sequence homology to the red cell cytoskeletal protein 4.1, and to other cytoskeletal proteins, *i.e.* talin and the PTK ezrin.

STIMULATION OF TYROSINE PHOSPHORYLATION IN PLATELETS BY AGONISTS AND ITS REGULATION BY SECOND MESSENGERS

Ferrell and Martin (1988) employed two methods to study tyrosine phosphorylation in intact washed platelets; *i.e.* western blotting with antiphosphotyrosine antibodies, and immunoprecipitation by those antibodies of proteins from intact platelets prelabelled with ^{32}P-phosphate. Thrombin was found to increase tyrosine phosphorylation within 20 sec at 37°C, concurrent with aggregation and secretion, but later than shape change. The rise in tyrosine phosphorylation occurred in 3 waves. The first wave, rapid and transient, involved primarily polypeptides of 70 and 68 KDa and lesser bands at 34 and 27 KDa. The second wave, which peaked in 1-3 min, was more sustained, and involved mainly a 130 KDa protein, a 60 Kda protein that comigrated with pp60^{c-src}, and proteins of 115 and 105 KDa. A 3rd wave, involving bands at 126, 108, 100 and 85 KDa coincided with aggregation. The average increase in tyrosine phosphorylation was 132-176% above basal levels. The 130 kDa protein increased the most, by 4-5 fold, and pp60^{c-src} tyrosine phosphorylation increased 25-75%. The maximum effect occurred with about 0.5U/ml thrombin, which was equivalent to the concentration required for maximal aggregation.

Golden and Brugge (1989) used an antiphosphotyrosine antibody to detect proteins of 170-180, 122, 66 and 59 KDa. Thrombin increased tyrosine phosphorylation most prominently in a 95-97 doublet, which occurred within 5 sec and was sustained for 15 min. Another antiphosphotyrosine antibody detected additional protein bands at 250, 170, 116, 83, 77, 38 and 34 KDa proteins in addition to the others mentioned above. With this antibody the 122 KDa band was the most prominent. One band at 66 KDa decreased after thrombin. PMA increased tyrosine phosphorylation of the 95-97 doublet to a lesser extent than thrombin. The pp60^{c-src} was found to be phosphorylated at Ser-12 and Tyr-527, but no change in the activity of the immunoprecipitated enzyme was measureable (Gould et al, 1985; Ferrell and Martin, 1988).

Thrombin stimulates tyrosine phosphorylation through its own receptor(s), independently of secondary mediators such as TXA$_2$ and ADP, as it remains fully effective in platelets that are treated with aspirin plus the ADP scavengers apyrase or CP/CP-kinase. The receptor responsible for this action appears to be the recently cloned thrombin receptor (Vu et al, 1991) that is a member of the G-protein linked family, and is detected by western blotting as a 66 KDa protein (Hung et al, 1992; Brass et al, 1992). This receptor is unique in that it is apparently *self-activated* by its own new N-terminal peptide sequence that is exposed by the proteolytic action of thrombin at the R41-S42 bond. This new N-terminal sequence is termed the *tethered ligand* (Vu et al, 1991). The peptides SFLLRN and SFLLRNPNDKYEPF, which represent the first 6 to 14 amino acids of the new N-terminal sequence, are known variously as the tethered ligand peptides, or the thrombin receptor peptides (TRP) (Vassallo et al, 1992). In agreement with Vassallo et al (1992) we find that these peptides stimulate tyrosine phosphorylation as effectively as α-thrombin. Furthermore, we observed that when the receptor was desensitized by prolonged exposure to SFLLRNPNDKYEPF the initial enhanced tyrosine phosphorylation fell back to virtually the pre-stimulation level after 20 min, at which time high concentrations of peptide no longer increased tyrosine phosphorylation.

Ferrell and Martin (1988) originally reported that the effect of thrombin was unique, as ADP, PAF, arachidonic acid, or epinephrine did *not* increase tyrosine phosphorylation. Subsequent studies have shown that tyrosine phosphorylation in human platelets *is* indeed stimulated by PAF (Dhar and Shukla, 1991, Dhar et al, 1991; Salari et al 1990), vasopressin (Granot et al, 1990) and the thromboxane A_2 analogs U46619 and STA2 (Nakashima et al, 1991). Using an antibody to aminobenzyl phosphonate Nakamura & Yamamura (1989) showed that collagen and thrombin caused tyrosine phosphorylation of 135 and 124 kDa platelet proteins, whereas thrombin affected an additional protein at 76 kDa. The magnitude of the effect was greatest with thrombin. The fact that these two agonists both increased tyrosine phosphorylation is of special interest because Smith et al (1992) report that these agonists differ substantially in their mode of signal transduction. However, since no effort was made to eliminate the secondary agonists (*i.e.* ADP, TXA_2) that are released from stimulated platelets the evidence is insufficient to conclude that collagen is able to *directly* induce tyrosine phosphorylation.

How is tyrosine phosphorylation stimulated by platelet agonists? None of the known platelet agonist receptors have intrinsic PTK activity like the growth factor and insulin receptors. Some of the agonists that increase tyrosine phosphorylation have G-protein linked receptors (*i.e.* thrombin, thromboxane A_2, PAF and vasopressin), which have no known *direct* linkage to tyrosine kinases. Ferrell and Martin (1989) suggested that platelet non-receptor PTKs may associate with membrane proteins that interact with receptors, by analogy with the lymphocyte membrane proteins CD4 and CD8 which interact with the T-cell receptor and the *lck* kinase. Evidence for the association of a platelet membrane protein with specific PTKs has been obtained by Huang et al (1991) by immunoprecipitation with polyclonal antibodies directed against the transmembrane glycoprotein GpIV (CD36), or against specific sequences of various *src*-related PTKs. PTKs were immunoprecipitated with antibodies specific for different *src*-family PTKs and incubated with ^{32}P-ATP to identify autophosphorylated PTKs. The nonreceptor PTKs *fyn, yes and lyn* were found to associate with GpIV (CD36). In contrast, the PTKs were not immunoprecipitated by anti-GpIV in platelets from a patient lacking GpIV. The relationship between these PTKs and GpIV is apparently unique and specific, because these associations occurred in the presence of much larger amounts of *src*, and antibodies to IIb-IIIa, Ia/IIa, Ib and the Fc receptor-II did not immunoprecipitate any PTKs. Furthermore, the same PTKs were co-immunoprecipitated with GpIV in C32 melanoma and HEL cells (Huang et al, 1991). It is not known if the putative PTK-GpIV complex influences the kinase activity, or if these associations are altered by the stimulation or aggregation of platelets. *Fyn, lyn* and *yes* also associate with p21^{ras}GAP in thrombin-stimulated platelets (Cichowski et al, 1992).

The G-protein family of receptors are not known to *directly* interact with non-receptor PTKs (Wahl et al, 1989a,b). Therefore, it is possible that tyrosine phosphorylation is stimulated by signals that are generated downstream from the receptor/G-protein interactions; *e.g.* by activated PKC and/or elevated cytosolic Ca^{2+}. Although Ferrell and Martin (1988) originally found little or no evidence that the effects of thrombin on tyrosine phosphorylation could be duplicated by elevated intracellular Ca^{2+} or PKC, others have shown that PMA and indolactam V (which activates PKC), and Ca^{2+} mobilizers (*i.e.* thapsigargin and Ca^{2+} ionophores) are effective stimulators of tyrosine phosphorylation (Golden and Brugge, 1989; Vostal et al, 1991), and the combination is synergistic (Takayama et al, 1991). We found that virtually all proteins affected by thrombin were also tyrosine phosphorylated when platelets were treated with A23187, or PMA, or a combination of both. In accord with this hypothesis we were able to eliminate tyrosine phosphorylation caused by thrombin by treating platelets with a combination of $1\mu M$ staurosporine plus BAPTA-AM, which respectively inhibited PKC and prevented the rise of intracellular Ca^{2+}. Bachelot et al (1992) also provided similar indirect evidence for a

PKC-mediated tyrosine phosphorylation in collagen and thrombin stimulated platelets.

Despite these findings the mechanisms responsible for increased tyrosine phosphorylation by platelet agonists remains to be established. It is likely that several different molecular mechanisms will emerge as we learn more about the interactions between different ligands and membrane proteins, including receptors and the various integrins such as GpIIb-IIIa and GpIV. The stimulation-induced tyrosine phosphorylation of pp125[fak] does not occur in platelets lacking GpIIb-IIIa, suggesting a role for that PTK in signal transduction emanating from a membrane integrin (Schaller et al, 1992). The possible role of GpIIb-IIIa in the regulation of tyrosine phosphorylation and its relationship with pp60[c-src] will be further discussed below. Another membrane protein that could be involved in transmembrane signalling is GpIb. The interaction of GpIb with vWF, which can be induced by adding the antibiotic ristocetin to platelets in the presence of vWF, causes agglutination of platelets, but we found no evidence of enhanced tyrosine phosphorylation accompanying this response.

Another way agonists may increase tyrosine phosphorylation of proteins would be to inhibit PTPases (*see below*), or to stimulate a PTPase which removes an inhibitory phosphotyrosine site on a PTK, as occurs in the case of the CD45 PTPase/p56[lck] kinase interaction in T- cells (Klausner and Samelson, 1991). In this regard it is important to note that PTPases homologous to CD45 have been cloned from a megakaryoblastic cDNA library (Gu et al, 1991).

Regardless of the exact mechanisms that promote tyrosine phosphorylation there must also be negative modulating signals. Cyclic AMP is a powerful inhibitor of receptor/G-protein mediated signal transduction in platelets (Zavoico et al, 1985), and agents that increase cAMP also affect tyrosine phosphorylation. Pumiglia et al (1990) first reported that the stimulation of tyrosine phosphorylation by a high concentration of thrombin (10nM) was strongly inhibited by prostacyclin, PGD_2 or $N^2,2'$-O-dibutyryl-cAMP, but not by sodium nitroprusside or 8-Br-cGMP. Prostacyclin inhibited tyrosine phosphorylation over the same concentration range that it antagonized $PtdIns(4,5)P_2$ hydrolysis (Zavoico et al, 1985). The agents that increase cAMP did not inhibit tyrosine phosphorylation caused by A23187, PMA or the PTPase inhibitor pervanadate, which is further evidence that the block produced by cAMP is directed against the very earliest receptor/G-protein mediated processes, and is not directly exerted on PTKs. Subsequently Oda et al (1992) showed that tyrosine phosphorylation evoked by low (*i.e.* 1nM), but not high (*i.e.* 10nM), concentrations of thrombin were inhibited by PGE_1 and agents that elevate cGMP; *i.e.* sodium nitroprusside, 8-Br-cGMP. It may be concuded that both cAMP and cGMP can inhibit the reactions that lead to enhanced tyrosine phosphorylation. The agents employed by Pumiglia et al (1990) exert a more powerful effect on cAMP levels than does PGE_1, which probably accounts for their ability to block even high concentrations of thrombin.

Forskolin, which also increases cAMP, blocked the tyrosine phosphorylation induced by thrombin and collagen (Golden et al, 1990). This finding contrasts with a recent study (Smith et al, 1992) which reported that *direct* responses to collagen (*i.e.* those not due to released ADP or generated TXA_2) are not inhibited by increasing cAMP. Further clarification of the action of collagen on tyrosine phosphorylation is needed.

ACTIVATION OF PLATELETS BY TYROSINE PHOSPHATASE INHIBITORS

Protein tyrosine phosphorylation in platelets must be under the control of tyrosine phosphatases (PTPases). Indeed, increased tyrosine phosphorylation accompanied by secretion was demonstrated using orthovanadate and molybdate in permeabilized platelets (Lerea et al, 1989), and pervanadate stimulated tyrosine phosphorylation and caused aggregation of intact platelets (Inazu et al, 1990). To further investigate the role of tyrosine

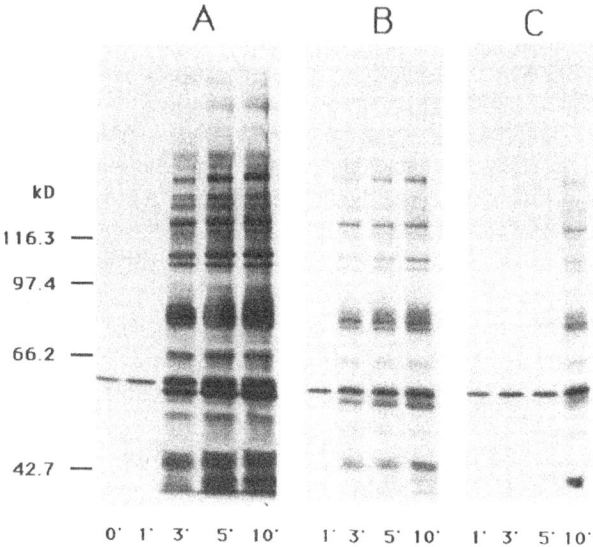

Figure 1   Stimulation of tyrosine phosphorylation by pervanadate. Washed platelets were incubated with pervanadate for 1,3,5 and 10 min at 37°C. Western blots were obtained using 70 μg of protein/lane and antiphosphotyrosine antibody plus [^{125}I]-protein A. Platelets/ml: (A) 3×10^8, (B) 6×10^8, (C) 1×10^9. The autorads were exposed optimally for the pervanadate effect; they are therefore underexposed for the control (0 min) and do not show the basal tyrosine phosphorylated proteins except for pp60. The rate and extent of tyrosine phosphorylation was decreased as the concentration of platelets was increased, because the cellular reducing agents (e.g. glutathione) inactivate the vanadyl hydroperoxide V^{4+}-OOH.

phosphatases in intact platelets we (Pumiglia et al, 1992) employed the reagent pervanadate - or vanadyl hydroperoxide - which is produced by the oxidation of orthovanadate by hydrogen peroxide, and permeates cells. Pervanadate is a powerful inhibitor of PTPases with little or no effect on Ser/Thr phosphatases.

Pervanadate stimulated tyrosine phosphorylation of platelet proteins about 30-fold greater than that obtained with several units/ml of thrombin (Figure 1). Pervanadate stimulated the tyrosine phosphorylation of the same protein bands as thrombin, as well as additional bands not seen with thrombin. Pervanadate caused all the aspects of platelet activation that were evoked by thrombin: *i.e.* hydrolysis of PtdIns(4,5)P$_2$, synthesis of PtdIns(3,4)P$_2$ and PtdOH, mobilization of intracellular Ca^{2+}, phosphorylation of PKC substrates (*e.g.* pleckstrin), secretion of the contents of dense and α-granules, increased polymerization of actin and cytoskeletal assembly (2/3 that of thrombin), activation of GpIIb-IIIa (measured with PAC-1 antibody obtained from Dr. S. Shattil), shape change, and platelet aggregation. Neither orthovanadate nor hydrogen peroxide alone had any significant effect. Furthermore, after pervanadate was formed the excess hydrogen peroxide could be destroyed by catalase without affecting the response to pervanadate.

The aggregation of platelets elicited by pervanadate was blocked by chelating extracellular Ca^{2+} or by RGDS, so it was GpIIb-IIIa and fibrinogen dependent. Preventing arachidonic acid metabolism with aspirin, and the effects of ADP with ADP-scavengers, had no effect on platelet aggregation caused by pervanadate. Tyrosine phosphorylation and all responses to pervanadate were blocked by PTK inhibitors; *i.e.* 90μM tyrphostin RG50864 (Yaish et al, 1988) and 5μM staurosporine (a higher concentration than necessary to block PKC). Tyrphostin-1 (Yaish et al, 1988), an inactive analog of RG50864, had no effect. The aggregation caused by pervanadate was abolished by preventing the elevation of intracellular Ca^{2+} with the cell permeant chelator BAPTA-AM, although shape change was unaffected. The release of arachidonic acid was also blocked by BAPTA-AM, indicating that it was probably caused by the elevation of Ca^{2+} owing to the pervanadate-induced activation of PLC to form $Ins(1,4,5)P_3$.

Staurosporine at a concentration of 1μM/10^9 platelets/ml totally blocked the PKC-mediated protein phosphorylation (^{32}P-phosphate incorporation) caused by pervanadate, with only a small ($\leq$ 20%) reduction in tyrosine phosphorylation. The increased protein phosphorylation caused by pervanadate probably resulted from the stimulation of a PLC to generate the PKC-activator diacylglycerol (and increased free Ca^{2+}). Under these conditions staurosporine strongly blocked dense granule secretion. Although the initial stimulation of platelets by pervanadate was due to its effect on tyrosine phosphorylation we conclude that phosphorylation of proteins by PKC was necessary for dense granule secretion, as was the case with other agonists (Watson and Hambleton, 1989). Secretion of α-granule proteins was only partially inhibited by RG50864 or BAPTA-AM. Later (*see below*) we will show that tyrosine phosphorylation was not necessary for thrombin-induced secretion.

Arachidonic acid release by pervanadate was blocked by the PTK inhibitor RG50864 (*see PTK inhibitors below*), and by high concentrations of staurosporine ($\geq$ 5μM) which concurrently prevent the stimulation of tyrosine phosphorylation. Staurosporine at 1μM, which preferentially blocked PKC, did not alter arachidonate release by pervanadate. Thus, PKC was not involved in arachidonate release by pervanadate, but tyrosine phosphorylation was necessary, probably because it activated a PLC that generated $Ins(1,4,5)P_3$ which released internal Ca^{2+}. In accord with this view we found that the release of arachidonic acid by pervanadate was blocked by chelating intracellular Ca^{2+} with BAPTA.

In order to further investigate how pervanadate stimulated phosphoinositide metabolism we utilized saponin-permeabilized platelets, and studied the synthesis of PtdOH as a marker for the hydrolysis of PtdIn(4,5)P_2 by PLC. In permeabilized platelets thrombin and the PTPase inhibitor pervanadate both stimulated PtdOH production, but only the former was blocked by GDPβS, suggesting the existence of separate G-protein dependent and tyrosine phosphorylation dependent PLCs. We conclude that pervanadate causes the inhibition of PTPases leading to the activation of a form of PLC (*e.g.* PLCγ-1) that requires tyrosine phosphorylation, and that all the subsequent biochemical and functional responses of the platelets caused by pervanadate are elicited by the second messengers produced by the hydrolysis of PtdIns(4,5)P_2.

EFFECTS OF TYROSINE KINASE INHIBITORS. WHAT RESPONSES ARE TYROSINE PHOSPHORYLATION-DEPENDENT ?

The tyrosine kinase inhibitors that have been used on platelets fall into 3 classes, genistein and erbstatin which are natural products, and the synthetic tyrphostins. Genistein, an isoflavone from *Pseudomonas* sp. was virtually equipotent against the EGF-receptor, *v-src* and pp110$^{gag\text{-}fes}$ PTK activities (Akiyama et al, 1987). Genistein is competitive with ATP and inhibits other protein kinases. Erbstatin, isolated from the culture medium of an actinomycete, also blocked the EGF-receptor tyrosine kinase activity. Yaish et al (1988) and Gazit et al (1989) produced synthetic PTK inhibitors based on the benzylidene malononitrile nucleus, that resemble the phenolic moeities of tyrosine and erbstatin. These compounds were found to be competitive with the substrate, and they were 1/10 as effective in inhibiting autophosphorylation of the receptor, and they inhibited EGF-dependent cell proliferation. RG50864 (Tyrphostin-12), the most active in the original series, had a K_1 of $0.85\mu M$ vs. the EGF-receptor tyrosine kinase activity with exogenous substrate. RG50864 was about 1000-fold less active against the insulin receptor tyrosine kinase, and was also much less effective against protein kinase A and some other kinases. These three classes of PTK inhibitors have been applied as tools to uncover the involvement of tyrosine phosphorylation in several aspects of platelet function, such as phosphoinositide metabolism, secretion and aggregation.

Several studies have noted very striking inhibition of platelet responses to agonists by PTK inhibitors (Dhar and Shukla, 1990; Salari et al, 1990). It has been common practice, therefore, to associate a platelet function with tyrosine phosphorylation if that function is modified by a PTK inhibitor. Three studies employing PTK inhibitors in rabbit platelets strongly implicated tyrosine phosphorylation in the regulation of platelet phosphoinositide metabolism and PLC activity, and placed tyrosine phosphorylation in a necessary central role for the transduction of stimuli into a wide range of responses. Gaudette and Holub (1990) showed that genistein inhibited the stimulation of phosphoinositide phosphorylation caused by U46619. Salari et al (1990) found that erbstatin blocked PAF-induced tyrosine phosphorylation, PI-hydrolysis, PKC activation, 5-HT secretion and aggregation of rabbit platelets. The tyrphostin RG50864 also inhibited many of these responses, whereas genistein was without effect. Platelet aggregation and 5-HT secretion caused by thrombin was also inhibited by erbstatin. Erbstatin's ability to prevent the formation of inositol phosphates by PAF suggested that PLC was affected. However, it should be note that the concentration of erbstatin necessary for these effects was 30 times greater than the IC_{50} for inhibition of the EGF-receptor PTK activity in vitro. In another study Dhar and Shukla (1990) observed that genistein blocked the tyrosine phosphorylation of 2 proteins of 50 and 60 kDa in rabbit platelets stimulated by PAF. They used a commercial (Boehringer Mannheim) antiphosphotyrosine antibody that detected only those two proteins. They further showed that a very high concentration of genistein (0.5 mM) blocked PAF-stimulated inositol phosphates formation, platelet aggregation and the phosphorylation of 20 Kda and 50 kDa proteins. The PMA-induced phosphorylation of a 40 kDa protein was also affected by genistein. They concluded that tyrosine phosphorylation was involved in the activation of PLC by PAF, although PLC activity measured in vitro was unaffected by genistein.

The association of tyrosine phosphorylation with secretion, noted in the studies of Salari et al (1990) was also suggested by two other findings: 1) Lerea et al (1989) found that orthovanadate and molybdate inhibited PTPase activity more than 97% in saponin permeabilized platelets and caused the tyrosine phosphorylation of a 50 kDa protein, which correlated closely with the secretion of 5-HT and PDGF; and 2) Rendu et al (1989) reported the presence of pp60[c-src] in the dense granule fraction of platelet homogenates.

The studies described above have shown PTK inhibitors to be effective inhibitors of virtually all responses of platelets to agonists, thereby implicating tyrosine phosphorylation in the earliest signalling events that lead to platelet responses. The information that can be derived from the use of PTK inhibitors is obviously dependent on their biochemical specificity, and unfortunately, each suffers from important limitations that greatly limit their usefulness. The work of Nakashima et al (1991) provided an important cautionary message. They also observed that genistein decreased aggregation, 5-HT secretion and protein tyrosine phosphorylation induced by collagen and the stable TXA_2 analogs U46619 and STA_2. In contrast, genistein did not inhibit responses to 0.1U/ml thrombin, or primary aggregation by ADP or epinephrine. However, Nakashima et al (1991) further showed that a causal relationship between tyrosine phosphorylation and the inhibitory effects of genistein on platelet responses was doubtful. Firstly, a related isoflavone, daidzin, lacked ability to inhibit tyrosine phosphorylation, but like genistein it inhibited the platelet responses to collagen and TXA_2. Secondly, genistein actually produced only a small inhibition of tyrosine phosphorylation, which is in accord with our own observations that genistein is a very poor inhibitor of PTKs in platelets. The effects of genistein were attributed to its inhibition of the binding of [³H]U46619 to washed platelets, indicating that it was affecting the TXA$_2$ receptor by a mechanism unrelated to tyrosine phosphorylation.

To further explore the functional roles of tyrosine phosphorylation our laboratory has employed several tyrphostins in human washed platelets stimulated by thrombin, PMA, A23187 and pervanadate. We found RG50864 to be a very effective inhibitor of the enhancement of tyrosine phosphorylation by these agents, as well as by PAF and vasopressin. Higher concentrations of RG50864 were necessary to block the stimulation of tyrosine phosphorylation by thrombin (or PMA) than were needed to counter the effect of

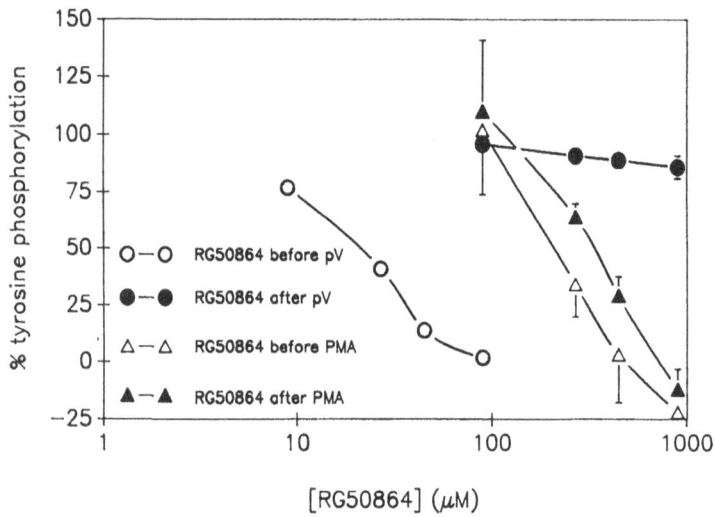

Figure 2   Effects of RG50864 on tyrosine phosphorylation in platelets stimulated with pervanadate (pV) or phorbol myristate actetate (PMA). RG50864 was added at various concentrations either 4 min before the stimulating agents, or 10 min after 80 nM PMA, or 15 min after pervanadate. The platelets were analyzed for tyrosine phosphorylated proteins after 5 min. The data is shown for a major substrate, pp128: 100% tyrosine phosphorylation is the amount without any RG50864.

pervanadate; *i.e.* maximum effects at 900 vs 90 µM (Figure 2). Basal tyrosine phosphorylation was only partially inhibited by 900 µM RG50864. We assessed the functional effects of tyrphostins mainly using thrombin as the stimulus. Our results differ substantially in many respects from the findings described above. We find that tyrosine phosphorylation is *not necessary* for the activation of PLC by thrombin, the secretion of dense granule constituents, or platelet aggregation.

RG50864 did inhibit secretion of 5-HT by low concentrations of thrombin, but the response to high concentrations of thrombin was unaffected, despite the total inhibition of the stimulated tyrosine phosphorylation. Similarly, the secretion induced by PMA was also resistant to RG50864, although PMA-stimulated tyrosine phosphorylation was blocked. The maximal secretion of α-granule proteins induced by thrombin or PMA was also not reduced by RG50864, although the rate was slowed. We conclude that the stimulation of tyrosine phosphorylation is not necessary for dense or α-granule secretion, but we have not ruled out a facilitory effect of tyrosine phosphorylation when the stimulus intensity is weak and Ca^{2+} mobilization is low.

RG50864 did effectively inhibit several responses to thrombin (10nM); *i.e.* the stimulation of tyrosine phosphorylation, aggregation, arachidonic acid release, $PtdIns(4,5)P_2$ resynthesis, and the production of PtdOH and $PtdIns(3,4)P_2$. The suppression of PtdOH production would normally be taken as evidence for the antagonism of PLC, but we find that the synthesis of $[^{32}P]PtdOH$ from diacylglycerol (DG) added to $[^{32}P]phosphate-loaded$ platelets was also effectively blocked by RG50864. Therefore, the tyrphostin appears to directly inhibit DG-kinase. Another finding that was consistent with this conclusion was that $5\mu M$ staurosporine blocked increased tyrosine phosphorylation by thrombin, but not the formation of PtdOH. The initial rapid fall in $[^{32}P]$-labelled $PtdIns(4,5)P_2$ normally caused by thrombin was not prevented by the tyrphostin, and in fact appeared to be somewhat larger because the ensuing rapid resynthesis phase was blocked. Similarly, King and Rittenhouse (1989) found no inhibition of $Ins(1,4,5)P_3$ production by 10 μM staurosporine, a concentration that we find to block the stimulation of tyrosine phosphorylation. Thus, we conclude that the activation of a PLC through the thrombin receptor in human platelets is not dependent on agonist-stimulated tyrosine phosphorylation. However, tyrosine phosphorylation may be important for the synthesis of $PtdIns(4,5)P_2$, as this was prevented by RG50864 or by 5 μM staurosporine. The effects of these inhibitors was consistent with the findings that tyrosine phosphorylation was implicated in the regulation of phosphoinositide kinase activities and the synthesis of $PtdIns(4)P$ and $PtdIns(4,5)P_2$ in A431 cells (Payastre et al, 1990), and the co-immunoprecipitation of PtdIns 3-kinase with $pp60^{c-src}$ and $pp59^{fyn}$ in platelets (Gutkind et al, 1990).

Nevertheless, we cannot yet exclude the possibility that the tyrphostin or staurosporine can *directly* inhibit these phosphoinositide kinases, independently of their effect on PTKs. An analog of RG50864 that does not inhibit PTKs, tyrphostin-1 (Yaish et al, 1988), produced none of the effects of RG50864 on platelets, but that may simply mean that only the active PTK inhibitors can also inhibit the other kinases. This is especially likely to occur in platelets which contain such very high PTK activity, making it necessary to use high concentrations of tyrphostins to block tyrosine phosphorylation. In this way the concentration of the tyrphostins used may extend into a concentration range that affects other enzymes.

Recently the synthesis of the novel phosphoinositide $PtdIns(3,4)P_2$ has been shown to occur in response to growth factors. Its synthesis, which requires a $PtdIns(4)P$ 3-kinase, has been shown to be on the pathway for the regulation of DNA synthesis by the PDGF receptor (Fanti et al, 1992). The synthesis of this lipid is also greatly stimulated in platelets by thrombin (Kucera and Rittenhouse, 1990; Sultan et al, 1990; Nolan and Lapetina, 1990). It is not known if this lipid plays any role in signal transduction in the anucleate platelet, or if it is merely a carryover from the megakaryocyte in which it may mediate growth factor controlled DNA synthesis. The equivalent production of $PtdIns(3,4)P_2$ in response to the PTPase inhibitor pervanadate suggests that tyrosine phosphorylation may be involved in the activation of the $PtdIns(4)P$ 3-kinase (Escobedo et al, 1991). This requires further study.

Another important response of platelets to thrombin is the release of arachidonic acid from phospholipids to provide the substrate for the synthesis of TXA_2 and several hydroxy fatty acids. We found that arachidonic acid release by thrombin was inhibited by tyrphostin compounds 8, 11, and RG50864; but their $IC_{50}s$ for the inhibition of arachidonate release did not agree with their IC_{50} *vs* tyrosine phosphorylation. The weaker PTK inhibitors were virtually equally good inhibitors of arachidonate release as RG50864. Another critical experiment tested the effect of high concentrations of staurosporine ($\geq 5\mu M$) on arachidonate release. The release of the fatty acid by thrombin was not inhibited by this high concentration of staurosporine which effectively blocked the thrombin-induced tyrosine phosphorylation. Indeed, arachidonate release was actually potentiated! Therefore, stimulated tyrosine phosphorylation is not necessary for arachidonate release by thrombin. One caveat, however, - staurosporine did not prevent the stimulated tyrosine phosphorylation of one protein band at 39 kDa, which was blocked by the tyrphostin.

Aggregation induced by thrombin, but not Ca^{2+} ionophore was blocked by RG50864. Shape change was not inhibited. RG50864 also inhibited aggregation caused by PMA, implicating tyrosine phosphorylation, an event that is downstream from the activation of PKC by PMA, in the aggregation induced by phorbol esters. The tyrphostin was capable of actually terminating platelet aggregation early on in its progression, when the inhibitor was added soon after PMA. This abrupt termination of aggregation was accompanied by a rapid dephosphorylation of tyrosine phosphorylated proteins, a result which suggested that aggregation is intimately associated with tyrosine phosphorylation, and may in fact be necessary for aggregation. This was a conclusion of the studies of Salari et al (1990) and Dhar and Shukla (1990) as well. However, there are two problems with this hypothesis: 1) the dose response for inhibition of aggregation elicited by thrombin or PMA was 10-fold higher than that necessary to prevent tyrosine phosphorylation by pervanadate. 2) But, most importantly over this 10-fold higher dose-response curve RG50864 also prevented aggregation in a situation that did not permit inhibition of tyrosine phosphorylation. In these experiments maximal tyrosine phosphorylation was produced by first exposing *unstirred* washed platelets to pervanadate for 15 min. When the platelets were then stirred they aggregated strongly. However, when RG50864 was added at the conclusion of the exposure to pervanadate, and the platelets were then stirred, aggregation was strongly suppressed, - *although the markedly enhanced tyrosine phosphorylation could not be reversed because the PTPases were inactivated.*

This result prompted us to further put to the test the thesis that tyrosine phosphorylation is necessary for platelet aggregation. One test involved the activation of GpIIb-IIIa with RGDS as described by Du et al (1991). Treatment of unstimulated platelets with RGDS, but not fibrinogen causes an activation of GpIIb-IIIa to a state that is competent to bind fibrinogen. This RGDS-induced state (*e.g.* measured by PAC-1 antibody binding) can be maintained by mild fixation with paraformaldehyde, so that upon subsequent washing to remove RGDS, and exposure to fibrinogen those platelets will aggregate when stirred. The aggregation occurred without the activation of the common intracellular signal transduction pathways, as it was not blocked by prostacyclin (Du et al, 1991), although it required fibrinogen (blocked by RGDS) and external Ca^{2+} (blocked by EGTA). We find that the original treatment with RGDS did not induce tyrosine phosphorylation, and the activating effect of RGDS was not altered by pretreatment of platelets with metabolic inhibitors (deoxyglucose + antimycin A). This mechanism to produce aggregation appears to be solely due to conformational changes in IIb-IIIa induced extracellularly (Du et al, 1991), and to be independent of ATP (energy metabolism). Nevertheless, aggregation under these conditions was blocked by tyrphostin RG50864 at the concentrations that inhibited aggregation by thrombin, PMA, and by pervanadate post-stimulation. This reveals that the tyrphostin possesses an action that is able to inhibit aggregation in a manner that is *independent* of tyrosine phosphorylation. Thus, the assumption of a requirement of tyrosine phosphorylation for aggregation that is based on inhibition of both by PTK inhibitors is unwarranted.

TYROSINE PHOSPHORYLATION SPECIFICALLY RELATED TO AGGREGATION AND THE FIBRINOGEN-GpIIb-IIIa INTERACTION

The aggregation of platelets by thrombin depends on at least 4 factors: external divalent cations (Ca^{2+}, Mg^{2+}), fibrinogen, the activation of GpIIb-IIIa to be competent to bind fibrinogen, and the frequency of platelet-platelet collisions. Several groups have observed that tyrosine phosphorylation of certain proteins is associated with some of these factors, but the existing studies are not in agreement on some fundamental points.

The binding of fibrinogen and platelet aggregation are blocked by RGDS (a peptide sequence of fibrinogen that binds to GpIIb-IIIa) or chelators of divalent cations such as EDTA. A link between tyrosine phosphorylation and these events was established when Ferrell & Martin (1989) reported that fibrinogen binding to GpIIB-IIIa was required for the third wave of tyrosine phosphorylation, but that aggregation *per se* was unnecessary, based on the fact that thrombin-induced tyrosine phosphorylation: *a*) occurred without aggregation (no stirring); *b*) was blocked by RGDS or EDTA, which prevent fibrinogen binding to GpIIb-IIIa; and *c*) did not occur in platelets from patients with Glanzmann's thrombasthenia. These platelets lack GpIIb-IIIa and do not bind fibrinogen or aggregate. Suprisingly, although their washed platelets aggregated when exposed to ADP + fibrinogen (or 10% plasma), no 3rd wave of tyrosine phosphorylation was detected. From these experiments Ferrell & Martin concluded that either fibrinogen binding to GpIIb-IIIa was not sufficient to elicit the RGDS-inhibitable tyrosine phosphorylation, - or alternatively, that the relevant ligand was not fibrinogen, vWF or fibronectin, all of which were present in plasma.

In contrast, Golden et al (1990) found that the tyrosine phosphorylation of 3 proteins depended both on fibrinogen binding to IIb-IIIa - *and subsequent aggregation*. They noted increased tyrosine phosphorylation of 97, 95 and 84 kDa proteins induced by thrombin, collagen, ADP/epinephrine, mastoparan (which stimulates G-protein) and PMA. It is not clear if these proteins are the same as those seen by Ferrell and Martin (i.e. 126, 108, 100 and 85 kDa). The important point of this study was that increased phosphorylation required *stirring and aggregation*. Furthermore, the aggregation-dependent tyrosine phosphorylation was inhibited by RGDS, by incubation of platelets with EGTA at 37°C to dissociate the GpIIb-IIIa complex, and by a monoclonal antibody that prevented fibrinogen binding. It was not blocked by aspirin, and was independent of secretion, and the intracellular protease calpain (which was antagonized by the cell permeant inhibitor E64D). As PMA also elicited aggregation-dependent (*i.e.* GpIIb-IIIa/fibrinogen-dependent) tyrosine phosphorylation it is evident that signals produced downstream from receptor/G-protein interactions may be responsible for aggregation-dependent tyrosine phosphorylation.

Golden et al [1990] attributed their differences in results from those of Ferrell and Martin to the likelihood that in the latter study microaggregates were actually formed when the washed platelets were stimulated by thrombin, even without stirring. Golden et al (1990) suggested that aggregation-dependent tyrosine phosphorylation may be related to the association of GpIIb-IIIa with the cytoskeleton, which only occurs when platelets aggregate (Phillips et al 1980). The controversy remains to be resolved. Exactly how the occupancy of GpIIb-IIIa by fibrinogen and the subsequent aggregation would stimulate tyrosine phosphorylation is not yet evident. Although GpIIb-IIIa can be tyrosine phosphorylated *in vitro* (Findik et al, 1990) this has not been demonstrated in intact platelets. Furthermore, no PTKs were found to associate with GpIIb-IIIa in immunoprecipitates of the latter (Ferrell and Martin, 1989).

Despite the described associations of enhanced tyrosine phosphorylation with aggregation, and the inhibition of aggregation by PTK-inhibitors (Salari et al, 1990; Dhar and Shukla, 1990), there are a number of experimental situations in which the two can be dissociated. One case involves the unique activation of GpIIb-IIIa by the ligand RGDS to a state that is competent to bind fibrinogen, which can then be fixed with paraformaldehyde to allow subsequent fibrinogen-dependent aggregation after RGDS is removed by washing (Du et al, 1991). We have confirmed this ability of RGDS to activate the integrin, but the extent of aggregation measured photometrically was significantly less than that caused by thrombin as the aggregates do not grow to the large size obtained with powerful agonists.

Nevertheless, visible aggregates do form, yet we found, as did Golden et al (1990) that tyrosine phosphorylation was not increased by RGDS. Furthermore, as we noted earlier, this effect of RGDS does not require ATP, as it is insensitive to metabolic inhibitors (deoxyglucose + antimycin A) which totally block tyrosine phosphorylation and aggregation of normal platelets. These experiments imply that tyrosine phosphorylation is *not necessary* for the activation of GpIIb-IIIa to a state competent to bind fibrinogen, nor is it necessary for any steps that might occur beyond the activation of GpIIb-IIIa that enable aggregation to occur. These experiments would not preclude the possibility that tyrosine phosphorylation is involved when GpIIb-IIIa is activated by *intracellular* signals (Shattil and Brass, 1987).

We obtained further evidence against a necessary role for tyrosine phosphorylation in aggregation by exposing unstirred platelets to 40μM SFLLRNPNDKYEPF for 20 min at 21°C to desensitize thrombin receptors. Under these conditions the platelets were rendered totally insensitive (no aggregation when stirred) to a 10-fold higher concentration of the peptide. The important point is that these platelets could be aggregated by thrombin although tyrosine phosphorylation was not stimulated. We conclude, therefore, that the binding of fibrinogen to GpIIb-IIIa is not sufficient to stimulate tyrosine phosphorylation, and the *enhanced tyrosine phosphorylation normally caused by thrombin is not necessary for the aggregation of platelets*. Another type of aggregation, that induced by ristocetin + vWF, which involves GpIb and was unaffected by RGDS, also occurred without enhanced tyrosine phosphorylation [Golden et al, 1990], and we found it to be unaffected by the PTK inhibitor RG50864. This implies that the action of the tyrphostin that inhibits aggregation is specifically directed against platelet interactions that are mediated by GpIIb-IIIa.

TYROSINE PHOSPHORYLATION AND THE PLATELET CYTOSKELETON

Many of the phenomena of platelet activation and aggregation may be directly or indirectly affected by the concomitant increased assembly of the cytoskeleton (CS) proteins. The CS from thrombin-aggregated platelets contains the bulk of the actin, myosin and actin-binding protein, and several other proteins that are fundamental for adhesion and aggregation (*e.g.* GPIIb-IIIa, GPIb, (Phillips et al 1980, Kouns et al, 1991) vinculin and talin, (Kouns et al, 1991, Asijee et al 1990). Additionally it has been reported that phospholipid metabolizing enzymes such as PI-kinases, diacylglycerol kinase and PLC (Grondin et al, 1991, Nayhas et al 1989, Zhang et al 1992) are present in the triton X-100 insoluble fraction of stimulated platelets, which is the operational designation of the CS. CS reorganization may also be involved in the activation of PLA_2 by collagen (Nakano et al 1989). The co-localization of some key enzymes with CS proteins demonstrates the intimate involvement of this structure in signal transduction, perhaps in direct response to cell adhesion or agreggation, or to make the latter events possible. Tyrosine phosphorylated proteins could play a role in platelet CS assembly/function. This was first suggested by the finding of a significant portion of the platelet $pp60^{c-src}$ in the CS under certain conditions (Grondin et al, 1991; Horvath et al, 1992).

Our laboratory investigated the presence of tyrosine phosphorylated proteins and PTKs in the CS. In these experiments we distinguished between unstimulated platelets, activated platelets (*i.e.* stimulated by agonist under non-aggregating conditions) and aggregated platelets. The CS obtained from unstimulated platelets was found to contain only a very small fraction of the total tyrosine phosphorylated proteins. After stimulation with thrombin, and when aggregation was prevented by the inclusion of EGTA or RGDS, there was an increase of total tyrosine phosphorylation and a small (1.4-fold) but detectable, increase in at least 6 proteins in the CS fraction. The level of CS tyrosine phosphorylation was low compared to that found in the triton soluble (non-CS) fraction. Two points are noteworthy about these findings: *1*) the large increase in tyrosine phosphorylation of non-CS proteins in the absence of platelet aggregation demonstrated that aggregation was not necessary for thrombin-induced tyrosine phosphorylation, in agreement with the original work of Ferrell and Martin, and in contrast to a more recent report which implied that aggregation was entirely responsible for the enhanced tyrosine phosphorylation caused by thrombin [Golden et al, 1990]; *2*) The increased incorporation of tyrosine phosphorylated proteins into the CS was specific as several major protein bands which were prominently phosphorylated (*e.g.* pp37, pp60^{c-src}) were excluded from the CS.

Under aggregating conditions there was a substantial (12.9-fold) increase in the tyrosine phosphorylated proteins associated with the CS, as measured with antiphosphotyrosine antibody and ^{125}I-protein A. It should be noted that the phosphotyrosine in the triton X-100 lysate was extremely labile, so that it was essential to add PTPase inhibitors to detect any tyrosine phosphorylated proteins. The same endogenous protein bands were more readily detectable with antiphosphotyrosine antibody when the isolated CS was incubated with ATP *in vitro*, as the PTPase activity was negligible under these conditions. The most prominent tyrosine phosphorylated proteins in the CS of aggregated platelets had molecular weights of 34, 50, 52, 55, 57, 60, and 130 kDa, while lesser bands were detected at 88 and 120 kDa. Several bands such as pp58 and pp52 were preexisting tyrosine phosphorylated proteins that were translocated from the triton soluble fraction into the triton-insoluble CS. Others such as pp130, pp34, pp120, and pp88 were newly tyrosine phosphorylated proteins, specific to aggregation, as they were absent or present at much lower levels in stimulated platelets under non-aggregating conditions. In fact several of these proteins (pp130, 120, 88) may correspond to those proteins previously reported to undergo aggregation dependent tyrosine phosphorylation (Ferrell and Martin, 1989; Golden et al, 1990).

We also measured PTK activity in the CS using enolase as an exogenous substrate. The PTK activity of CS increased 1.4-fold by the activation (unstirred) of platelets with thrombin, and a small amount of p59^{lck} was also detected by Western blotting in the CS with a specific antibody. The PTK activity measured with enolase increased to 14.9-fold in the CS of platelets that were fully aggregated by thrombin.

Using mAb327 to measure pp60^{c-src} by immunoprecipitation we found very low levels of the *src* kinase in the resting CS, representing approximately 4.4% of the total pp60^{c-src} in the platelet. We could detect no appreciable change after stimulation with 1 U/ml α-thrombin if aggregation was blocked by preincubation with either RGDS (1mM) or EGTA (2mM). Likewise, stimulation in the absence of shaking resulted in little or no change in the CS pp60^{c-src}. However, under fully aggregating conditions (vigorous shaking, 37°C, absence of EGTA or RGDS) we measured a 13.9-fold increase in the level of detectable *src* in the triton insoluble CS, which amounted to 45-60% of the total platelet pp60^{c-src}. The remainder of the total pp60^{c-src} was quantitatively accounted for in the triton soluble fraction. The increase in the mass of pp60^{c-src} in the CS after aggregation corresponded closely to the increased rate of enolase phosphorylating activity, which suggests that the activity of the *src* kinase was not affected by its association with the CS.

The tethered ligand agonist peptide (TRAP, TRP) for the thrombin receptor TRP also produced an aggregation dependent translocation of pp60^{c-src} to the CS, that was indistinguishable from thrombin itself. Additionally, we incubated platelets with exogenous fibrinogen (240μg/ml) and 1 U/ml thrombin under non-stirring conditions. Under these conditions appreciable insoluble fibrin was generated as evidenced by the increase in the protein pelleted after lysis with triton X-100. However there was no increase in the CS association of pp60^{c-src}, which demonstrates that: *1*) activation of the tethered ligand thrombin receptor causes pp60^{c-src} translocation to the CS, and *2*) that the generation of fibrin by thrombin is neither necessary nor sufficient for this effect.

The association of *src* with the CS does not take place under any condition in which platelets undergo homologous cross-linking. Using ristocetin and cryoprecipitate as a source of VWF we induced GPIb/vWF-mediated agglutination. RGDS was included, as it does not effect the agglutination process, but does prevent vWF and fibrinogen (present in high amounts in the cryoprecipitate) from binding to GPIIb/IIIa. Under these conditions ristocetin induced agglutination produced similar changes in light transmittance to thrombin mediated aggregation, indicative of a similar degree of platelet clumping. However, only the thrombin aggregated platelets showed significant amounts of pp60$^{c\text{-src}}$ associated with the CS. This experiment additionally demonstrates that the translocation of the *src* protein was not an artifact due to incomplete solubilization of proteins from platelet aggregates. We conclude that the translocation of pp60$^{c\text{-src}}$ to the CS is a specific aggregation dependent event mediated by GPIIb/IIIa, rather than a more generalized phenomenon resulting from platelet-platelet contact.

We then asked if the aggregation-dependent incorporation of pp60$^{c\text{-src}}$ into the CS occurred in parallel with IIb-IIa, since it is known that upon aggregation of platelets there is a marked and specific incorporation of GPIIb/IIIa into the platelet cytoskeleton (Phillips et al 1980). Although both appeared to occur concurrently in thrombin-treated platelets, we found that the two events to be separable. Concanavalin A also causes association of GpIIb-IIIa with the cytoskeleton (Painter and Ginsberg, 1982) in non-aggregating platelets. Using antibody to IIb we observed marked translocation of the integrin to the CS under these conditions, but no pp60$^{c\text{-src}}$ was associated with the CS. Therefore, there is no obligatory linkage of GPIIb-IIIa to pp60$^{c\text{-src}}$, and the translocation of the integrin is insufficient for the incorporation of pp60$^{c\text{-src}}$ into the CS.

From these results we conclude that when washed platelets are stimulated with thrombin, without stirring, GpIIb-IIIa activation occurs, as measured by PAC-1 binding, and there is an increase of tyrosine phosphorylation, but very little incorporation of pp60$^{c\text{-src}}$ into the cytoskeleton. Therefore, activation of platelets *via* a G-protein linked receptor to produce activation of GpIIb-IIIa, and the other normal intracellular signals, is insufficient to cause the translocation of pp60$^{c\text{-src}}$ into the CS. The conformational change in GpIIb-IIIa caused by binding RGDS was also insufficient to elicit the translocation of the *src* protein.

Our results linked the association of pp60$^{c\text{-src}}$ with the platelet CS to aggregation and the tyrosine phosphorylation of multiple CS proteins. We next sought to better understand the temporal relationship between these events and aggregation. We utilized a platelet concentration of 3×10^8/ml, mild shaking (or rotating) at room temperature in order to slow the aggregation process. After stimulation with 10nM thrombin duplicate samples were subjected to CS isolation, or were fixed to examine by bright field microscopy for the presence of platelet aggregates. Treatment with thrombin under these conditions led to the rapid disappearance of single platelets, indicating that the initial aggregation event was rapid. By 1.5 min the majority of platelets were associated with an aggregate complex, and the disappearance of single platelets had nearly reached a plateau. The *src* protein was undetectable in the CS of those platelets. When the process was studied out to 7 minutes after the addition of thrombin aggregates were clearly visible to the eye, although they appeared to be of a finer or smaller nature than seen with maximal aggregation. Parallel analysis with GPIIb antibody demonstrated that under these "low aggregating" conditions, GPIIb/IIIa was also excluded from the cytoskeleton. However, when the platelets were subjected to more vigorous shaking at 37°C, large aggregates were formed and both pp60$^{c\text{-src}}$ and GPIIb were substantially associated with the CS. Thus, it appears that the formation of *large aggregates* is a prerequisite for the translocation of pp60$^{c\text{-src}}$ and IIb/IIIa to the CS. It may be that the shear forces that are exerted on large aggregates require the formation of a more stable structure of membrane integrins linked to the CS, which may provide enhanced stability and load-bearing capacity. Such a mechanism might also be operative in the adhesion of platelet plugs to damaged vessel walls. The association of pp125^{fak} with focal adhesion sites may be indicative of the involvement of PTKs with the structural stability of large aggregates. The role that tyrosine phosphorylation of proteins plays in such a process has yet to be established.

REFERENCES

Akiyama, T., Kadowaki, T., Nishida, E., Kadooka, T., Ogawara, H., Fukami, Y.,Sakai, H., Takaku, F. and Kasuga, M., 1986, Substrate specificities of tyrosine-specific protein kinase toward cytoskeletal proteins in vitro, *J. Biol. Chem.* 261:14797.

Akiyama, T., Ishida, J., Nakagawa, S., Ogawara, H., Watanabe, S-I., Itoh, N., Shibuya, M. and Fukam, Y., 1987, Genistein, a specific inhibitor of tyrosine-specific protein kinases, *J. Biol. Chem.* 262:5592.

Anderson, D., Koch, C.A., Grey, L., Ellis, C., Moran, M.F. and Pawson, T., 1990, Binding of SH2 domains of phospholipase Cγ1, GAP, and Src to activated growth factor receptors, *Science* 250:979.

Asijee, G.M., Sturk, A., Bruin, T., Wilkinson, J.M. and Ten Cate, J.W., 1990, Vinculin is a permanent component of the membrane skeleton and is incorporated into the (re)organising cytoskeleton upon platelet activation, *Eur. J. Biochem.* 189:131.

Bachelot, C., Cano, E., Grelac, F., Saleun, S., Druker, B.J., Levy-Toledano, S., Fischer, S. and Rendu, F., 1992, Functional implications of tyrosine protein phosphorylation in platelets, *Biochem. J.* 284:923.

Brass, L.F., Vassallo, R.R., Jr., Belmonte, E., Ahuja, M., Cichowski, K. and Hoxie, J.A., 1992, Structure and function of the human platelet thrombin receptor, *J. Biol. Chem.* 267:13795.

Cichowski, K., McCormick, F. and Brugge, J.S., 1992, p21^{ras}GAP association with Fyn, Lyn, and Yes in thrombin-activated platelets, *J. Biol. Chem.* 267:5025.

Collett, M.S., Purchio, A.F. and Erikson, R.L., 1980, Avian sarcoma virus-transforming protein, pp60^{src} shows protein kinase activity specific for tyrosine, *Nature* 285:167.

Dhar, A. and Shukla, S.D., 1991, Involvement of pp60^{c-src} in platelet-activating factor-stimulated platelets, *J. Biol. Chem.* 266:18797.

Dhar, A., Paul, A.K. and Shukla, S.D., 1990, Platelet-activating factor stimulation of tyrosine kinase and its relationship to phospholipase C in rabbit platelets: studies with genistein and monoclonal antibody to phosphotyrosine, *Mol. Pharm.* 37:519.

Du, X., Plow, D.F., Frelinger, A.L., III, O'Toole, T.E., Loftus, J.C. and Ginsberg, M.H., 1991, Ligands "activate" integrin $\alpha_{IIb}\beta_3$ (platelet GPIIb-IIIa), *Cell* 65:409.

Eckhart, W., Hutchinson, M.A. and Hunter, T., 1979, An activity phosphorylating tyrosine in polyoma T antigen immunoprecipitates, *Cell* 18:925.

Elmore, M.A., Anand, R., Horvath, A.R. and Kellie, S., 1990, Tyrosine-specific phosphorylation of gpIIIa in platelet membranes, *FEBS Lett.* 269:283.

Escobedo, J.A., Navankasattusas, S., Kavanaugh, W.M., Miltay, D., Fried, V.A. and Williams, L.T., 1991, cDNA cloning of a novel 85 kd protein that has SH2 domains and regulates binding of PI3-kinase to the PDGF ß-receptor, *Cell* 65:75.

Fanti, W.J., Escobedo, J.A., Martin, G.A., Turck, C.W., del Rosario, M., McCormick, F. and Williams, L.T., 1992, Distinct phosphotyrosines on a growth factor receptor bind to specific molecules that mediate different signaling pathways, *Cell* 69:413.

Feder, D. and Bishop, J.M., 1990, Purification and enzymatic characterization of pp60^{c-src} from human platelets, *J. Biol. Chem.* 265:8205.

Feder, D. and Bishop, J.M., 1991, Identification of platelet membrane proteins that interact with amino-terminal peptides of pp60^{c-src}, *J. Biol. Chem.* 266:19040.

Ferrell, J.E., Jr. and Martin, G.S., 1988, Platelet tyrosine-specific protein phosphorylation is regulated by thrombin, *Mol. Cell. Biol.* 8:3603.

Ferrell, J.E., Jr. and Martin, G.S., 1989, Tyrosine-specific protein phosphorylation is regulated by glycoprotein IIb-IIIa in platelets, *Proc. Natl. Acad. Sci. USA* 86:2234.

Ferrell, J.E., Noble, J.A., Martin, G.S., Jacques, Y.V., Bainton, D.F., 1990, Intracellular localization of pp60^{c-src} in human platelets, *Oncogene* 5:1033.

Findik, D., Reuter, C. and Presek, P., 1990, Platelet membrane glycoproteins IIb and IIIa are substrates of purified pp60^{c-src} protein tyrosine kinase, *FEBS Lett.* 262:1.

Gaudette, D.C. and Holub, B.J., 1990, Effect of genistein, a tyrosine kinase inhibitor, on U46619-induced phosphoinositide phosphorylation in human platelets, *Biochem. Biophys. Res. Commun.* 170:238.

Gazit, A., Yaish, P., Gilon, C and Levitzki, A., 1989, Tyrphostins I: synthesis and biological activity of protein tyrosine kinase, *J. Med. Chem.* 32:2344.

Glover, C.J., Goddard, C. and Felsted, R.L., 1988, N-myristoylation of p60^{src}, *Biochem. J.* 250:485.

Goddard, C., Arnold, S.T. and Felsted, R.L., 1989, High affinity binding of an N-terminal myristoylated p60^{src} peptide, *J. Biol. Chem.* 264:15173.

Golden, A., Nemeth, S.P. and Brugge, J.S., 1986, Blood platelets express high levels of the pp60^{c-src}-specific tyrosine kinase activity, *Proc. Natl. Acad. Sci. USA* 83:852.

Golden, A. and Brugge, J.S., 1989, Thrombin treatment induces rapid changes in tyrosine phosphorylation in platelets, *Proc. Natl. Acad. Sci. USA* 86:901.

Golden, A., Brugge, J.S. and Shattil, S.J., 1990, Role of platelet membrane glycoprotein IIb-IIIa in agonist-induced tyrosine phosphorylation of platelet proteins, *J. Cell Biol.* 111:3117.

Gould, K.L., Woodgett, J.R., Cooper, J.A., Buss, J.E., Shalloway, D. and Hunter, T., 1985, Protein kinase C phosphorylates pp60^{src} at a novel site, *Cell* 42:849.

Granot, Y., Van Putten, V. and Schrier, R.W., 1990, Vasopressin dependent tyrosine phosphorylation of a 38 kDa protein in human platelets, *Biochem. Biophys. Res. Commun.* 168:566.

Grondin, P., Plantavid, M., Sultan, C., Breton, M., Mauco, G. and Chap, H., 1991, Interaction of pp60^{c-src}, phospholipase C, inositol-lipid, and diacylglycerol kinases with the cytoskeletons of thrombin-stimulated platelets, *J. Biol. Chem.* 266:15705.

Gu, M., York, J.D., Warshawsky, I. and Majerus, P.W. 1991, Identification, cloning, and expression of a cytosolic megakaryocyte protein tyrosine phosphatase with sequence homology to cytoskeletal protein 4.1, *Proc. Natl. Acad. Sci. USA* 88:5867.

Gutkind, J.S., Lacal, P.M. and Robbins, K.C., 1990, Thrombin-dependent association of phosphatidylinositol-3 kinase with p60^{c-src} and p59^{fyn} in human platelets, *Mol. Cell. Biol.* 10:3806.

Horak, I.D., Corcoran, M.L., Thompson, P.A., Wahl, L.M. and Bolen, J.B., 1990, Expression of p60^{fyn} in human platelets, *Oncogene* 5:597.

Horvath, A.R., Muszbek, L. and Kellie, S., 1992, Translocation of pp60^{c-src} to the cytoskeleton during platelet aggregation, *EMBO J.* 11:855.

Huang, M-M., Bolen, J.B., Barnwell, J.W., Shattil, S.J. and Brugge, J.S., 1991, Membrane glycoprotein IV (CD36) is physically associated with the Fyn, Lyn, and Yes protein-tyrosine kinases in human platelets, *Proc. Natl. Acad. Sci. USA* 88:7844.

Hung, D.T., Vu, T-K.H., Wheaton, V.I., Ishii, K. and Coughlin, S.R., 1992, Cloned platelet thrombin receptor is necessary for thrombin-induced platelet activation, *J. Clin. Invest.* 89:1350.

Hunter, T. and Cooper, J.A., 1985, Protein tyrosine kinases, *Ann. Rev. Biochem.* 54:897.

Inazu, T., Taniguchi, T., Yanagi, S. and Yamamura, H., 1990, Protein tyrosine phosphorylation and aggregation of intact human platelets by vanadate with H_2O_2, *Biochem. Biophys. Res. Commun.* 170:259.

King, W.G. and Rittenhouse, S.E., 1989, Inhibition of protein kinase C by staurosporine promotes elevated accumulations of inositol trisphosphates and tetrakisphosphate in human platelets exposed to thrombin, *J. Biol. Chem.* 264:6070.

Klausner, R.D. and Samelson, L.E., 1991, T cell antigen receptor activation pathways: the tyrosine kinase connection, *Cell* 64:875.

Koch, C.A., Anderson, D., Moran, M.F., Ellis, C. and Pawson, T., 1991, SH2 and SH3 domains: elements that control interactions of cytoplasmic signaling proteins, *Science* 252:668.

Kouns, W.C., Fox, C.F., Lamoreaux, W.J., Coons, L.B. and Jennings, L.K., 1991, The effect of glycoprotein IIb-IIIa receptor occupancy on the cytoskeleton of resting and activated platelets, *J. Biol. Chem.* 266:13891.

Kucera, G.L. and Rittenhouse, S.E., 1990, Human platelets form 3-phosphorylated phosphoinositides in response to α-thrombin, U46619, or GTPγS, *J. Biol. Chem.* 265:5345.

Lerea, K.M., Tonks, N.K., Krebs, E.G., Fischer, E.H. and Glomset, J.A., 1989, Vanadate and molybdate increase tyrosine phosphorylation in a 50-kilodalton protein and stimulate secretion in electropermeabilized platelets, *Biochemistry* 28:9286.

Mustelin, T., Coggeshall, K.M., Isakov, N. and Altman, A., 1990, T cell antigen receptor-mediated activation of phospholipase C requires tyrosine phosphorylation, *Science* 247:1584.

Nahas, N., Plantavid, M., Mauco, G. and Chap, H., 1989, Association of phosphatidylinositol kinase and phosphatidylinositol 4-phosphate kinase activities with the cytoskeleton in human platelets, *FEBS Lett.* 246:30.

Nakamura, S-I. and Yamamura, H., 1989, Thrombin and collagen induce rapid phosphorylation of a common set of cellular proteins on tyrosine in human platelets, *J. Biol. Chem.* 264:7089.

Nakano,T., Hanasaki,K. and Arita H., 1989, Possible involvement of cytoskeleton in collagen-stimulated activation of phospholipases in human platelets. *J. Biol. Chem.* 264: 5400.

Nakashima, S., Koike, T. and Nozawa, Y., 1991, Genistein, a protein tyrosine kinase inhibitor, inhibits thromboxane A_2-mediated human platelet responses, *Mol. Pharm.* 39:475.

Nolan, R.D. and Lapetina, E.G., 1990, Thrombin stimulates the production of a novel polyphosphoinositide in human platelets, *J. Biol. Chem.* 265:2441.

Oda, A., Druker, B.J., Smith, M. and Salzman, E.W., 1992, Inhibition by sodium nitroprusside or PGE_1 of tyrosine phosphorylation induced in platelets by thrombin or ADP, *Am. J. Physiol*, 262:C701.

Ohta, S., Taniguchi, T., Asahi, M., Kato, Y., Nakagawara, G. and Yamamura, H., 1992, Protein-tyrosine kinase p72^{syk} is activated by wheat germ agglutinin in platelets, *Biochem. Biophys. Res. Commun.* 185:1128.

Okano, Y., Sugimoto, Y., Fukuoka, M., Matsui, A., Nagata, K-I., and Nozawa, Y., 1991, Identification of rat cDNA encoding hck tyrosine kinase from megakaryocytes, *Biochem. Biophy. Res. Comm.* 181:1137.

Painter, R.G. and Ginsberg, M, 1982, Concanavalin a induces interactions between surface glycoproteins and the platelet cytoskeleton, *J. Cell Biol.* 92:565.

Payrastre, B., Plantavid, M., Breton, M., Chambaz, E. and Chap, H., 1990, Relationship between phosphoinositide kinase activities and protein tyrosine phosphorylation in plasma membranes from A431 cells, *Biochem. J.* 272:665.

Phillips, D.R., Jennings, L.K. and Edwards, H.H., 1980, Identification of membrane proteins mediating the interaction of human platelets, *J. Cell Biol.* 86:77.

Pumiglia, K.M., Huang, C-K. and Feinstein, M.B., 1990, Elevation of cAMP, but not cGMP, inhibits thrombin-stimulated tyrosine phosphorylation in human platelets, *Biochem. Biophys. Res. Commun.* 171:738.

Pumiglia, K.M., Lau, L-F., Huang, C-K., Burroughs, S. and Feinstein, M.B., 1992, Activation of signal transduction in platelets by the tyrosine phosphatase inhibitor pervanadate (vanadyl hydroperoxide), *Biochem. J.* 286:1.

Rendu, F., Lebret, M., Danielian, S., Fagard, R., Levy-Toledano, S. and Fischer, S., 1989, High pp60^{c-src} level in human platelet dense bodies, *Blood* 73:1545.

Salari, H., Duronio, V., Howard, S.L., Demos, M., Jones, K., Reany, A. Hudson, A.T. and Pelech, S.L., 1990, Erbstatin blocks platelet activating factor-induced protein-tyrosine phosphorylation, polyphosphoinositide hydrolysis, protein kinase C activation, serotonin secretion and aggregation of rabbit platelets, *FEBS Lett.* 263:104.

Schaller, M.D., Borgman, C.A., Cobb, B.S., Vines, R.R., Reynolds, A.B. and Parsons, J.T., 1992, pp125^{fak}, a structurally distinctive protein-tyrosine kinase associated with focal adhesions, *Proc. Natl. Acad. Sci. USA* 89:5192.

Shattil, S.J. and Brass, L.F., 1987, Induction of the fibrinogen receptor on human platelets by intracellular mediators, *J. Biol. Chem.* 262:992.

Smith, J.B., Dangelmaier, C., Selak, M.A., Ashby, B. and Daniel, J., 1992, Cyclic AMP does not inhibit collagen-induced platelet signal transduction, *Biochem. J.* 283:889.

Sorisky, A., Lages, B., Weiss, H.J. and Rittenhouse, S.E., 1992, Human platelets deficient in dense granules contain normal amounts of pp60^{c-src}, *Thrombosis Res.* 65:77.

Sultan, C., Breton, M., Mauco, G., Grondin, P., Plantavid, M. and Chap, H., 1990, The novel inositol lipid phosphatidylinositol 3,4-bisphosphate is produced by human blood platelets upon thrombin stimulation, *Biochem. J.* 269:831.

Takayama, H., Nakamura, T., Yanagi, S., Taniguchi, T., Nakamura, S-I. and Yamamura, H., 1991, Ionophore A23187-induced protein-tyrosine phosphorylation of human platelets: possible synergism between Ca^{2+} mobilization and protein kinase C activation, *Biochem. Biophys. Res. Commun.* 174:922.

Tuy, F.P.D., Henry, J., Rosenfeld, C. and Kahn, A., 1983, High tyrosine kinase activity in normal nonproliferating cells, *Nature* 305:435.

Ullrich, A. and Schlessinger, J., 1990, Signal transduction by receptors with tyrosine kinase activity, *Cell* 61:203.

Vassallo, R.R., Jr., Kieber-Emmons, T., Cichowski, K. and Brass, L.F., Structure-function relationships in the activation of platelet thrombin receptors by receptor-derived peptides, *J. Biol. Chem.* 267:6081.

Vostal, J.G., Jackson, W.L. and Shulman, N.R., 1991, Cytosolic and stored calcium antagonistically control tyrosine phosphorylation of specific platelet proteins, *J. Biol. Chem.* 266:16911.

Vu, T-KH., Hung, D.T., Wheaton, V.I. and Coughlin, S.R., 1991, Molecular cloning of a functional thrombin receptor reveals a novel proteolytic mechanism of receptor activation, *Cell* 64:1057.

Wahl, M.I., Nishibe, S., Suh, P-G., Rhee, S.G. and Carpenter, G., 1989a, Epidermal Growth factor stimulates tyrosine phosphorylation of phospholipase C-II independently of receptor internalization and extracellular calcium, *Proc. Natl. Acad. Sci. USA* 86:1568.

Wahl, M.I., Olashaw, N.E., Nishibe, S., Rhee, S.G., Pledger, W.J. and Carpenter, G., 1989b, Platelet-derived growth factor induces rapid and sustained tyrosine phosphorylation of phospholipase C-γ in quiescent BALB/c 3T3 cells, *Mol. Cell. Biol.* 9:2934.

Watson, S.P. and Hambleton, S., 1989, Phosphorylation-dependent and -independent pathways of platelet aggregation, *Biochem. J.* 258:479.

Yaish, P, Gazit, A., Gilon, C and Levitzki, A., 1988, Blocking of EGF-dependent cell proliferation by EGF receptor kinase inhibitors, *Science* 242:933.

Zavoico, G.B., Halenda, S.P., Chester, D and Feinstein, M.B., 1985, Control of Ca^{2+} mobilization and polyphosphoinositide metabolism in platelets by prostacyclin, *in* Prostaglandins, Leukotrienes, and Lipoxins, pp. 345-356, (Bailey, J.M., ed.), *Plenum Publ. Corp.*

Zhang, J., Fry, M.J., Waterfield, M.D., Jaken, S., Liao, L., Fox, J.E.B. and Rittenhouse, S.E., 1992, Activated phosphoinositide 3-kinase associates with membrane skeleton in thrombin-exposed platelets, *J. Biol. Chem.* 267:4686.

EVIDENCE THAT ACTIVATION OF PHOSPHOLIPASE D CAN MEDIATE

SECRETION FROM PERMEABILIZED PLATELETS

Richard J. Haslam and Jens R. Coorssen

Department of Pathology
McMaster University
Hamilton, Ontario, Canada

INTRODUCTION

Blood platelets possess three types of secretory organelle, namely dense granules that contain serotonin (5-HT), ADP, ATP and Ca^{2+}, α-granules that store a wide variety of proteins, including β-thromboglobulin (βTG), fibrinogen, thrombospondin and various growth factors, and lysosomes that contain acid hydrolases (Zucker and Nachmias, 1985). Secretion of the contents of platelet dense and α-granules in response to stimuli such as thrombin or thromboxane A_2 plays an important role in both physiology and pathology, contributing to normal hemostasis and to the development of arterial thrombi. Many studies on the signal transduction pathways that initiate these secretory responses have emphasized the receptor and G protein-dependent activation of phosphoinositide-specific phospholipase C (PLC) and the consequent mobilization of Ca^{2+} ions by inositol 1,4,5-trisphosphate (IP_3) and activation of protein kinase C (PKC) by 1,2-diacylglycerol (DAG) (Nishizuka, 1984; Siess, 1989). However, it is now clear that the process of platelet activation is much more complex than this, involving rapid parallel and sequential changes in the activities of several effector enzymes and ion channels, with many positive and negative feedback loops. As a result, the relative importance of Ca^{2+} ions and PKC, the contributions made by other factors, and the molecular mechanisms responsible for exocytosis of granule contents have proved difficult to analyse in intact platelets. Several groups have therefore used platelets permeabilized by saponin or by high-voltage electric discharges to study specific aspects of signal transduction and to identify factors involved in the secretion of granule constituents.

Electropermeabilization has proved particularly useful with platelets and has several major advantages (Knight and Scrutton, 1986a). First, stable access of small molecules (M_r < 2,000) to the cytosol can be maintained sufficiently long for equilibration of the intracellular space with the extracellular medium, so that the effects of defined concentrations of added compounds can be tested. Second, soluble proteins involved in signal transduction and the capacity of platelets to secrete granule constituents by exocytosis are preserved, though secretion is appreciably slower than in intact platelets. In the modification of this method developed in this laboratory (Haslam and Davidson, 1984a,b), the electropermeabilized platelets are isolated on a column of Sepharose CL-4B to remove endogenous low-M_r compounds as completely as possible, and are then stored for up to 2 h at 0°C in a glutamate medium containing MgATP. This method permits a wide range of biochemical studies to be carried out using a homogeneous preparation of permeabilized platelets without loss of sensitivity to stimuli that cause secretion. Equilibration of these permeabilized platelets with low-M_r compounds requires 15 min at 0°C and is usually followed by incubation for a further 10 min at 25°C, during which secretion can occur. Use of higher incubation temperatures leads to premature resealing of the platelets.

In the present paper, we will review progress in understanding the regulation of secretion of granule constituents arising from experiments with electropermeabilized platelets. The original concept, which emphasized the roles in platelets of Ca^{2+} ions (Knight and Scrutton, 1980) and of the guanine nucleotide-dependent activation of PLC and PKC (Haslam and Davidson, 1984a,b,c) must now be modified. Thus, in platelets, as in neutrophils (Barrowman et al., 1986) and insulin-secreting RINm5F cells (Vallar et al., 1987), guanine nucleotides have now been shown to promote the Ca^{2+}-independent secretion of granule constituents by a mechanism that does not involve activation of phosphoinositide-specific PLC (Coorssen et al., 1990). Gomperts has argued that guanine nucleotides may exert PLC and PKC-independent effects on secretion through an unidentified GTP-binding protein ('G_E'), which is directly involved in the exocytotic methanism (Gomperts, 1990). Our recent experiments on permeabilized platelets are consistent with this view and suggest that phospholipase D (PLD) is one possible target of 'G_E'.

FACTORS REGULATING SECRETION FROM PERMEABILIZED PLATELETS

Ca^{2+}-Dependent Secretion

As first shown by Knight and Scrutton (1980), addition of Ca^{2+} buffers giving pCa values below 6 (i.e. $> 1 \mu M$) induce secretion of dense granule 5-HT (Figure 1A). Very similar results were obtained with respect to the secretion of α-granule βTG (Figure 1B), though the maximum secretion (about 50%) was less, probably because much of the βTG is bound to the α-granule membranes (Coorssen et al., 1990). Lysosomal enzymes are also secreted in response to Ca^{2+} ions, though in this case the maximum response appears to be restricted by Ca^{2+}-dependent proteolysis (Athayde and Scrutton, 1990). In each case, the effective Ca^{2+} concentrations exceeded those associated with agonist-induced secretion from intact platelets by about an order of magnitude (Rink and Sage, 1990), suggesting that Ca^{2+} is not the only regulatory factor involved.

The first indication of a role for PKC in secretion from permeabilized platelets was provided by the demonstration that phorbol 12-myristate 13-acetate (PMA), 1-oleoyl-2-acetylglycerol and thrombin all caused a leftward displacement of the Ca^{2+} log-concentration response curve for secretion of 5-HT (Knight and Scrutton, 1984). This action of thrombin was associated with the formation of DAG and the phosphorylation of pleckstrin (Haslam and Davidson, 1984a). The latter is the major PKC substrate in platelets and serves as a convenient indicator of PKC activity, though its function has not yet been determined (Tyers et al., 1988). An optimal concentration of PMA (100 nM) caused five to tenfold decreases in the Ca^{2+} concentrations required for secretion of 5-HT or βTG, together with small increases in the maximum secretion obtained (Figure 1). Thus, simultaneous activation of PKC can account, at least in part, for the discrepancy between the Ca^{2+} concentrations associated with secretion from permeabilized and intact

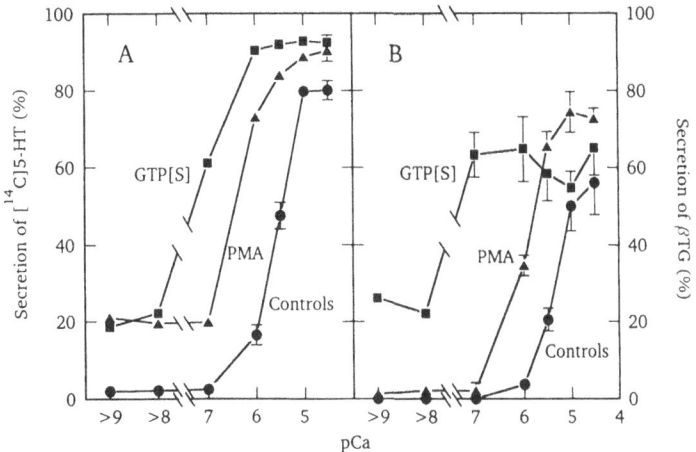

Figure 1   Effects of PMA and of GTP[S] on the secretion of [^{14}C]5-HT (A) and βTG (B) from permeabilized human platelets in the presence of various buffered Ca^{2+} concentrations. Platelets were labelled with [^{14}C]5-HT, electropermeabilized and transferred to a glutamate-based medium (0°C) containing ATP (Haslam and Davidson, 1984a). Samples were equilibrated for 15 min at 0°C with the indicated Ca^{2+} buffers and either no other additions (●), 100 nM PMA (▲) or 100 μM GTP[S] (■); pCa values of >8 and >9 represent the calculated maximum Ca^{2+} concentrations with no added Ca^{2+} in the presence of 2.5 mM and 12.5 mM EGTA, respectively. Equilibrated samples were incubated for 10 min at 25°C before measurement of the secretion of dense granule [^{14}C]5-HT and α-granule βTG. Values are means $\pm$ S.E. from triplicate samples. This figure is reproduced with permission from Coorssen et al. (1990), in which full experimental details are given.

platelets. Phosphorylation of pleckstrin (presumably by PKC) was also observed when Ca^{2+} alone was added to permeabilized platelets. However, whether thrombin was present or not, maximal phosphorylation of pleckstrin occurred under conditions in which there was relatively little secretion (Haslam and Davidson, 1984a). Thus, activation of PKC cannot represent a final common pathway leading to secretion, and Ca^{2+} must be assumed to play an additional role. These early results suggested that the presence of both Ca^{2+} ions and activated PKC was required for optimal secretion from permeabilized platelets. This conclusion is supported by some recent results with protein kinase inhibitors. Thus, addition of staurosporine prevented phosphorylation of pleckstrin and secretion induced by addition of Ca^{2+} buffers alone (Table 1). Similar results have been obtained with more specific PKC inhibitors, including pseudosubstrate peptides. These experiments with permeabilized platelets have confirmed and refined the concept, developed initially from observation of synergistic effects of Ca^{2+} ionophore and phorbol ester on intact platelets, that Ca^{2+} and PKC can act together to promote secretion (Nishizuka, 1984).

Further progress has depended on studies of the effects of exogenous guanine nucleotides on secretion from permeabilized platelets, which necessarily have no parallel in experiments with intact platelets. The initial investigations showed that 100 μM GTP enhanced the Ca^{2+} sensitivity of 5-HT secretion to about the same extent as did thrombin and that lower GTP concentrations potentiated the effect of thrombin (Haslam and Davidson, 1984b,c). Metabolically stable analogues of GTP, such as guanosine 5'-O-(3-thiotriphosphate) (GTP[S]), exerted much more powerful effects, causing over 100-fold increases in the Ca^{2+} sensitivity of secretion at 100 μM (Figure 1). Secretion of βTG (Coorssen et al., 1990) and of lysosomal enzymes (Athayde and Scrutton, 1990) were similarly potentiated by guanine nucleotides. Investigation of the mechanisms of action of

Table 1. Inhibition by staurosporine of secretion from permeabilized platelets induced by Ca^{2+} alone but not of secretion induced by Ca^{2+} and GTP[S]

Additions	Secretion of [^{14}C]5-HT (%)	Secretion of βTG (%)	Phosphorylation of pleckstrin (pmol ^{32}P/10^9 platelets)
Ca^{2+}	75 $\pm$ 2	45 $\pm$ 2	347 $\pm$ 64
Ca^{2+} + staurosporine	22 $\pm$ 2	7 $\pm$ 1	52 $\pm$ 6
Ca^{2+} + GTP[S]	93 $\pm$ 1	54 $\pm$ 3	336 $\pm$ 31
Ca^{2+} + GTP[S] + staurosporine	78 $\pm$ 1	60 $\pm$ 6	50 $\pm$ 8

Samples of permeabilized platelets containing dense granule [^{14}C]5-HT and, when required, [γ-^{32}P]ATP were equilibrated (15 min at 0°C) and incubated (10 min at 25°C) with Ca^{2+} buffer (pCa 4.5) and with staurosporine (2 μM) and GTP[S] (100 μM), as indicated. Mg^{2+}$_{free}$ was 6.4 mM. Secretion of [^{14}C]5-HT and βTG (triplicate samples) and the phosphorylation of pleckstrin (duplicate samples) were determined (see legends to Figures 1 and 2); values are means $\pm$ S.E. or means $\pm$ range, respectively.

these guanine nucleotides showed that GTP, and particularly GTP[S], stimulated phosphoinositide-specific PLC activity in permeabilized platelets, as shown by enhanced DAG and inositol phosphate accumulation (Haslam and Davidson, 1984b; Culty et al., 1988). The ability of thrombin to activate PLC in permeabilized platelets was also enhanced in the presence of these guanine nucleotides. Guanosine 5'-O-(2-thiodiphosphate) (GDP[S]) inhibited the actions of guanine nucleotides on both secretion and PLC activity, but was without effect on secretion induced by Ca^{2+} alone or by Ca^{2+} and thrombin (Haslam and Davidson, 1984b,c; Knight and Scrutton, 1986b). Potentiation of secretion by guanine nucleotides was associated with increases in PKC activity, consistent with the concept that Ca^{2+} and PKC act together to mediate secretion. However, several observations showed that GTP[S] exerted effects on secretion that were not mediated by PKC. First, the ability of this compound to enhance the Ca^{2+} sensitivity of 5-HT and βTG secretion greatly exceeded that of a concentration of PMA that was capable of inducing maximal phosphorylation of pleckstrin (Coorssen et al., 1990; Figure 1). Second, in the presence of high concentrations of Ca^{2+} and GTP[S], staurosporine did not block secretion, despite almost complete inhibition of protein phosphorylation caused by PKC and most other protein kinases (Table 1). The latter observation implies that in the presence of Ca^{2+}, an unidentified GTP[S]-stimulated reaction can replace PKC activity. This is reminiscent of results obtained with mast cells in which Ca^{2+} and guanine nucleotide, rather than PKC, appear to be the proximal inducers of exocytosis (Gomperts, 1990).

Ca^{2+}-Independent Secretion

Early experiments suggested that secretion from permeabilized platelets was largely Ca^{2+}-dependent, in that the relatively low residual secretion of 5-HT observed on addition of PMA or GTP[S] in the absence of added Ca^{2+} seemed likely to represent incomplete buffering of endogenous sources of Ca^{2+} by EGTA (Knight and Scrutton, 1984; Haslam and Davidson, 1984c). To settle this question, we have compared the effects of buffering platelet Ca^{2+} with 2.5 mM and 12.5 mM EGTA which, even if all the platelet Ca^{2+} was released into the medium, would give limiting pCa values of >8 and >9, respectively (Coorssen et al., 1990). As shown in Figure 1A, the residual secretion of 5-HT caused by

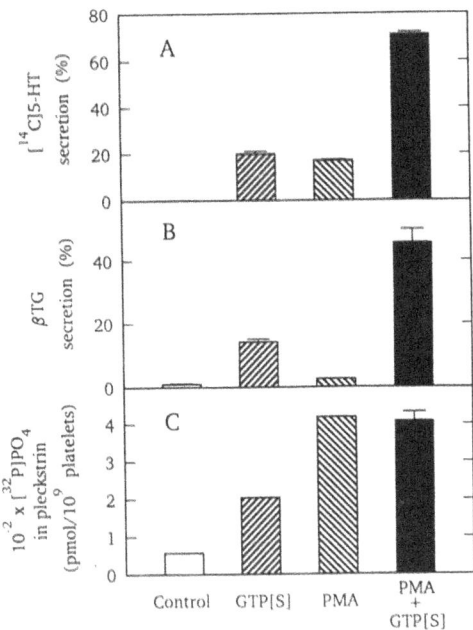

Figure 2   Synergistic actions of GTP[S] and PMA on secretion from permeabilized platelets in the absence of Ca^{2+}; relationship to pleckstrin phosphorylation. Samples of permeabilized platelets containing dense granule [^{14}C]5-HT and when required, [γ-^{32}P]ATP, were equilibrated (15 min at 0°C) and incubated (10 min at 25°C) at pCa >9 with variously, no other additions (control), GTP[S] (100 μM), PMA (100 nM) or both GTP[S] and PMA. Mg^{2+}_{free} was 2.7 mM. Secretion of [^{14}C]5-HT (A) and βTG (B) and the phosphorylation of pleckstrin (C) were then determined as described by Coorssen et al. (1990); values are means $\pm$ S.E. from triplicate samples ([^{14}C]5-HT, βTG) or means $\pm$ range from duplicate samples (phosphorylation).

PMA or GTP[S] was not affected by increasing the pCa from >8 to >9, and can therefore be regarded as Ca^{2+}-independent. The results also showed that GTP[S], but not PMA, caused Ca^{2+}-independent secretion of α-granule βTG (Figure 1B). In general, GTP[S] alone stimulated the secretion of a larger fraction of the releasable α-granule contents than of the dense granule constituents, whereas the converse applied with PMA. Thus, these stimuli exerted partially selective effects in the absence of Ca^{2+} (Coorssen et al., 1990).

GTP[S] and PMA had much more marked supra-additive effects on the secretion of both dense granule 5-HT and α-granule βTG when added together to permeabilized platelets at pCa >9 (Figures 2A and 2B). These Ca^{2+}-independent effects of GTP[S] and PMA did not correlate with the simultaneous increases in PKC activity, as determined by the phosphorylation of pleckstrin (Figure 2C). Thus, GTP[S] caused about half the phosphorylation seen with an optimal concentration of PMA, but was much more effective than PMA in stimulating secretion of βTG. More importantly, addition of GTP[S] with PMA had no effect on PMA-induced pleckstrin phosphorylation but greatly potentiated secretion of both 5-HT and βTG. The latter observation provides a further unequivocal demonstration that GTP[S] enhances secretion from permeabilized platelets by mechanisms additional to the activation of PKC. Nevertheless, in the absence of Ca^{2+}, activation of PKC was essential if not sufficient for secretion to occur, since the effects of GTP[S], of PMA, or of both stimuli together, were blocked by staurosporine or PKC pseudosubstrate peptides (Coorssen et al., in preparation).

At pCa >9, GTP[S] did not stimulate the formation of [^{3}H]inositol phosphates in permeabilized platelets containing ^{3}H-labelled phosphoinositides, though it readily did so at pCa 7 (Coorssen et al., 1990). PMA inhibited rather than stimulated [^{3}H]inositol phosphate formation, when this was detectable. Thus, activation of phosphoinositide-specific PLC is not involved in Ca^{2+}-independent secretion from platelets. The same conclusion has been drawn from studies with other permeabilized cells in which Ca^{2+}-independent secretion has been observed (Barrowman et al., 1986; Vallar et al., 1987). In permeabilized mast cells, in which both Ca^{2+} and GTP[S] are required for exocytosis, GTP[S] stimulates PLC activity and also acts at a downstream site ('G_E'), which could be more directly involved in the exocytotic mechanism (Cockcroft et al., 1987; Gomperts, 1990).

Implications for the Regulation of Exocytosis

The results cited above show that near maximal secretion from permeabilized platelets of both dense and α-granule constituents can be induced by Ca^{2+} in combination with PKC (in the absence of exogenous guanine nucleotide), by Ca^{2+} in combination with GTP[S] (in the absence of PKC activity), and by GTP[S] in combination with PKC (in the absence of Ca^{2+}). This suggests that none of the three stimulatory factors studied is an essential component of the exocytotic process, though all may act to promote secretion through synergistic effects on the same effector mechanism. The identification of any common target of Ca^{2+}, PKC and GTP[S] is thus of particular interest. In this connection, studies in this laboratory have shown that these three factors regulate the PLD activity of rabbit platelet membranes (Van der Meulen and Haslam, 1990) and we have suggested that PLD could therefore play a role in exocytosis from platelets (Coorssen et al., 1990). The ability of GTP[S] to stimulate PKC in permeabilized platelets in the absence of Ca^{2+} or PLC activity is consistent with this possibility, since it implies the formation of a lipid mediator by another phospholipase. We have therefore studied the activation of PLD in permeabilized platelets.

CORRELATIONS BETWEEN SECRETION AND PLD ACTIVITY

The immediate hydrolysis products of the action of PLD on its most likely substrate, phosphatidylcholine, are phosphatidic acid (PA) and choline (Exton, 1990). Of these, only PA and its metabolites are plausible second messengers. However, since PA is readily formed by the action of DAG kinase, the accumulation of PA alone does not provide a valid measure of PLD activity. We have therefore exploited the ability of PLD to catalyse a transphosphatidylation reaction in which ethanol is substituted for water with the formation of the metabolically stable phospholipid, phosphatidylethanol (PEt), instead of PA (Kobayashi and Kanfer, 1987). In these experiments, platelets were labelled with [^{3}H]arachidonate before permeabilization and [^{3}H]PEt accumulation was then determined during incubations in the presence of 100-400 mM ethanol. Similar results were obtained after labelling platelets with [^{3}H]glycerol.

Effects of GTP[S], PMA and Ca^{2+} on PLD activity

Addition of either 100 μM GTP[S] or 100 nM PMA to permeabilized platelets at pCa >9 caused only modest accumulations of [^{3}H]PEt, but together these stimuli had a potent synergistic effect on [^{3}H]PEt formation that correlated with their synergistic action on 5-HT secretion (Figure 3A). In the presence of Ca^{2+} (pCa 6), GTP[S] alone was a much more potent stimulus to both PLD and dense granule secretion (Figure 3B). PMA

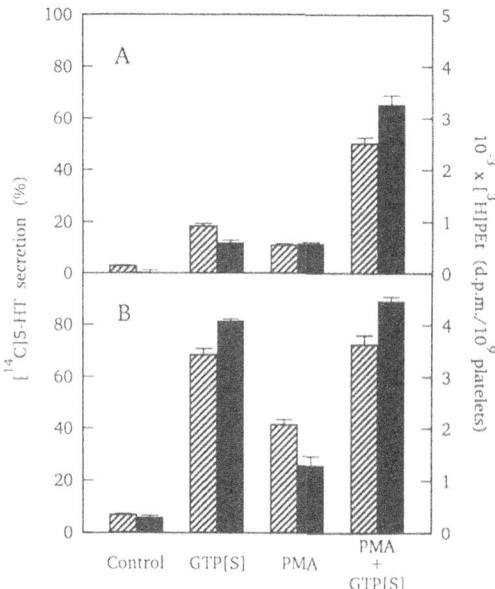

Figure 3   Correlation between [^{14}C]5-HT secretion from permeabilized platelets and PLD activity. Samples of permeabilized platelets containing [^{14}C]5-HT and phospholipids labelled with [^{3}H]arachidonate were equilibrated (15 min at 0°C) and incubated (10 min at 25°C) with variously, no stimulus (control), GTP[S] (100 μM), PMA (100 nM) or both GTP[S] and PMA, each in the absence of Ca^{2+} (A, pCa >9) and the presence of Ca^{2+} (B, pCa 6). All samples contained 5 mM Mg$^{2+}_{free}$ and 100 mM ethanol. Secretion of [^{14}C]5-HT (▨) and the formation of [^{3}H]PEt (■) were determined, the latter as described by Van der Meulen and Haslam (1990). Values are means ± S.E. from triplicate samples.

was less effective than GTP[S], but nevertheless potentiated PLD activity and secretion to a greater extent in the presence than in the absence of Ca^{2+}. At pCa 6, GTP[S] and PMA did not show synergism with respect to either PLD activity or secretion. This could reflect the ability of Ca^{2+} and GTP[S] to activate PKC more effectively than GTP[S] alone or, possibly, the inhibition of PLC activity by PMA (Coorssen et al., 1990). In either case, there was an excellent correlation between PLD activity and dense granule secretion under all the conditions tested (Figure 3). Comparison of Figures 1B, 2B and 3 indicates that there was also a correlation between PLD activity and α-granule secretion under most of the conditions studied.

Inhibitory Effects of Ethanol

The experiment shown in Figure 3 was carried out in the presence of only 100 mM ethanol to perturb PLD and secretion as little as possible, while still permitting measurement of the former. Similar studies have also been performed in the presence and absence of 400 mM ethanol to determine whether replacement of the [^{3}H]PA generated by PLD with [^{3}H]PEt is associated with changes in platelet function (Table 2). In the absence of Ca^{2+}, this concentration of ethanol inhibited the increase in 5-HT secretion caused by GTP[S] alone by about 80% and that caused by GTP[S] and PMA by about 50%. The [^{3}H]PA accumulation caused by these stimuli in the absence of ethanol paralleled that of [^{3}H]PEt in the presence of ethanol, and was decreased 70% on average by ethanol. However, [^{3}H]PEt accumulation often exceeded the associated decrease in [^{3}H]PA by a factor of up to two, suggesting that some metabolism of PA occurred. The results indicate

Table 2. Effects of ethanol on secretion and on the formation of phospholipid metabolites in permeabilized platelets

Additions		Secretion of [^{14}C]5-HT	Formation of [^{3}H]phospholipid metabolites (10^{-3} x d.p.m./10^9 platelets)		
Stimuli	Ethanol	(%)	[^{3}H]PA	[^{3}H]PEt	[^{3}H]DAG
None	-	3±1	5.7±0.2	---	-4.6±0.4
	+	5±1	5.4±0.4	0.3±0.1	-3.7±0.5
GTP[S]	-	18±2	7.1±0.3	---	-3.2±0.2
	+	8±1	5.7±0.3	1.4±0.1	-2.5±0.3
GTP[S]+PMA	-	44±2	13.6±0.4	---	-2.4±0.9
	+	25±1	8.1±0.2	9.0±0.3	-3.1±0.3
Ca^{2+}	-	12±0	12.1±0.4	---	-0.6±0.3
	+	7±1	7.1±0.4	1.6±0.2	-2.9±0.2
Ca^{2+}+GTP[S]	-	66±2	25.1±0.2	---	55.3±2.7
	+	62±3	13.8±0.5	13.5±0.8	37.7±1.9
Ca^{2+}+GTP[S]+PMA	-	67±3	26.6±1.6	---	20.6±1.8
	+	64±1	13.8±0.6	15.2±0.4	3.7±1.3

Samples of permeabilized platelets containing dense granule [^{14}C]5-HT and phospholipids labelled with [^{3}H]arachidonate were equilibrated (15 min at 0°C) and incubated (10 min at 25°C) with the indicated combinations of GTP[S] (100 μM), PMA (100 nM) and ethanol (400 mM), in both the absence and presence of Ca^{2+} (pCa >9 and 6). Mg$^{2+}_{free}$ was 5 mM. Secretion of [^{14}C]5-HT and the formation of [^{3}H]phospholipid metabolites were determined; values are means ± S.E. from triplicate samples.

that at pCa >9 most of the [^{3}H]PA formed in response to GTP[S] or to GTP[S] and PMA was generated by PLD activity and that secretion correlated well with the amount of this [^{3}H]PA that accumulated. Some [^{3}H]PA was also formed in control incubations, apparently from endogenous [^{3}H]DAG (Table 2), but this PA was never associated with secretion. Increases in [^{3}H]DAG were rarely observed in incubations at pCa >9, indicating not only the absence of PLC activity, but also that net conversion of the [^{3}H]PA formed by PLD to [^{3}H]DAG did not occur.

Markedly different results were obtained at pCa 6 (Table 2). Large amounts of [^{3}H]DAG, as well as [^{3}H]PA, were formed on stimulation of the permeabilized platelets with GTP[S]. Ethanol partly inhibited the accumulation of both but did not affect secretion of 5-HT. The decrease in [^{3}H]PA could be accounted for by [^{3}H]PEt formation but that in [^{3}H]DAG is more likely to represent inhibition of PLC by the ethanol. Thus PMA, which has been shown to inhibit PLC in permeabilized platelets (Coorssen et al., 1990), inhibited [^{3}H]DAG formation but not that of [^{3}H]PA. The results show that relatively little of the [^{3}H]DAG formed at pCa 6 was converted to [^{3}H]PA. It is apparent that although secretion of 5-HT at pCa 6 correlated with PLD activity, as represented by the formation of [^{3}H]PEt, it did not in this instance correlate well with [^{3}H]PA formation. However, it is possible that the total accumulation of phospholipid breakdown products in the presence of GTP[S] at pCa 6 so far exceeded that required for a maximal secretory response that an inhibitory effect of ethanol on secretion was difficult to detect. The results also demonstrate that secretion did not correlate with [³H]DAG accumulation, which indicates that this lipid is unlikely to promote exocytosis through a direct membrane-fusogenic effect.

Inhibition by BAPTA

Several studies have indicated that 1,2-bis(o-aminophenoxy)ethane-N,N,N',N'-tetraacetic acid (BAPTA), a pH-independent Ca^{2+}-chelating agent (Tsien, 1980), inhibits Ca^{2+}-independent secretion (Knight et al., 1989). For example, in Sendai virus-permeabilized neutrophils, EGTA supported a higher level of GTP[S]-induced secretion of β-glucuronidase than BAPTA (Barrowman et al., 1986), whereas in electropermeabilized platelets, GTP[S]-induced secretion of dense granule and lysosomal constituents was observed in the presence of 20 mM EGTA but not of 20 mM BAPTA (Athayde and Scrutton, 1990). Finally, Penner and Neher (1988) found that dibromo-BAPTA inhibited GTP[S]-induced exocytosis from mast cells even in the presence of 530 nM Ca^{2+}. These observations have suggested the possibility that BAPTA could have an unidentified 'toxic effect' on secretion (Knight et al., 1989).

We have therefore investigated the effects of BAPTA on secretion and PLD activity in permeabilized platelets in which Ca^{2+} was already buffered with EGTA (Figure 4). At

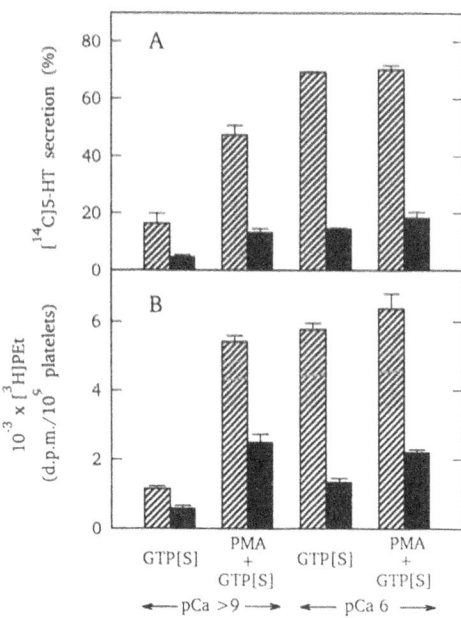

Figure 4   Inhibition of [¹⁴C]5-HT secretion and PLD by BAPTA in permeabilized platelets. Samples of permeabilized platelets containing [¹⁴C]5-HT and phospholipids labelled with [³H]arachidonate were equilibrated (15 min at 0°C) and incubated (10 min at 25°C) with GTP[S] (100 μM) or with GTP[S] (100 μM) and PMA (100 nM), each in the absence (▨) and presence (■) of BAPTA (10 mM). Samples were adjusted to give final pCa values of >9 or 6; all contained 5 mM Mg^{2+}_{free} and 200 mM ethanol. Secretion of [¹⁴C]5-HT (A) and [³H]PEt formation (B) were determined; values are means $\pm$ S.E. from triplicate samples.

pCa >9, BAPTA (10 mM) inhibited 5-HT secretion and [^{3}H]PA or [^{3}H]PEt formation by over 50%, whether GTP[S] or GTP[S] and PMA were used to stimulate secretion. In general, secretion was inhibited slightly more potently than PLD activity and the actions of GTP[S] were inhibited more effectively when PMA was absent. Under all conditions, addition of 20-30 mM BAPTA was sufficient to abolish both secretion and PLD activity. Measurement of pleckstrin phosphorylation showed that BAPTA inhibited the stimulation of PKC by GTP[S] alone at pCa >9, but had relatively little effect on PKC activity when PMA was also present. These results suggest that BAPTA may exert most if not all of its effects on Ca^{2+}-independent secretion by inhibiting PLD activity and that in the absence of PMA, activation of PLD by GTP[S] may be responsible for the observed increase in PKC activity.

At present it is not clear whether BAPTA inhibits PLD directly or indirectly. BAPTA binds Ca^{2+} much more rapidly than EGTA (Tsien, 1980) and it has been suggested that this accounts for its ability to inhibit neurotransmitter release (Adler et al., 1991). However, it is not plausible that Ca^{2+} chelation is responsible for the ability of BAPTA to inhibit secretion or PLD activity in permeabilized platelets. In the latter system, secretion is very slow (Coorssen et al., 1990) and therefore very unlikely to involve Ca^{2+} transients that cannot be buffered by EGTA. Moreover, the ability of BAPTA to inhibit secretion and PLD activation in the platelet system was retained in the presence of Ca^{2+} buffered to a final pCa of 6 or lower (Figure 4). Since BAPTA is known to bind to various proteins (Chiancone et al., 1986), possible targets of BAPTA include the GTP-binding protein mediating the effects of GTP[S] on secretion ('G$_E$'), any accessory proteins that affect guanine nucleotide exchange, and PLD itself.

POSSIBLE ROLES FOR PLD IN SECRETION

The results discussed above show that striking correlations exist between the activation of PLD and secretion from permeabilized platelets, particularly in the absence of Ca^{2+}. Relationships between PLD activity and secretion have also been observed in other secretory cells, including pancreatic islet cells (Dunlop and Metz, 1989), mast cells (Gruchalla et al., 1990) and neutrophils (Kanaho et al., 1991). However, to establish cause and effect, it is necessary to define the mechanism of action of the PA formed by PLD. Although conversion of this PA to DAG has been well-documented in neutrophils (Billah et al., 1989), there is evidence that PA may be more important than DAG in secretion from these cells (Kanaho et al., 1991). Consistent with this view, there appeared to be little conversion to DAG of the PA generated by PLD in permeabilized platelets (Table 2), though some metabolism of PA to other unidentified products is likely. One of the best documented effects of PA itself is on protein phosphorylation. Thus, PA is known to be able to replace phosphatidylserine in the activation of PKC, though little activity has been seen in the absence of Ca^{2+}, except when DAG was also present (Kaibuchi et al., 1981; Epand and Stafford, 1990). Recently, evidence has been obtained that PA can induce distinctive patterns of protein phosphorylation in supernatant fractions from some tissues in either the presence or absence of Ca^{2+}, suggesting that this phospholipid may stimulate protein kinases in addition to PKC (Bocckino et al., 1991). We have therefore investigated the effects of PA and related phospholipids on the Ca^{2+}-independent phosphorylation of platelet supernatant proteins.

As shown in Figure 5, addition of dioleoyl-PA alone to platelet supernatant markedly stimulated the phosphorylation of pleckstrin and very slightly increased that of the 20 kDa light-chain of myosin. This result corresponds closely to what was seen in permeabilized platelets incubated with GTP[S] at pCa >9 (Coorssen et al., 1990). It is difficult to exclude the possibility that the effects of dioleoyl-PA on platelet supernatant

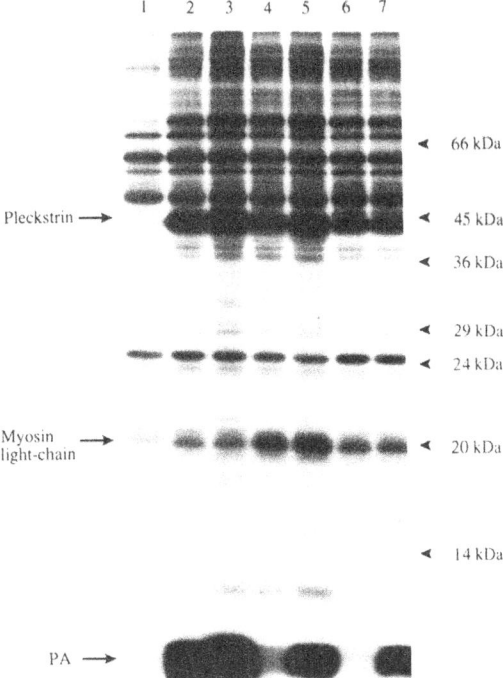

| | 1 | 2 | 3 | 4 | 5 | 6 | 7 | |

Pleckstrin →

◄ 66 kDa

◄ 45 kDa

◄ 36 kDa

◄ 29 kDa

◄ 24 kDa

Myosin light-chain →

◄ 20 kDa

◄ 14 kDa

PA →

Figure 5   Phosphorylation of proteins in platelet supernatant fraction caused by PA and analogues. Supernatant protein (60 μg) was incubated for 4 min at 37°C in medium containing 50 mM Hepes, pH 7.2, 5 mM EGTA, 2 mM MgCl$_2$, 200 μM [γ-^{32}P]ATP and the additions listed below.   Proteins were resolved by SDS-polyacrylamide gel electrophoresis; an autoradiograph is shown.   Other additions were as follows: lane 1, none; lane 2, 30 μM dioleoyl-PA; lane 3, 30 μM dioleoyl-PA + 10 μM diolein; lane 4, 30 μM dioleoyl-thiophosphatidic acid; lane 5, 30 μM dioleoylthiophosphatidic acid + 10 μM diolein; lane 6, 30 μM dioleoyl-PEt; lane 7, 30 μM dioleoyl-PEt + 10 μM diolein.   The positions of protein standards are shown on the right.

were potentiated by the trace formation of diolein, as implied by the appearance of ^{32}P labelled PA on the autoradiograph (Figure 5).   Further addition of diolein certainly increased the phosphorylation of pleckstrin and of many minor protein kinase substrates. However, we found that dioleoylthiophosphatidic acid, an analogue of dioleoyl-PA that is not readily hydrolysed to diolein (Bonnel et al., 1989), was also a potent activator of pleckstrin phosphorylation.   Unexpectedly, this PA analogue stimulated myosin light-chain phosphorylation much more effectively than did PA itself (Figure 5).   These different patterns of protein phosphorylation suggest the presence of more than one PA-activated protein kinase.   Finally, we investigated the effects on platelet protein phosphorylation of PEt, which has been reported to activate the brain-specific PKC isozyme, PKC-γ (Asaoka et al., 1988).   Dioleoyl-PEt not only stimulated the phosphorylation of pleckstrin in the absence of Ca^{2+}, but also caused partial phosphorylation of myosin light-chain.   This observation suggests that PEt may not be biologically inert in the platelet and could help to explain why ethanol inhibited secretion from permeabilized platelets less effectively than it inhibited PA accumulation.   These Ca^{2+}-independent effects of PA, thiophosphatidic acid and PEt on protein phosphorylation in platelet supernatant fraction differed from each other, not only with respect to the relative extents of phosphorylation of pleckstrin and myosin light-chain, but also with respect to the phosphorylation of minor substrates and the potentiating action of added diolein (Figure 5). All three compounds differed in their effects from that of phosphatidylserine, which caused no phosphorylation of myosin light-chain in the absence of Ca^{2+}. These observations are insufficient to establish that PA alone can act as a second messenger that stimulates protein phosphorylation in platelets, but are consistent with this possibility and indicate that platelets may contain two or more protein kinases that are activated by this phospholipid. Protein kinases stimulated by PA include PKC-α and PKC-β (Sekiguchi et al., 1988). However, platelets also contain Ca^{2+}-insensitive PKC isozymes, including PKC-δ (Grabarek et al., 1992) and apparently, PKC-ζ (Crabos et al., 1991), which may also be activated by PA.

Although activation of PKC is essential for GTP[S]-induced secretion from permeabilized platelets at pCa >9 and is most easily accounted for by direct or indirect effects of the PA generated by PLD, our results show that GTP[S] also exerts major effects on secretion that are not mediated by PKC. Since secretion correlates with PLD activity, this implies that PA or metabolites may stimulate secretion by other mechanisms. Activation by PA of an unidentified protein kinase (Boeckino et al., 1991) or phosphoprotein phosphatase (Gomperts, 1990) must be considered. In addition, PA has been shown to inhibit the activities of some of the factors that regulate low-M_r GTP-binding proteins, including the *ras* p21 GTPase-activating protein (Tsai et al., 1989) and the *rap* 1B GDP-dissociation inhibitor (Itoh et al., 1991). Finally, the possibility of a localized fusogenic action of PA on platelet membranes cannot be discounted (Leventis et al., 1986). Much further work will be required to determine which, if any, of these effects of PA have a role in mediating the stimulation of exocytosis by guanine nucleotides.

EXTRAPOLATION TO INTACT PLATELETS

Our results have shown that in electropermeabilized platelets, activation of PLD accounts for much, or in the absence of Ca^{2+}, all of the PA formed in response to stimuli that cause secretion. In this system, DAG generated by PLC activity in the presence of Ca^{2+} tended to accumulate and was only slowly converted to PA (Table 2). This finding contrasts with the situation in intact platelets, in which most of the PA accumulating during activation of the platelets by thrombin is derived from DAG formed by the action of PLC on phosphoinositides (Broekman et al., 1981; Tysnes et al., 1988; Huang et al., 1991). However, thrombin has been shown to stimulate [³H]PEt formation in intact human platelets that have been labelled with [³H]arachidonate and then incubated in the presence of ethanol (Rubin, 1988; Nakashima et al., 1991). In addition, thrombin increased the formation of labelled PA in platelets in which phosphatidylcholine had been selectively labelled by pre-incubation of the platelets with 1-alkyl-2-lysophosphatidylcholine containing [³H]hexadecyl- and [³²P]phosphoryl-residues (Huang et al., 1991). Thus, PLD is certainly activated on stimulation of intact platelets.

To obtain estimates of the relative importance of PLD and PLC in PA formation in intact platelets, we studied the effects of ethanol on [³H]PA and [³H]PEt accumulation after labelling with [³H]arachidonate or [³H]glycerol. In five minute incubations of [³H]arachidonate-labelled human platelets with a high thrombin concentration (2 units/ml), [³H]PEt formation in the presence of 200 mM ethanol amounted to about 12% of [³H]PA formation in the absence of ethanol (Table 3). Moreover, ethanol caused a significant decrease in [³H]PA accumulation that was similar in magnitude to the increase in [³H]PEt. The same methods also permitted detection of significant accumulations of PEt in either one minute incubations with 2 units of thrombin/ml or in five minute incubations with 0.1 unit of thrombin/ml. In contrast, PMA alone caused little [³H]PEt formation, but increased that observed with thrombin (Table 3). The latter effect was particularly marked with low thrombin concentrations, which may not have activated PKC fully, and was reminiscent of

Table 3. Effects of thrombin and PMA on the formation of [³H]PA and [³H]PEt in intact human platelets labelled with [³H]arachidonate.

Additions		Formation of [³H]phospholipids (10^{-3} x d.p.m./10^9 platelets)	
Stimuli	Ethanol	[³H]PA	[³H]PEt
None	-	-1.0 ± 0.5	---
	+	-0.9 ± 0.4	0.1 ± 0.2
Thrombin	-	51.4 ± 2.3	---
	+	43.1 ± 0.6	6.2 ± 0.2
PMA	-	0.3 ± 0.6	---
	+	0.3 ± 0.6	0.4 ± 0.1
Thrombin + PMA	-	34.6 ± 1.9	---
	+	29.0 ± 1.1	7.6 ± 0.2

Platelets labelled with [³H]arachidonate were washed and resuspended in Tyrode's solution containing 0.35% BSA and 5 mM Hepes, pH 7.4. Samples (2.7×10^8 platelets in 0.5 ml) were incubated for 5 min at 37°C with the indicated combinations of thrombin (2 units/ml), PMA (100 nM) and ethanol (200 mM). Increases in [³H]PA and [³H]PEt were then determined as described by Van der Meulen and Haslam, 1990; values are means ± S.E. from triplicate samples.

the synergism between GTP[S] and PMA observed in permeabilized platelets incubated at pCa > 9. A comparable synergistic activation of PLD by platelet-activating factor and PMA has also been reported in U937 cells (Balsinde and Mollinedo, 1991). In platelets incubated with PMA as well as thrombin, PLC activity and the total [³H]PA formation were inhibited, with the result that PLD activity then accounted for 20-25% of the [³H]PA that accumulated (Table 3). When [³H]glycerol-labelled platelets were stimulated with thrombin, [³H]PEt formation in the presence of ethanol was equivalent to about 20% of the [³H]PA that accumulated in the absence of ethanol, even in the absence of PMA. This probably results from a higher ratio of phosphatidylcholine to phosphatidylinositol labelling by [³H]glycerol than by [³H]arachidonate (Mahadevappa et al., 1986), and emphasizes the fact that prelabelling methods may not accurately indicate mass changes in phospholipids. It should be noted that estimates of the contribution of PLD activity to PA formation based upon PEt formation are also subject to at least two other sources of error. First, biologically acceptable ethanol concentrations may not completely prevent PA formation by PLD. Second, PA undergoes further metabolism, whereas PEt does not. Despite these difficulties, our results indicate that PLD makes a larger contribution to PA formation in intact platelets than a previous study in which the same methodology was used (Nakashima et al., 1991) and are consistent with indirect determinations based upon the specific activity of PA in platelets labelled with [³²P]P$_i$ (Huang et al., 1991). In summary, we conclude that PLD activity is likely to account for 10-20% of the total PA formation in thrombin-stimulated platelets.

Ethanol (25-150 mM) has been reported to inhibit secretion of 5-HT from intact platelets caused by low but not high thrombin concentrations (Benistant and Rubin, 1990). Although diversion of PA formed by PLD to the synthesis of PEt could have played a role in this effect, no inhibition of PA formation or of PKC activity was detected. Even if the results obtained with permeabilized platelets are correctly interpreted as indicating that PA generated by PLD activity can induce secretion, two questions must be answered before the same mechanism can be invoked in intact platelets. First, it is necessary to know whether PA itself or a metabolite is the relevant second messenger. In permeabilized platelets at pCa >9, little or no DAG accumulates and DAG is unlikely to mediate the activation of PKC. However, at higher pCa values or in intact platelets, DAG is formed by activation of PLC and it is unnecessary to invoke any role for PA or breakdown products in the stimulation of PKC activity. A role for PA or a metabolite in intact platelets can therefore only be considered in relation to the PKC-independent effects of guanine nucleotides on secretion. The second important question is that of whether PA generated by PLD differs in any substantive way, such as subcellular localization or fatty acid composition, from that formed by the combined actions of PLC and DAG kinase. Only if such a difference exists could PLD have an important role in intact platelets, since at least 80% of the PA that accumulates is a product of the activation of PLC. Resolution of these questions will be facilitated by the development of inhibitors of PLD that are effective in intact cells and are more selective and potent than ethanol or BAPTA.

SUMMARY

Studies on electropermeabilized human platelets indicated that any two of three distinct factors must be present for marked secretion of dense or α-granule constituents to occur. These factors are Ca^{2+}, activation of protein kinase C (PKC) and activation of an unidentified GTP-binding protein ('G_E'). Thus, in the absence of Ca^{2+}, phorbol ester and GTP[S] acted synergistically to promote secretion, whereas in the presence of Ca^{2+}, *either* activation of PKC *or* addition of GTP[S] was sufficient. In all cases, secretion correlated with the activation of phospholipase D (PLD), as detected by the formation of [^{3}H]phosphatidic acid (PA) in the absence of ethanol or of [^{3}H]phosphatidylethanol (PEt) in the presence of ethanol. Secretion did not correlate with phospholipase C (PLC) activity or with the accumulation of 1,2-diacylglycerol (DAG), both of which required Ca^{2+} and were inhibited by phorbol ester. Ethanol partially inhibited secretion in the absence of Ca^{2+}. BAPTA, a known inhibitor of Ca^{2+}-independent secretion in permeabilized cells, caused parallel inhibitions of secretion and PLD activity. GTP[S] enhanced PKC activity, as indicated by pleckstrin phosphorylation, apparently by stimulating the formation of PA in the absence of Ca^{2+}, as well as of DAG in the presence of Ca^{2+}. PA and stable analogues, including PEt, stimulated the Ca^{2+}-independent phosphorylation of pleckstrin and other proteins in platelet supernatant fraction. The results suggest that PA formed by activation of PLD may mediate secretion from permeabilized platelets by PKC-dependent and independent mechanisms. However, in intact platelets stimulated by thrombin, PLD accounted for only 10-20% of the total PA formed and can only play a major role in secretion if this PA fraction is distinct from that formed by the combined actions of PLC and DAG kinase.

Acknowledgements

The authors wish to thank Monica Davidson and Xiao-Tang Fan for their expert assistance. This work was supported by the Medical Research Council of Canada (Grant MT-5626). J.R.C. held an M.R.C. Studentship.

REFERENCES

Adler, E.M., Augustine, G.J., Duffy, S.N, and Charlton, M.P., 1991, Alien intracellular calcium chelators attenuate neurotransmitter release at the squid giant synapse, *J. Neurosci.* 11:1496.

Asaoka, Y., Kikkawa, U., Sekiguchi, K., Shearman, M.S., Kosaka, Y., Nakano, Y., Satoh, T., and Nishizuka, Y., 1988, Activation of a brain-specific protein kinase C subspecies in the presence of phosphatidylethanol, *FEBS Lett.* 231:221.

Athayde, C.M., and Scrutton, M.C., 1990, Guanine nucleotides and Ca^{2+}-dependent lysosomal secretion in electropermeabilised human platelets, *Eur. J. Biochem.* 189:647.

Balsinde, J., and Mollinedo, F., 1991, Platelet-activating factor synergizes with phorbol myristate acetate in activating phospholipase D in the human promonocytic cell line U937. Evidence for different mechanisms of activation, *J. Biol. Chem.* 266:18726.

Barrowman, M.M., Cockcroft, S., and Gomperts, B.D., 1986, Two roles for guanine nucleotides in the stimulus-secretion sequence of neutrophils, *Nature* 319:504.

Benistant, C., and Rubin, R., 1990, Ethanol inhibits thrombin-induced secretion by human platelets at a site distinct from phospholipase C or protein kinase C, *Biochem. J.* 269:489.

Billah, M.M., Eckel, S., Mullmann, T.J., Egan, R.W., and Siegel, M.I., 1989, Phosphatidylcholine hydrolysis by phospholipase D determines phosphatidate and diglyceride levels in chemotactic peptide-stimulated human neutrophils. Involvement of phosphatidate phosphohydrolase in signal transduction, *J. Biol. Chem.* 264:17069.

Bocckino, S.B., Wilson, P.B., and Exton, J.H., 1991, Phosphatidate-dependent protein phosphorylation, *Proc. Natl. Acad. Sci. U.S.A.* 88:6210.

Bonnel, S.I., Lin, Y.-P, Kelley, M.J., Carman, G.M., and Eichberg, J., 1989, Interactions of thio-phosphatidic acid with enzymes which metabolize phosphatidic acid. Inhibition of phosphatidic acid phosphatase and utilization by CDP-diacylglycerol synthase, *Biochim. Biophys. Acta* 1005:289.

Broekman, M.J., Ward, J.W., and Marcus, A.J., 1981, Fatty acid composition of phosphatidylinositol and phosphatidic acid in stimulated platelets, *J. Biol. Chem.* 256:8271.

Chiancone, E., Thulin, E., Boffi, A., Forsen, S., and Brunori, M., 1986, Evidence for the interaction between the calcium indicator 1,2-bis(*o*-aminophenoxy)ethane-*N,N,N',N'*-tetraacetic acid and calcium-binding proteins, *J. Biol. Chem.* 261:16306.

Cockcroft, S., Howell, T.W., and Gomperts, B.D., 1987, Two G-proteins act in series to control stimulus-secretion coupling in mast cells: use of neomycin to distinguish between G-proteins controlling polyphosphoinositide phosphodiesterase and exocytosis, *J. Cell Biol.* 105:2745.

Coorssen, J.R., Davidson, M.M.L., and Haslam, R.J., 1990, Factors affecting dense and α-granule secretion from electropermeabilized human platelets: Ca^{2+}-independent actions of phorbol ester and GTPγS, *Cell Regulation,* 1:1027.

Crabos, M., Imber, R., Woodtli, T., Fabbro, D., and Erne, P., 1991, Different translocation of three distinct PKC isoforms with tumor-promoting phorbol ester in human platelets, *Biochem. Biophys. Res. Commun.* 178:878.

Culty, M., Davidson, M.M.L., and Haslam, R.J., 1988, Effects of guanosine 5'-[γ-thio]triphosphate and thrombin on the phosphoinositide metabolism of electropermeabilized human platelets, *Eur. J. Biochem.* 171:523.

Dunlop, M., and Metz, S.A., 1989, A phospholipase D-like mechanism in pancreatic islet cells: stimulation by calcium ionophore, phorbol ester and sodium fluoride, *Biochem. Biophys. Res. Commun.* 163:922.

Epand, R.M., and Stafford, A.R., 1990, Counter-regulatory effects of phosphatidic acid on protein kinase C activity in the presence of calcium and diolein. *Biochem. Biophys. Res. Commun.* 171:487.

Exton, J.H., 1990, Signaling through phosphatidylcholine breakdown, *J. Biol. Chem.* 265:1.

Gomperts, B.D., 1990, G_E: a GTP-binding protein mediating exocytosis, *Annu. Rev. Physiol.* 52:591.

Grabarek, J., Raychowdhury, M., Ravid, K., Kent, K.C., Newman, P.J., and Ware, J.A., 1992, Identification and functional characterization of protein kinase C isozymes in platelets and HEL cells, *J. Biol. Chem.* 267:10011.

Gruchalla, R.S., Dinh, T.T., and Kennerly, D.A., 1990, An indirect pathway of receptor-mediated 1,2-diacylglycerol formation in mast cells. I. IgE receptor-mediated activation of phospholipase D, *J. Immunol.* 144:2334.

Haslam, R.J., and Davidson, M.M.L., 1984a, Potentiation by thrombin of the secretion of serotonin from permeabilized platelets equilibrated with Ca^{2+} buffers. Relationship to protein phosphorylation and diacylglycerol formation, *Biochem. J.* 222:351.

Haslam, R.J., and Davidson, M.M.L., 1984b, Receptor-induced diacylglycerol formation in permeabilized platelets; possible role for a GTP-binding protein, *J. Receptor Res.* 4:605.

Haslam, R.J., and Davidson, M.M.L., 1984c, Guanine nucleotides decrease the free [Ca^{2+}] required for secretion of serotonin from permeabilized blood platelets. Evidence of a role for a GTP-binding protein in platelet activation, *FEBS Lett.* 174:90.

Huang, R., Kucera, G.L., and Rittenhouse, S.E., 1991, Elevated cytosolic Ca^{2+} activates phospholipase D in human platelets, *J. Biol. Chem.* 266:1652.

Itoh, T., Kaibuchi, K., Sasaki, T., and Takai, Y., 1991, The *smg* GDS-induced activation of *smg* p21 is initiated by cyclic AMP-dependent protein kinase-catalyzed phosphorylation of *smg* p21, *Biochem. Biophys. Res. Commun.* 177:1319.

Kaibuchi, K., Takai, Y., and Nishizuka, Y., 1981, Cooperative roles of various membrane phospholipids in the activation of calcium-activated, phospholipid-dependent protein kinase, *J. Biol. Chem.* 256:7146.

Kanaho, Y., Kanoh, H., Saitoh, K., and Nozawa, Y., 1991, Phospholipase D activation by platelet-activating factor, leukotriene B_4, and formyl-methionyl-leucyl-phenylalanine in rabbit neutrophils. Phospholipase D activation is involved in enzyme release, *J. Immunol.* 146:3536.

Knight, D.E., and Scrutton, M.C., 1980, Direct evidence for a role for Ca^{2+} in amine storage granule secretion by human platelets, *Thromb. Res.* 20:437.

Knight, D.E., and Scrutton, M.C., 1984, Cyclic nucleotides control a system which regulates Ca^{2+} sensitivity of platelet secretion, *Nature* 309:66.

Knight, D.E., and Scrutton, M.C., 1986a, Gaining access to the cytosol: the technique and some applications of electropermeabilization, *Biochem. J.* 234:497.

Knight, D.E., and Scrutton, M.C., 1986b, Effects of guanine nucleotides on the properties of 5-hydroxytryptamine secretion from electropermeabilised human platelets, *Eur. J. Biochem.* 160:183.

Knight, D.E., von Grafenstein, H., and Athayde, C.M., 1989, Calcium-dependent and calcium-independent exocytosis, *TINS* 12:451.

Kobayashi, M., and Kanfer, J.N., 1987, Phosphatidylethanol formation via transphosphatidylation by rat brain synaptosomal phospholipase D, *J. Neurochem.* 48:1597.

Leventis, R., Gagne, J., Fuller, N., Rand, R.P., and Silvius, J.R., 1986, Divalent cation induced fusion and lipid lateral segregation in phosphatidylcholine-phosphatidic acid vesicles, *Biochemistry* 25:6978.

Mahadevappa, V.G., Belkhode, M.L., and Holub, B.J., 1986, Degradation of plasma membrane phospholipids in thrombin-stimulated human platelets following pre-labelling with different lipid precursors, *Progr. Lipid Res.* 25:87.

Metz, S.A., and Dunlop, M., 1990, Stimulation of insulin release by phospholipase D. A potential role for endogenous phosphatidic acid in pancreatic islet function, *Biochem. J.* 270:427.

Nakashima, S., Suganuma, A., Matsui, A., and Nozawa, Y., 1991, Thrombin induces a biphasic 1,2-diacylglycerol production in human platelets, *Biochem. J.* 275:355.

Nishizuka, Y., 1984, The role of protein kinase C in cell surface signal transduction and tumour promotion, *Nature* 308:693.

Penner, R., and Neher, E., 1988, Secretory responses of rat peritoneal mast cells to high intracellular calcium, *FEBS Lett.* 226:307.

Rink, T.J., and Sage, S.O., 1990, Calcium signaling in human platelets, *Annu. Rev. Physiol.* 52:431.

Rubin, R., 1988, Phosphatidylethanol formation in human platelets: evidence for thrombin-induced activation of phospholipase D, *Biochem. Biophys. Res. Commun.* 156:1090.

Sekiguchi, K., Tsukuda, M., Ase, K., Kikkawa, U., and Nishizuka, Y., 1988, Mode of activation and kinetic properties of three distinct forms of protein kinase C from rat brain, *J. Biochem.* 103:759.

Siess, W., 1989, Molecular mechanisms of platelet activation, *Physiol. Rev.* 69:58.

Tsai, M.-H., Yu, C.-L., Wei, F.-S., and Stacey, D.W., 1989, The effect of GTPase activating protein upon ras is inhibited by mitogenically responsive lipids, *Science* 243:522.

Tsien, R.Y., 1980, New calcium indicators and buffers with high selectivity against magnesium and protons: design, synthesis and properties of prototype structures, *Biochemistry* 19:2396.

Tyers, M., Rachubinski, R.A., Stewart, M.I., Varrichio, A.M., Shorr, R.G.L., Haslam, R.J., and Harley, C.B., 1988, Molecular cloning and expression of the major protein kinase C substrate of platelets, *Nature* 333:470.

Tysnes, O.-B., Verhoeven, A.J.M., and Holmsen, H., 1988, Rates of production and consumption of phosphatidic acid upon thrombin stimulation of human platelets, *Eur. J. Biochem.* 174:75.

Vallar, L., Biden, T.J., and Wollheim, C.B., 1987, Guanine nucleotides induce Ca^{2+}-independent insulin secretion from permeabilized RINm5F cells, *J. Biol. Chem.* 262:5049.

Van der Meulen, J., and Haslam, R.J., 1990, Phorbol ester treatment of intact rabbit platelets greatly enhances both the basal and guanosine 5'-[γ-thio]triphosphate-stimulated phospholipase D activities of isolated platelet membranes, *Biochem. J.* 271:693.

Zucker, M.B., and Nachmias, V.T., 1985, Platelet activation, *Arteriosclerosis* 5:2.

INOSITOL LIPID METABOLISM, THE CYTOSKELETON, GLYCOPROTEIN

IIB IIIA AND PLATELETS

Gérard P. Mauco, Claire Sultan, Bernard Payrastre, Monique Plantavid, Monique Breton and Hugues Chap

INSERM 326, Phospholipides Membranaires, Signalisation Cellulaire et Lipoprotéines, Hôpital Purpan 31 059 Toulouse Cédex, France

INTRODUCTION

Blood platelets, like many other cells respond to a wide range of agonists by modifications of their lipid metabolism (Lloyd et al., 1973; Lloyd and Mustard, 1974; Mauco et al., 1978, 1983, 1984; Billah and Lapetina 1982a,b; Agranoff et al., 1983; Broekman *et al.*, 1981 etc...) including a transient fall in $[^{32}P]PtdIns4,5P_2$ (1) labelling, increases in diacylglycerol (DAG) production (Rittenhouse-Simmons, 1979; Mauco et al., 1984) and phosphatidic acid (PtdOH) and at later times increases in the mass of $PtdIns4,5P_2$ (Perret et al., 1983). However these lipid modifications are not alone and a full range of metabolic responses have been reported, including GTP binding to a full set of G-proteins, serine/threonine phosphorylations via Ca^{2+}/calmodulin-, cyclic AMP-dependent protein kinases or protein kinases C, tyrosine protein phosphorylations. The cellular responses are also extremely diverse, ranging from cytoplasm i.e. pH changes and free calcium concentration increases to cytoskeletal modifications and secretion. Although many of the cellular events have been studied up to the end of aggregation (i.e. 1 to 5 minutes after stimulation), it is important to keep in mind that the aggregated platelets are still able to react to stimulating agonists (ADP, TXA_2, adrenaline etc...) or physiological antagonists (PGI_2...) as shown by experiments on the platelet-strip (Salganicoff et al., 1983). For instance, in vivo, the contractile properties are responsible for bringing together the lips of a wound.

We shall briefly review here some of the general data on inositol lipid metabolism in stimulated platelets and focus on results on the links between inositol lipids, the cytoskeleton and integrins.

The "classical" phospholipid effect

Since the pioneering work by Lloyd et al., (1973) and Lloyd and Mustard (1974), it is well established that platelets undergo a typical "phospholipid effect", involving the action of phospholipases C (Mauco et al., 1979; Rittenhouse-Simmons, 1979, 1983; Baldassare and Fisher, 1986; Banno et al., 1986; for a recent review, see Nozawa et al, 1991), action of diacylglycerol kinases, and PtdIns resynthesis followed by activation by subsequent phosphorylation into PtdInsP and PtdInsP$_2$ (Kaulen and Gross, 1976). However, as in other cells, it has been shown that the main changes upon stimulation are observed in the equilibria between the phosphorylated derivatives, PtdIns4P and PtdIns4,5P$_2$, the latter being assumed to be the main substrate for hydrolysis by specific phospholipases C, giving rise to the now classical second messengers inositol 1,4,5 trisphosphate (Agranoff et al., 1983) and diacylglycerol. These two messengers participate in elevating cytosolic calcium concentration and activating protein kinase(s) C.

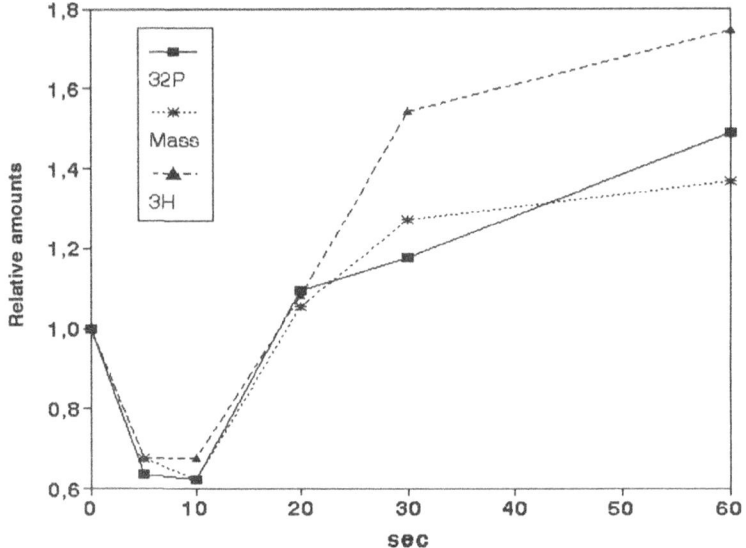

Figure 1   Relative changes in PtdIns4,5P$_2$ in thrombin stimulated platelets. Gel-filtered platelets were prelabelled with either [^{32}P]orthophosphate or [^{3}H] arachidonate, stimulated with 1 NIHU thrombin, lipids were analysed by thin layer chromatography or gas-liquid chromatography (Mauco et al., 1984).

Perret et al, (1983) and Mauco et al, (1984), demonstrated an increase in PtdIns4,5P$_2$ 1 min after stimulation by thrombin (1U/ml) (Figure 1). This was observed either by direct mass measurement, or after [^{32}P]orthophosphate or [^{3}H]-arachidonate prelabelling.   Therefore a specific activation of the synthesis of this lipid has been proposed, occurring at late times of activation (see below). This activation has also been reported to occur very early on in platelets stimulated by low concentrations of thrombin (Lassing and Lindberg, 1990), and in platelets loaded with Quin2 (Sultan et al., manuscript in preparation). In such cases an increase in labelling of PtdIns4,5P$_2$ is observed before activation of phospholipase C (Lassing and Lindberg, 1990) or in the absence of phospholipase C activity due to extremely low calcium concentrations obtained with Quin2 (our unpublished results). Other agonists are also able to increase PtdIns4,5P$_2$ in A431 cells (Payrastre et al., 1990). These observations point out to a more complex role for inositol lipids than previously thought; the amount of PtdIns4,5P$_2$ could be important by itself, not only for the production of the now classical second messengers, DAG and Ins1,4,5P$_3$.

It appears that under certain circumstances and in some cells, inositol lipid metabolism is compartmentalised (Monaco and Woods 1982; Koreh and Monaco, 1983). Preliminary data from Billah et al., 1983, suggest that phosphatidylinositol metabolism is also compartmentalised. Nevertheless, a more precise knowledge of the chemistry of inositol lipids and of their subcellular compartmentation is needed. Different authors have demonstrated that most of the PtdIns in platelets contains stearic and arachidonic acids (Broekman et al., 1981; Mauco et al., 1984; Mahadevappa and Holub; 1987). Figure 2 reports the composition of the inositol lipids and their derivatives found in human blood platelets. It is clear that almost all the inositol phospholipids contain these two fatty acids, suggesting a common origin. This is less clear for PtdOH and DAG. Indeed other fatty acids are also found, indicating either another source for these lipids or a worse "signal to noise" ratio (Mauco et al., 1984).

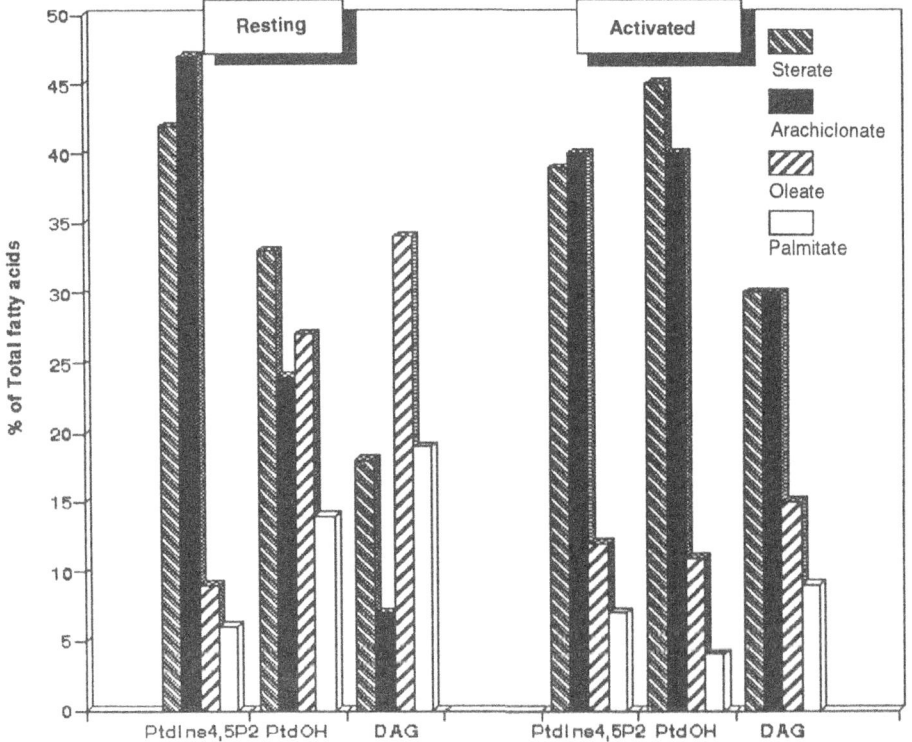

Figure 2   Changes in fatty acids upon thrombin stimulation. Platelet phospholipids were assayed before and 10 sec after addition of thrombin (calculated from Mauco et al., 1984).

The subcellular localization of inositol lipids in platelets has seldom been addressed. We observed that when platelets were labelled with [^{32}P] orthophosphate, all the PtdInsP and PtdInsP$_2$ were present in plasma membranes. Labelling with [^{3}H]inositol revealed that about 60 % of the total label incorporated in PtdIns was in internal membranes compared to less than 10 % of PtdInsP. Only traces of PtdInsP$_2$ were recovered in these internal organelles (Mauco et al., 1987). Furthermore we obtained results demonstrating that the plasma membranes were able to recycle the DAG moiety into PtdIns. Therefore all the inositol lipid metabolism could occur in the platelet plasma membrane. These unpublished data confirm those obtained by Imai and Gershengorn (1986) in GH3 cells. In our opinion it is important to the physiology of such active cells as platelets that all the machinery is present at, or near, the sites of modification, making the regulation of the enzymes by receptor occupancy more efficient and faster.

Platelets produce inositol lipids phosphorylated on the D-3 position of inositol

Whitman et al., 1987, 1988 first reported the occurence of novel phospholipids, bearing a phosphate on the D-3 hydroxyl of myo-inositol. These new compounds were identified to be $PtdIns3P$, $PtdIns3,4P_2$ and $PtdIns3,4,5P_3$. They represented at most 1 % of total phospholipids. For some time these lipids were believed to be exclusively linked to mitosis, since they were found in dividing cells, for instance upon stimulation by PDGF or EGF (Whitman et al., 1987 ; Kaplan et al., 1987; Auger et al, 1989a,b). The production of these novel lipids was linked to associations of lipid kinases and tyrosine kinases like the PDGF receptor, the oncogene product $pp60^{v-src}$, and others (Courtneidge and Heber, 1987 ; Kaplan et al., 1987; Whitman et al., 1988; Gutkind et al., 1990). Platelets have very high contents of the cellular counterpart of $pp60^{v-src}$, known as $pp60^{c-src}$ (Golden et al., 1986; Golden and Brugge, 1989). Even if it is still not clear that this protein actually is a tyrosine kinase in intact cells, it has been demonstrated to display such an activity in acellular preparations. It was then possible to hypothesize that platelets, of course devoid of any mitotic capability, could synthesize these novel lipids. Indeed, during the same period three reports documented this issue : Kucera and Rittenhouse, 1990; Nolan and Lapetina, 1990 and Sultan et al., 1990, demonstrated that platelets contain $PtdIns3P$ under resting conditions. Quiescent platelets did not contain appreciable amounts of either of the other two novel phospholipids. However they produced $PtdIns3,4P_2$ in a dose dependent manner when stimulated by thrombin with a lag of about 10 sec. This lipid peaked after 5 min of stimulation by 1 U of thrombin and decreased thereafter. Masses of the lipids correlated well with radioactivity changes (Mauco and Mayr manuscript in preparation). The time course of formation of $PtdIns3,4,5P_3$, was totally different : this lipid peaked after 10 sec stimulation and decreased rapidly thereafter. Very recently, Huang et al. (1991) demonstrated that the cloned thrombin receptor (Vu et al., 1991) was able to elicit all of the effects described above.

These data draw attention to a pleiotropic role of the novel inositol lipids in cellular biology. Indeed these 3-phosphorylated inositol lipids are likely to have a role in non dividing cells. For instance Traynor-Kaplan et al., 1989 reported their occurence in neutrophils, which are also devoid of mitosis capacity. One may wonder what kind of common mechanism would be the link between mitosis and terminally differentiated cell activation. One possibility could be a role in the cytoskeletal changes which accompany both cell division and activation.

Interactions between the Cytoskeletal Proteins and Phospholipids

Cytoskeleton assembly is a complex phenomenom occurring in platelets upon stimulation. As a dedicated chapter is included in this book, we will restrict our commentaries to the specific aspects of interactions between the cytoskeletal proteins and phospholipids. It has been proposed that Triton X100 insoluble fractions of blood platelets consist mainly of cytoskeletal proteins assembled together. We will refer to this preparation as the cytoskeleton (for a discussion see the chapter by J. Fox in this book). Analysis of cytoskeletons from platelets and other cells revealed the presence of phospholipids (Schick et al., 1983; Tuszinsky et al., 1984) as well as DAG (Burn et al., 1985). Moreover the linkage between cytoskeletal assembly and phospholipid metabolism became an intriguing issue with the hypothesis of Lassing and Lindberg who proposed a specific interaction between PtdIns4,5P_2 and profilactin (Lassing and Lindberg, 1985, 1988). Briefly the actin capping protein profilin could be bound either to PtdIns4,5P_2 or G-actin. The relative amounts of the lipid to actin could then promote association or dissociation of the profilactin complex. Therefore, PtdIns4,5P_2 amounts would be a putative means for regulating the equilibrium between G and F actin. Indeed platelet contents of this phospholipid change during the time of activation, first decreasing (10-20 sec) then rising above resting levels (1 min and after), see figure 1. A more complex interaction appears to exist, however, can even been proposed since at least another regulating protein can become associated with PtdInsP_2 : gelsolin, an actin severing protein (Janmey et al. 1987).

Another possible link between cytoskeletal proteins and inositol lipid metabolism has been proposed by Goldschmidt-Clermont et al., 1990. Since the concentrations of profilin and PtdIns4,5P_2 are close, it would be possible that profilin binds to all the lipid present in the cell, therefore hiding it from the action of phospholipase C, in a manner not very different from that of annexins. Indeed, the same team reported an inhibition of unphosphorylated form of PLCγ1 by profilin. Furthermore they suggested that profilin could have a rapid and transient activating action on F-actin assembly, and not an inhibitory one (Goldschmidt-Clermont et al., 1991).

Some molecular structures have also to be reviewed briefly here. Indeed some of the enzymes of the lipid metabolism, i.e. PLCγ1 and the regulatory subunit of PtdIns 3-kinase, p85 possess a peptide domain called SH3 (pp60src homology, region 3) which is suggested to be a domain responsible for the association of some proteins to cytoskeleton. Interestingly, a second region SH2 is also present in these molecules and is supposed to be the site of linkage to phosphorylated tyrosine residues (Koch et al., 1991). These two sites are of course present on the putative tyrosine kinase pp60^{c-src}. These data could be an indication of a specific association of inositol lipid metabolizing enzymes to the cytoskeleton on one side and to phosphotyrosine proteins on the other.

Inositol Lipid Metabolism is Associated to Cytoskeleton in Platelets

Nahas et al. (1989) have reported an association of almost 60 % of the PtdIns kinase and 50 % of PtdIns$P$ kinase to platelet cytoskeletons. A study by Grondin et al. (1991) have extended these results to kinases and phospholipases : PtdIns 3- and 4-kinase, as well than PtdIns4P 5-kinase, DAG kinase and PLC. All of these enzymes were detectable in cytoskeletons from resting platelets, but their specific activities were increased in cytoskeletons from thrombin-activated platelets. The kinetics of association were slightly different, some enzymes becoming associated very shortly after thrombin addition (PLC and PtdIns4P kinase), others being slightly more delayed. Furthermore we demonstrated a translocation of pp60^{c-src} from Triton-soluble to a Triton—insoluble fraction upon thrombin addition. Interestingly, no phosphorylation of PtdIns4P into PtdIns3,4P_2 was detected, the only way to obtain this lipid was to add PtdIns to the reaction mixture indicating the presence of A PtdIns 3-kinase acting on PtdIns. This could be a confirmation of results by Cunningham et al., 1990 and Yamamoto et al., 1990, on the sequence of phosphorylations leading to the production of novel phospholipids in platelets :

PtdIns --- > PtdIns3P ---- > PtdIns3,4P2

instead of a direct conversion from PtdIns4P. A similar association of PtdIns 3-kinase has been very recently reported by Zhang et al. (1992), suggesting a specific interaction with the membrane cytoskeleton. This confirms our hypothesis, proposed by Nahas et al. (1989), on the localization of the enzymes of the lipid metabolism to the anchoring points of the cytoskeleton to the membrane.

Gp IIb-IIIa is Involved in Platelet Production of PtdIns3,4P_2

The production of the novel inositol lipids has been linked to tyrosine phosphorylation in cells stimulated by PDGF or EGF, and indeed the PtdIns 3-kinase regulatory subunit p85 is proposed to be a site of phosphorylation on tyrosine residues in platelets (Gutkind et al., 1990). On the other hand tyrosine phosphorylation is also a regulatory means proposed for PLC and PtdIns-4-kinase (Kaplan et al., 1986; Wahl et al., 1989; Kanoh et al., 1990; Payrastre et al., 1990). Platelets are known to contain exceptionally high levels of pp60^{c-src}, a putative in vivo tyrosine kinase as well as other tyrosine kinases i.e. p59fyn. Upon stimulation they undergo at least three waves of tyrosine phosphorylations, the last one being strongly inhibited if aggregation is impaired, either by the addition of RGDS or lack of external calcium and agitation (Ferrell and Martin, 1989, Golden and Brugge, 1990). We decided to investigate whether inhibition of fibrinogen binding to GPIIb-IIIa (αIIbβ3 integrin) could have some inhibitory effect upon PtdIns3,4P_2 production. Indeed, as reported in Table 1 RGDS had a strong inhibitory effect on production of this lipid, this inhibition was strictly parallel to the effect on aggregation. Other inhibitors of the binding of fibrinogen were effective in inhibiting this lipid synthesis, in a manner totally parallel to their aggregation inhibitory capacity. Furthermore Glanzman's thrombasthenic patients were shown to have only a basal level of PtdIns3,4P_2, after thrombin-stimulation and this was not affected by RGDS (Sultan et al., 1991).

Table 1.   Inhibition of thrombin-induced production of [^{32}P]PtdIns3,4P_2 in platelets by RGD peptides

Peptide	Conc. (mM)	% of inhibition
RGDS	0.25	58$\pm$5
	1.00	63$\pm$7
RGD	0.25	17$\pm$13
	1.00	56$\pm$10

This is the first time that an effect of inhibition of ligand binding to an integrin is shown to have an effect on a putative second messenger. However Sinigaglia et al., (1989) have suggested the possibility of an inhibitory effect of GPIIb-IIIa on PLC. They also demonstrated the association of a GTP binding protein of low molecular weight to this same integrin (Torti et al., 1991). Taken together these data could open a very exciting new field for signal transduction research. Moreover Zhang et al. (1992), in their very recent studies on association of PtdIns 3- kinase to cytoskeleton, suggest an association of this enzyme to the αIIbβ3 integrin, in the membrane skeleton.

CONCLUSION

This brief review summarises some of the most important data on inositol lipid metabolism in platelets. However, these observations extend beyond the platelet field, and could help understand the interactions between metabolism stimulated by "primary" agonists, morphological changes and modulation of the responses by a "secondary" agonist. A hypothetical scheme, taking into account the diverse data obtained so far, is illustrated in Figure 3.

In resting cells, most of the PtdInsP_2 is linked to profilin and therefore unaccessible to PLC. The amount of profilactin is stable. Upon stimulation, PLC is activated (targetted to the membrane ?) while some synthesis of PtdInsP_3 and PtdInsP_2 occurs, the latter giving the opportunity for PLC to produce DAG and inositol1,4,5P_3. In some way profilin is then released into the cytoplasm, regulating the assembly of F-actin, by increasing ADP/ATP exchange and binding of divalent cations (Goldschmidt-Clermont et al., 1991). At the same early time, tyrosine phosphorylations start to appear. Later on the membrane cytoskeleton gets associated with whole cell cytoskeleton. GpIIb-IIIa is then associated to this total cytoskeleton and could serve as a substratum for aggregation of different molecules into "signal particles", which can be extracted with Triton X100. Of course, other integrins or

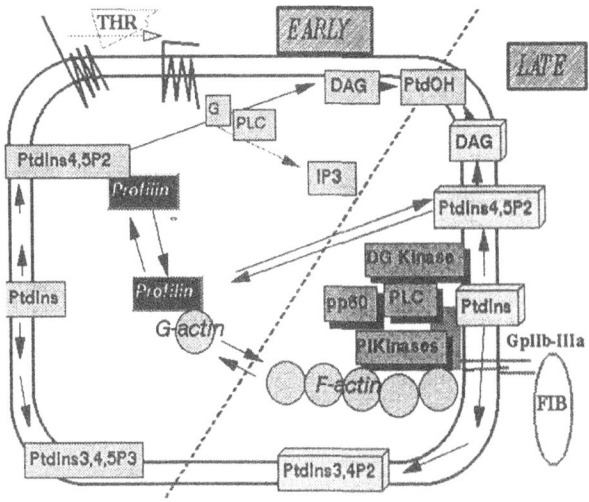

Figure 3   The interactions between inositol lipid metabolism and a number of platelet responses. This scheme has to be understood as a two part cartoon : all the metabolic steps drawn to the left of the dotted diagonal line are more or less stimulated by thrombin addition itself, the ones drawn to the right are late events, linked either to cytoskeleton aggregation or/and integrin occupancy. G stands for guanine nucleotide binding protein, PIKinases for all the enzymes able to phosphorylate the inositol lipids, FIB is fibrinogen and THR, thrombin, IP3 inositol 1,4,5 trisphosphate.

receptors could serve as a substratum for this "macromolecular crowding" as has been reported for other cells : EGF or NGF receptors for instance. The participants in the "signal particles" could be associated either via SH2 or SH3 regions to phosphorylated tyrosines of other proteins or to actin itself. This could also provide a means of regulating PtdIns4,5P_2 hydrolysis (as well as other phospholipids) and synthesis in discrete areas of the cell, analogue to focal contacts in adherent cells.   Thereafter the specific 3 - kinase pathway becomes activated, provided fibrinogen binding is allowed to occur, and leads to a net production of PtdIns3,4P_2. This scheme, of course does not tell much about the role of these new lipids, however their appearance is linked to late events during platelet activation, and is regulated by integrin occupancy. Indeed GpIIb-IIIa regulates 60% of their total production, and this extends the role of this integrin towards regulation of metabolic events. This new field needs more data and also new approaches. What is responsible for activation of PtdIns 3-kinase pathway? Is GpIIb-IIIa regulating the enzyme by itself? If yes it would be necessary to find a new means of regulation, this integrin being presumably devoid of enzymatic capabilities. However it has been described as a calcium channel. Is it possible that its occupancy by fibrinogen gives local increases in free calcium, at the site of signal particles? What are then the necessary preliminaries, priming the cell for production of PtdIns3,4P_2? Are all the agonists leading to aggregation, able to induce an activation of the 3-kinase pathway? What other molecule is responsible for the remaining levels of PtdIns3,4P_2 after RGDS addition and in thrombasthenic patients?

REFERENCES

Agranoff, B.W., Murphy, P. and Seguin, E.B. (1983) Thrombin-induced phosphodiesteratic cleavage of phosphatidylinositol bisphosphate in human platelets. *J. Biol. Chem.* 258: 2076.

Auger, K.R., Serunian, L.A., Soltoff, S.P., Libby, P., Cantley, L.C and (1989a) PDGF-dependent tyrosine phosphorylation stimulates production of novel polyphosphoinositides in intact cells. *Cell* 57: 167.

Auger, K.R., Carpenter, C.L., Cantley, L.C. and Varticovski, L. (1989b) Phosphatidylinositol 3-Kinase and its novel product, phosphatidyl-inositol 3-phosphate, are present in Saccharomyces-Cerevisiae. *J. Biol. Chem.* 264: 20181.

Baldassare, J.J. and Fisher, G.J. (1986) Regulation of membrane-associated and cytosolic phospholipase C activities in human platelets by guanosine triphosphate. *J. Biol. Chem.* 261: 11942.

Banno, Y., Nakashima, S. and Nozawa, Y. (1986) Partial purification of phosphoinositide phospholipase C from human platelet cytosol; characterization of its three forms. *Biochem. Biophys. Res. Commun.* 136: 713.

Billah, M.M. and Lapetina, E.G. (1982a) Evidence for multiple metabolic pools of phosphatidylinositol in stimulated platelets. *J. Blol. Chem.* 257: 11856.

Billah, M.M. and Lapetina, E.G. (1982b) Rapid decrease of phosphatidylinositol 4,5-bisphosphate in thrombin-stimulated platelets. *J. Biol. Chem.* 257: 12705.

Broekman, M.J., Ward, J.W. and Marcus, A.J. (1981) Fatty acid composition of phosphatidylinositol and phosphatidic acid in stimulated platelets: Persistence of arachidonyl-stearyl structure. *J. Biol. Chem.* 256: 8271.

Burn, P., Rotman, A., Meyer, R.K. and Burger, M.M. (1985) Diacylglycerol in large α-actinin/actin complexes and in the cytoskeleton of activated platelets. *Nature* 314: 469.

Courtneidge, S.A. and Heber, A. (1987) An 81kd protein complexed with middle T antigen and pp60 c-src: a possible phosphatidylinositol kinase. *Cell* 50: 1031.

Cunningham, T.W., Lips, D.L., Bansal, V.S., Caldwell, K.K., Mitchell, C.A. and Majerus, P.W. (1990) Pathway for the formation of D-3 phosphate containing inositol phospholipids in intact human platelets. *J. Biol. Chem.* 265: 21676.

Ferrell, J.E. and Martin, G.S. (1989) Tyrosine-specific protein phosphorylation is regulated by glycoprotein-IIb-IIIa in platelets. *Proc. Natl. Acad. Sci.* 86: 2234.

Golden, A. and Brugge, J.S. (1989) Thrombin treatment induces rapid changes in tyrosine phosphorylation in platelets. *Proc. Natl. Acad. Sci.* 86: 901.

Golden, A., Nemeth, S.P. and Brugge, J.S. (1986) Blood platelets express high levels of the pp60c-src-specific tyrosine kinase activity. *Proc. Natl. Acad. Sci.* 83: 852.

Goldschmid-Clermont,P.J., Machesky, L.M., Baldassare, J.J. and Pollard, T.D. (1990) The actin-binding protein profilin binds to PIP2 and inhibits its hydrolysis by phospholipase-C. *Science* 247: 1575.

Goldschmidt-Clermont,P.J., Machesky, L.M., Doberstein, S.K. and Pollard, T.D. (1991) Mechanism of the interaction of human platelet profilin with actin. *J. Cell Biol.* 113: 1081.

Grondin, P., Plantavid, M., Sultan, C., Breton, M., Mauco, G. and Chap, H. (1991) Interaction of pp60c-src, phospholipase-C, inositol-Lipid, and diacylglycerol kinases with the cytoskeletons of thrombin-stimulated platelets. *J. Biol. Chem.* 266: 15705.

Gutkind, J.S., Lacal, P.M. and Robbins, K.C. (1990) Thrombin-dependent association of phosphatidylinositol-3 kinase with P60-c-src and P59Fyn in human platelets. *Mol. Cell. Biol.*. 10: 3806.

Huang, R.S., Sorisky, A., Church, W.R., Simons, E.R. and Rittenhouse, S.E. (1991) Thrombin receptor-directed ligand accounts for activation by thrombin of platelet phospholipase C and accumulation of 3-phosphorylated phosphoinositides. *J. Biol. Chem.* 266: 18435.

Imai, A. and Gershengorn, M.C. (1987) Independent phosphatidylinositol synthesis in pituitary plasma membrane and endoplasmic reticulum. *Nature* 325: 726.

Janmey, P.A., Iida, K., Yin, H.L. and Stossel, T.P. (1987) Polyphosphoinositide micelles and polyphosphoinositide-containing vesicles dissociate endogenous gelsolin-actin complexes and promote actin assembly from the fast-growing end of actin filaments blocked by gelsolin. *J. Biol. Chem.* 262: 12228.

Kanoh, H., Banno, Y., Hirata, M. and Nozawa, Y. (1990) Partial purification and characterization of phosphatidylinositol kinases from human platelets. *Biochim. Biophys. Acta* 1046: 120.

Kaplan, D.R., Whitman, M., Schaffhausen, B., Raptis, L., Garcea, R.L., Pallas, D., Roberts, T.M. and Cantley, L. (1986) Phosphatidylinositol metabolism and polyoma-mediated transformation. *Proc. Natl. Acad. Sci.* 83: 3624.

Kaplan, D.R., Whitman, M., Schauffhausen, B., Pallas, D.C., White, M., Cantley, L. and Roberts, T.M. (1987) Common elements in growth factor stimulation and oncogenic transformation: 85 kd phosphoprotein and phosphatidylinositol kinase activity. *Cell* 50: 1021.

Kaulen, H.D. and Gross, R. (1976) Metabolic properties of human platelet membranes II. Thrombin-induced phosphorylation of membrane lipids and demonstration of phosphorylating enzymes in the platelet membrane. *Thromb. Haemost.* 35: 364.

Koch, C.A., Anderson, D., Moran, M.F., Ellis, C. and Pawson, T. (1991) SH2 and SH3 domains - Elements that control interactions of cytoplasmic signaling proteins. *Science* 252: 668.

Koreh, K. and Monaco, M.E. (1986) The relationship of hormone-sensitive and hormone-insensitive phosphatidylinositol to phosphatidylinositol 4,5-bisphosphate in the WRK-1 cell. *J. Biol. Chem.* 261: 88.

Kucera, G.L. and Rittenhouse, S.E. (1990) Human platelets form 3-phosphorylated phosphoinositides in response to alpha-thrombin, U46619, or GTP-gamma-S. *J. Biol. Chem.* 265: 5345.

Lassing, I. and Lindberg, U. (1985) Specific interaction between phosphatidylinositol 4,5-bisphosphate and profilactin. *Nature* 314: 472.

Lassing, I. and Lindberg, U. (1990) Polyphosphoinositides synthesis in platelets stimulated with low concentrations of thrombin is enhanced before activation of phospholipase C. *FEBS Lett.* 262: 231.

Lassing, I. and Lindberg, U. (1988) Evidence that the phosphatidylinositol cycle is linked to cell motility. *Cell Res.* 174: 1.

Lloyd, J.V., Nishizawa, E.E. and Mustard, J.F. (1973) Effect of ADP-induced shape change on incorporation of ^{32}P into platelet phosphatidic acid and mono-, di- and triphosphatidyl inositol. *Br. J. Haematol.* 25: 77.

Lloyd, J.V. and Mustard, J.F. (1974) Changes in ^{32}P-content of phosphatidic acid and the phosphoinositides of rabbit platelets during aggregation induced by collagen or thrombin. *Br. J. Haematol.* 26: 243.

Mahadevappa, V.G. and Holub, B.J. (1987) Quantitative loss of individual eicosapentaenoyl-relative to arachidonoyl- containing phospholipids in thrombin-stimulated human platelets. *J. Lipid. Res.* 28: 1275.

Mauco, G., Chap, H., Simon, M.F. and Douste-Blazy, L. (1978) Phosphatidic and lysophosphatidic acid production in phospholipase C-and thrombin-treated platelets. Possible involvement of a platelet lipase. *Biochimie* 60: 653.

Mauco, G., Chap, H. and Douste-Blazy, L. (1979) Characterization and properties of a phosphatidylinositol phosphodiesterase (phospholipase C) from platelet cytosol. *FEBS Lett.* 100: 367.

Mauco, G., Chap, H. and Douste-Blazy, L. (1983) Platelet activating factor (PAF-acether) promotes an early degradation of phosphatidylinositol-4,5-biphosphate in rabbit platelet. *FEBS Lett.* 153: 361.

Mauco, G., Dangelmaier, C.A. and Smith, J.B. (1984) Inositol lipids, phosphatidate and diacylglycerol share stearoylarachidonoylglycerol as a common backbone in thrombin-stimulated human platelets. *Biochem. J.* 224: 933.

Mauco, G., Dajeans, P., Chap, H. and Douste-Blazy, L. (1987) Subcellular localization of inositol lipids in blood platelets as deduced from the use of labelled precursors. *Biochem. J.* 244: 757.

Monaco, M.E. and Woods, D. (1983) Characterization of the hormone-sensitive phosphatidylinositol pool in WRK-1 cells. *J. Biol. Chem.* 258: 15125.

Nahas, N., Plantavid, M., Mauco, G. and Chap, H. (1989) Association of phosphatidylinositol kinase and phosphatidylinositol 4-phosphate kinase activities with the cytoskeleton in human platelets. *FEBS Lett.* 246: 30.

Nolan,R.D., and Lapetina, E.G. (1990) Thrombin stimulates the production of a novel polyphosphoinositide in human platelets. *J. Biol. Chem.* 265; 2441.

Nozawa, Y., Nakashima, S. and Nagata, K. (1991) Phospholipid-mediated signaling in receptor activation of human platelets. *Biochem. Biophys. Acta.* 1082: 219.

Payrastre, B., Plantavid, M., Breton, M., Chambaz, E and Chap, H. (1990) Relationship between phosphoinositide kinase activities and protein tyrosine phosphorylation in plasma membranes from A431 cells. *Biochem. J.* 272: 665.

Perret, B., Levy-Toledano, S., Plantavid, M., Bredous, R., Chap, H., Tobelem, G., Douste-Blazy, L. and Caen, J.P. (1983) Abnormal phospholipid organization in Bernard-Soulier platelets. *Thromb. Res.* 31: 529.

Rana, R.S. and Hokin, L.E. (1990) Role of phosphoinositides in transmembrane signaling. *Physiol. Rev.* 70: 115.

Rittenhouse, S.E. (1983) Human platelets contain phospholipase C that hydrolyzes polyphosphoinositides. *Proc. Natl. Acad. Sci.* 80: 5417.

Rittenhouse-Simmons, S. (1979) Production of diglyceride from phosphatidylinositol in activated human platelets. *J. Clin. Invest.* 63: 580.

Salganicoff, L., Loughnane, M.H., Sevy, R.W. and Russo M. (1985) The platelet-strip. I A low-fibrin contractile model of thrombin-activated platelets. *Amer. J. Physiol.* 249: C279

Schick, P.K., Tuszynski, G.P. and Vander Voort, P.W. (1983) Human platelet cytoskeletons: Specific content of glycolipids and phospholipids. *Blood* 61: 163.

Sinigaglia,F., Torti M., Ramashi, G. and Bauluini, C. (1989) The occupancy of glyprotein IIb-IIIa complex modulates thrombin activation of human platelets. *Biochim. Biophys. Acta.* 984; 225.

Stossel, T.P. (1990) How cells crawl. *American Scientist* 78: 408.

Sultan, C., Breton, M., Mauco, G., Grondin, P., Plantavid, M. and Chap, H. (1990) The novel inositol lipid phosphatidylinositol 3,4- bisphosphate is produced by human blood platelets upon thrombin stimulation. *Biochem. J.* 269: 831.

Sultan,C., Plantavid, M., Bachelot, C., Grondin, P., Breton, M., Mauco, G., Levy-Toledano, S., Caen, J.P. and Chap, H. (1991) Involvement of platelet glycoprotein-IIb-IIIa (alphaIIb-beta3 integrin) in thrombin-induced synthesis of phosphatidylinositol 3',4'-bisphosphate. *J. Biol. Chem.* 266: 23554.

Torti, M., Sinigaglia, F., Ramashi, G. and Balduini, C. (1991) Platelet glycoprotein IIb-IIIa is associated with 21 kDa GTP-binding protein. *Biochim. Biophys. Acta:* 1070, 20

Traynor-Kaplan, A.E.Thompson, B.L., Harris, A.L., Taylor, P., Omann, G.M. and Klar, L.A. (1989) Transient increase in phosphatidylinositol 3,4-bisphosphate and phosphatidylinositol trisphosphate during activation of human neutrophils. *J. Biol. Chem.* 264: 15668.

Tuszynski, G.P., Mauco, G., Koshy, A., Sckick, P.K. and Walsh, P.N. (1984) The platelet cytoskeleton contains elements of the prothrombinase complex. *J. Biol. Chem.* 259: 6947.

Vu, T-K., Hung, D.T., Wheaton, V.I. and Coughlin, S.R. (1991) Molecular cloning of a functional thrombin receptor reveals a novel proteolytic mechanism of receptor activation. *Cell.* 64: 1057.

Wahl, M.I., Olashaw, N.E., Nishibe, S., Rhee, S.G., Pledger, W.J. and Carpenter, G. (1989) Platelet-derived growth factor induces rapid and sustained tyrosine phosphorylation of phospholipase C-gamma in quiescent BALB/C-3T3-Cells. *Mol. Cell. Biol.* 9: 2934.

Whitman, M., Kaplan, D., Roberts, T. and Cantley, L. (1987) Evidence for two distinct phosphatidylinositol kinases in fibroblasts. Implications for cellular regulation. *Biochem. J.* 247: 165.

Whitman, M., Downes, C.P., Keeler, M., Keller, T. and Cantley, L. (1988) Type I phosphatidylinositol kinase makes a novel inositol phospholipid, phosphatidylinositol-3-phosphate. *Nature* 332: 644.

Yamamoto, K., Graziani, A., Carpenter, C., Cantley, L.C. and Lapetina, E.G. (1990) A novel pathway for the formation of phosphatidylinositol 3,4- bisphosphate - Phosphorylation of phosphatidylinositol 3- monophosphate by phosphatidylinositol-3-monophosphate 4-kinase. *J. Biol. Chem.* 265: 22086.

Zhang, J., Fry, M.J., Waterfield, M.D., Jaken, S., Liao, L., Fox, J.E.B. and Rittenhouse, S.E. (1992) Activated phosphoinositide 3-kinase associates with membrane skeleton in thrombin-exposed platelets. *J. Biol. Chem.* 267: 4686.   .

REGULATION OF PLATELET FUNCTION BY THE CYTOSKELETON

Joan E.B. Fox

Gladstone Institute of Cardiovascular Disease
Department of Pathology and Cardiovascular Research Institute
University of California, San Francisco
San Francisco, California 94140

and

Children's Hospital Oakland Research Institute
747 52nd Street
Oakland, California 94609

INTRODUCTION

The platelet cytoplasm is filled with a network of actin filaments. The organization of these filaments is regulated by their association with proteins such as α-actinin, tropomyosin, and actin-binding protein (Fox, 1987). When platelets are activated, myosin associates with these cytoplasmic actin filaments, thus generating the tension required for the centralization of granules (Fox and Phillips, 1982; Painter and Ginsberg, 1984). New filaments are formed by the polymerization of monomeric actin (Carlsson et al, 1979; Jennings et al, 1981); polymerization of these new filaments induces the extension of filopodia (Casella et al, 1981).

In recent years, it has become apparent that platelets contain a membrane skeleton in addition to the cytoplasmic actin filaments. As illustrated schematically in Fig. 1, the membrane skeleton coats the plasma membrane and is linked to the cytoplasmic domains of certain membrane glycoproteins. It is becoming increasingly apparent that the membrane skeleton is critically important in regulating various aspects of platelet function both in unstimulated and activated cells. This article will briefly review the evidence that platelets contain a membrane skeleton and will discuss potential functions of this structure. It will focus primarily on the possibility that the membrane skeleton may be involved in regulating the adhesive function of the glycoprotein (GP) IIb-IIIa complex and the transmembrane signalling that results from binding of adhesive ligand to this integrin.

Mechanisms of Platelet Activation and Control, Edited by
K.S. Authi *et al.*, Plenum Press, New York, 1993

The Platelet Membrane Skeleton

The initial identification of a membrane skeleton in platelets came from experiments in which platelets were lysed with Triton X-100. This detergent solubilizes most platelet proteins but does not solubilize actin filaments. As diagrammed in Fig. 2, cytoplasmic actin filaments could be sedimented from the detergent lysates at low g-forces. However, a small pool of actin filaments in the detergent lysates required higher g-forces to be sedimented. These filaments were recovered in association with specific membrane glycoproteins (Fox, 1985a).

Morphological experiments subsequently confirmed the idea that platelets contain a membrane skeleton. When platelets were lysed with Triton X-100, the actin filaments could be seen to be present throughout the cytoplasm and as a continuous layer external to the microtubule coil (Fox et al, 1988). However, an additional, more amorphous layer was visualized at the periphery of the cytoplasmic filaments. When Ca^{2+} was included in the lysis buffer, essentially all of the visible actin filaments were lost, presumably as a result of depolymerization by gelsolin, but the amorphous layer remained (Fox et al, 1988). The amorphous layer followed the outlines of the platelet and was shown immunocytochemically to contain the membrane GP Ib-IX complex (Fox et al, 1988). Recently, Hartwig and coworkers utilized a high=resolution quick-freeze deep-etch approach to show that the platelet membrane skeleton has an appearance very similar to that of red blood cell membrane skeletons (Hartwig and DeSisto, 1991). Isolation of platelet skeletons by centrifugation, followed by analysis of their components on sodium dodecyl sulfate-polyacrylamide (SDS) gels has revealed that the platelet membrane skeleton contains short actin filaments cross-linked by fodrin (Fox et al, 1987), actin-binding protein (Fox, 1985b) and other unidentified proteins (Fox et al, 1988).

Because the membrane skeleton lines the plasma membrane, it appeared likely that it may play an important role in regulating properties of the plasma membrane. For example, by analogy to the red blood cell, it appeared probable that the platelet membrane skeleton regulates the contours of the plasma membrane. Evidence for this possibility came from the observation that the platelet membrane skeleton retains the contours of the platelet for many hours after solubilization of the rest of the platelet proteins (Fox et al, 1988), indicating that the skeleton is a self-supporting structure. Further evidence for the idea that the membrane skeleton regulates the contours of the plasma membrane comes from the observation that platelets from patients with Bernard-Soulier syndrome, in which the major site of attachment for the membrane skeleton on the plasma membrane (the GP Ib-IX

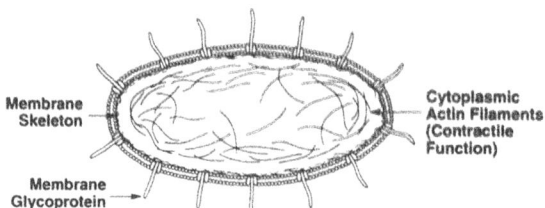

Figure 1   Schematic representation of the organization of actin filaments in an unstimulated platelet. Platelets contain two systems of actin filaments. Many of the filaments are organized into a three-dimensional network throughout the cytoplasm. These perform the contractile functions of the cytoskeleton. Other filaments are part of a lattice-like structure that coats the inner surface of the plasma membrane. This structure, known as the membrane skeleton, associates with the cytoplasmic domain of membrane glycoproteins and regulates properties of the plasma membrane.

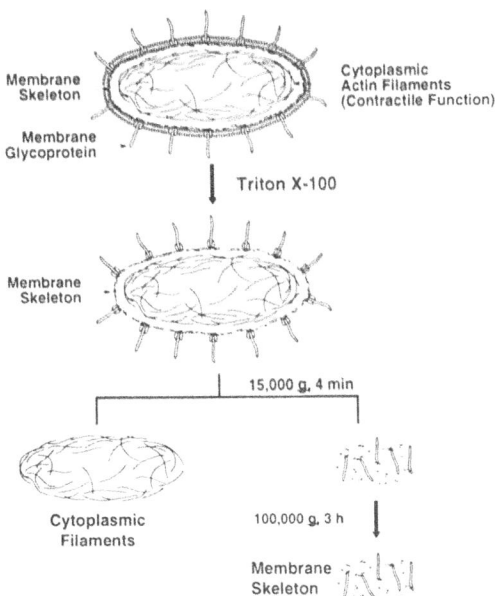

Figure 2   Schematic representation of the separation of the membrane skeleton and cytoplasmic actin filaments in detergent-lysed platelets. When platelets are lysed with Triton X-100, many of the cytoplasmic actin filaments remain sufficiently cross-linked into networks that they can be sedimented by centrifugation of the lysates at low g-forces. However, the membrane skeleton fragments (Hartwig and DeSisto, 1991) and requires higher g-forces to be sedimented (Fox, 1985a).

complex) is missing, often have bizarre shapes and are readily deformable (Bernard and Soulier, 1948). A second function that has been suggested for the platelet membrane skeleton is that it may regulate the lateral distribution of membrane glycoproteins to which it is attached. Evidence for this possibility comes from the observation that the GP Ib-IX complex is present over the entire surface of the plasma membrane in unstimulated platelets, but clusters when the membrane skeleton is disrupted (Boyles et al, 1987). A third suggestion is that the platelet membrane skeleton may stabilize the plasma membrane, preventing it from fragmenting. In support of this idea, disruption of the skeleton (for example by exposure of platelets to agonists or other agents, such as dibucaine or ionophore A23187, that activate calpain within platelets) results in the formation of small membrane vesicles that bud and break away from the platelets (Fox et al, 1990; Fox et al, 1991).

Each of these functions for the platelet membrane skeleton was suggested by analogy to the known functions of the red blood cell membrane skeleton. However, unlike the red blood cell, the platelet is a very dynamic cell, responding rapidly to a variety of different agonists. Thus, we have considered the possibility that the membrane skeleton in platelets might have functions in addition to those of the membrane skeleton in the red blood cell. When platelets are activated, agonists activate a variety of intracellular events. We reasoned that the lattice-like skeleton that coats the plasma membrane is likely to play an important role in regulating these transmembrane signaling events. For example, it may act as a framework for various regulatory molecules, perhaps recruiting important signaling molecules to appropriate submembranous locations. It also appeared possible to us that activation-induced changes in the composition or organization of the skeleton may modulate the structure or function of plasma membrane components in activated cells.

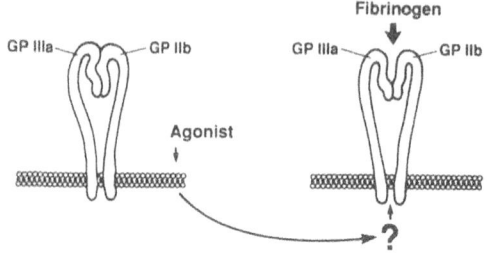

Figure 3   Schematic representation of the agonist-induced activation of the GP IIb-IIIa complex. When platelets are activated, an unidentified intracellular event acts on the cytoplasmic domain of the GP IIb-IIIa complex. This in turn induces a conformational change in the extracellular domain of the complex that allows adhesive ligands, including fibrinogen, von Willebrand factor, and fibronectin, to bind.

One membrane protein that is altered upon platelet activation is the GP IIb-IIIa complex. The GP IIb-IIIa complex is a member of the integrin family of adhesive receptors (Phillips et al, 1988). As shown in Fig. 3, the complex is not active in unstimulated platelets. Upon addition of an agonist, an unidentified intracellular event causes a change in the cytoplasmic domain of the GP IIb-IIIa complex; this causes the extracellular domain to undergo a conformational change that allows it to bind its adhesive ligand, fibrinogen. The binding of fibrinogen in turn causes signaling events in the other direction such that intracellular events, including activation of tyrosine kinases (Ferrell and Martin, 1988; Golden et al, 1990), activation of calpain (Fox et al, 1993), and activation of Na^+/H^+ exchange are induced (Banga et al, 1986). It has often been suggested that the cytoskeleton regulates the function of integrins. The best evidence for this comes from the study of focal-contacts of cells such as cultured chick embryo fibroblasts (see Burridge et al, 1988 for a review). Focal contacts are sites at which the plasma membrane of adherent cells is in closest contact with the extracellular matrix. When cells are plated on dishes coated with an adhesive ligand, the integrin that binds that particular ligand clusters. Bundles of actin filaments (known as stress fibers) form within the cell and associate (via unidentified linkage proteins) with the cytoplasmic domain of the integrins in the resulting focal contacts. Thus, the integrins mediate a transmembrane linkage between the cytoskeleton and the extracellular matrix at these sites. Antibodies that reverse the integrin-matrix interaction cause disruption of the stress fibers. Cytochalasins, which disrupt the stress fibers, cause the integrin to dissociate from its adhesive ligand. Thus, as with the GP IIb-IIIa in platelets, there appears to be a two-way communication across the membrane; in the case of focal contacts of cultured cells, the cytoskeleton regulates the adhesive properties of the integrin and the binding of extracellular adhesive proteins regulates the organization of the cytoskeleton. As a first step in addressing the possibility that the cytoskeleton plays a similar role in the two-way communication across the GP IIb-IIIa complex in platelets, we examined whether the GP IIb-IIIa complex is associated with the cytoskeleton in platelets.

ASSOCIATION OF THE GP IIb-IIIa COMPLEX WITH THE PLATELET MEMBRANE SKELETON

In the past, there has been some controversy as to whether the GP IIb-IIIa complex is associated with the cytoskeleton in platelets. Painter and coworkers (Painter et al, 1985) suggested that a small amount of the GP IIb-IIIa complex is associated with the cytoskeleton in unstimulated platelets. This suggestion was based on the observation that approximately 10% of the GP IIb-IIIa complex in isolated platelet membranes could be recovered in association with actin upon solubilization of the membranes. In contrast, Phillips and coworkers (Phillips et al, 1980) were unable to detect association of GP IIb-IIIa with the cytoskeleton when unstimulated platelets were lysed with Triton X-100-containing buffers. However, they found that GP IIb-IIIa became associated with the cytoskeleton in aggregating platelets. We reasoned that if GP IIb-IIIa was associated with the cytoskeleton, it would be associated with the membrane skeleton that coats the inner surface of the plasma membrane, rather than with the cytoplasmic actin filaments. As diagrammed in Fig. 2, upon lysis and centrifugation of platelets, the membrane skeleton separates from the underlying cytoplasmic actin and fragments. Thus, whereas the cytoplasmic actin filaments can be isolated by centrifugation at low g-forces, the fragments of membrane skeleton require higher g-forces to be sedimented. To determine whether the GP IIb-IIIa complex was associated with the membrane skeleton in unstimulated platelets, we lysed platelets in Triton X-100 and isolated the cytoplasmic actin filaments and membrane skeleton by differential centrifugation. As shown in Fig. 4A, essentially no GP IIb-IIIa was associated with the cytoplasmic actin filaments but approximately 20% of the total platelet GP IIb-IIIa sedimented with the membrane skeleton, suggesting that a subpopulation of GP IIb-IIIa is indeed associated with the cytoskeleton in unstimulated

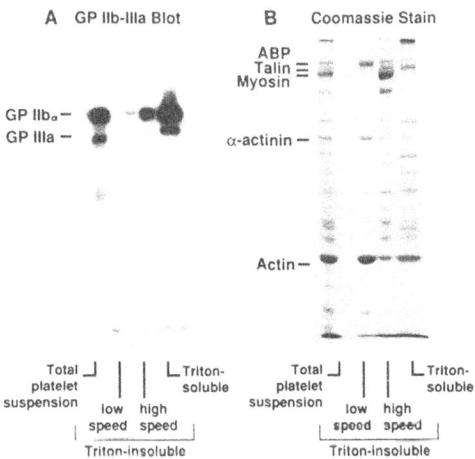

Figure 4   Recovery of the GP IIb-IIIa complex and other proteins with soluble and insoluble fractions from Triton X-100-lysed platelets. Platelets were lysed with Triton X-100 in the presence of EGTA. Cytoplasmic actin filaments were sedimented at low g-forces and the membrane skeleton subsequently sedimented by centrifugation of the resulting low-speed supernatant at high speeds. The low-speed pellet, high-speed pellet, and high-speed supernatant (Triton-soluble fraction) were electrophoresed through SDS-polyacrylamide gels. The GP IIb-IIIa complex was detected on a Western blot; other proteins were detected with Coomassie Brilliant Blue. ABP, actin-binding protein.

platelets. The reason that this GP IIb-IIIa went undetected in previous studies on detergent-lysed platelets was presumably that the existence of the membrane skeleton had not yet been recognized at the time of the early studies. Thus, only the low-speed pellet (i.e., the cytoplasmic actin filaments) had been examined for GP IIb-IIIa.

As shown in Fig. 4B, many proteins sediment with the membrane skeleton at high g-forces. Not all of these (for example, thrombospondin) are components of the membrane skeleton; many of them are simply inherently insoluble in Triton X-100. Thus, it is essential to demonstrate criteria other than insolubility in detergent before concluding that a protein is associated with the cytoskeleton. In the case of GP IIb-IIIa, we have utilized a number of techniques, including sedimentation with actin on sucrose density gradients and decreased sedimentation under conditions in which calpain (which cleaves several components of the membrane skeleton) is active. It appears, therefore, that in washed platelet suspensions, a subpopulation of GP IIb-IIIa is associated with the membrane skeleton.

ROLE OF THE GP IIb-IIIa-CYTOSKELETON INTERACTION IN REGULATING LIGAND-INDUCED TRANSMEMBRANE SIGNALING

By analogy to focal contacts of cultured cells, we reasoned that binding of adhesive ligand to the extracellular domain of the GP IIb-IIIa complex in platelets might drive the skeletal-associated GP IIb-IIIa into focal contact-like structures. Some evidence for such

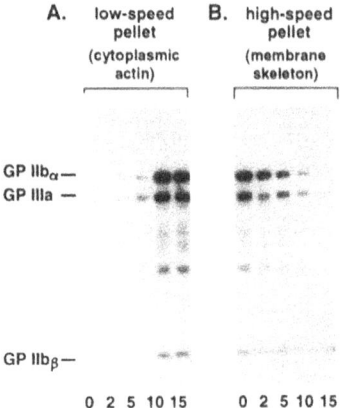

Figure 5 Western blots of SDS-polyacrylamide gels showing the thrombin-induced redistribution of GP IIb-IIIa in platelets. Platelets were incubated for increasing lengths of time with thrombin, lysed with Triton X-100, and the low-speed and high-speed pellets isolated by centrifugation as described in the legend to Fig. 4. Sedimented material was solubilized in an SDS-containing buffer and electrophoresed through SDS-polyacrylamide gels. The GP IIb-IIIa complex was detected on Western blots.

a possibility has been presented by Isenberg and coworkers, who showed that the binding of adhesive ligand to the GP IIb-IIIa complex results in a clustering of the complex (Isenberg et al, 1987). By analogy to focal contacts of cultured cells, we reasoned that this clustering might result because of ligand-induced association of the membrane skeleton and associated GP IIb-IIIa complex with cytoplasmic actin filaments. We predicted that if this was the case, the GP IIb-IIIa complex that requires high g-forces to be sedimented from unstimulated platelets might be sedimented at low g-forces when platelets are activated and bind their adhesive ligand. As shown in Fig. 5, the Triton X-100-insoluble GP IIb-IIIa did indeed shift from the high-speed pellet to the low-speed pellet when platelets were activated. It appears probable that this shift resulted from binding of adhesive ligand (secreted from the α-granules during platelet activation) to the GP IIb-IIIa complex. In support of this idea, we found that addition of RGDS, which binds to the fibrinogen-binding site on the GP IIb-IIIa complex, accelerated the shift from the low- to high-speed pellet. Fig. 6 is a schematic illustration of the formation of these focal contact-like structures in activated platelets.

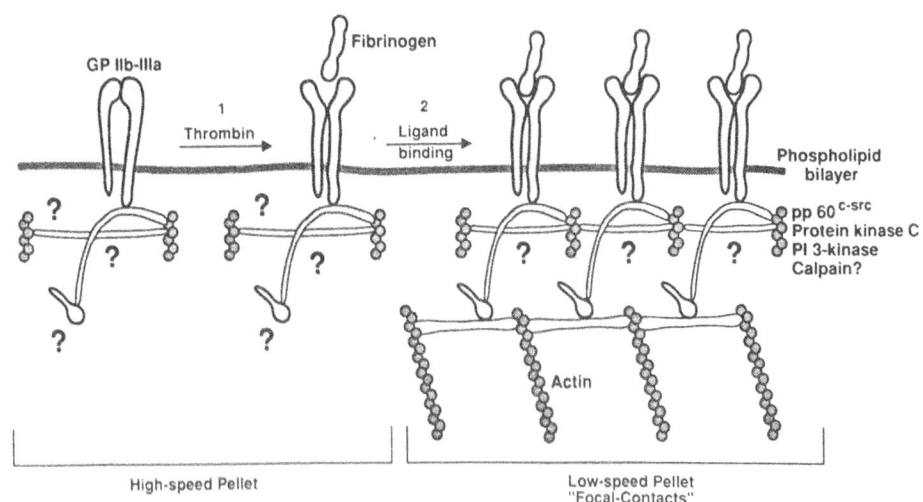

Figure 6   Schematic representation of the formation of focal contact-like structures in activated platelets. In unstimulated platelets, the GP IIb-IIIa complex is associated with unidentified components of the membrane skeleton (indicated with a question mark). Only a subpopulation of the GP IIb-IIIa complex is recovered in association with the membrane skeleton, perhaps because the affinity of the interaction is such that much of the complex dissociates upon cell lysis. When platelets are activated, the GP IIb-IIIa complex binds fibrinogen. Binding of fibrinogen occurs over a period of 15 to 30 min; binding induces clustering of the GP IIb-IIIa complex, and an association of the membrane skeleton with underlying actin filaments (Fig. 5). Proteins that have been recovered in association with these focal contact-like structures include pp60[c-src], protein kinase C, and phosphoinositide 3-kinase (PI 3-kinase), making these potential enzymes involved in this ligand-induced reorganization of the cytoskeleton. Although calpain has not yet been localized to focal contacts in platelets, it is concentrated in focal contacts of cultured cells and is activated as a consequence of ligand binding to the GP IIb-IIIa complex in platelets. Thus, calpain is another enzyme likely to play an important role in mediating the ligand-induced cytoskeleton reorganizations in platelets.

In platelets and other adherent cells there is considerable interest in identifying the intracellular enzymes that are activated as a consequence of binding of adhesive ligands to integrins and in understanding how activation of these enzymes leads to a reorganization of the cytoskeleton. In focal contacts of adherent cells, it has been difficult to elucidate the molecular mechanisms involved in the ligand-induced reorganization of cytoskeletons. One of the problems is that the focal-contacts represent such a small percentage of the total cellular protein that it has been difficult to determine whether any proteins are becoming selectively modified at these sites or whether any enzymes are becoming selectively activated as a consequence of the binding of adhesive ligand to the integrin. In contrast, platelets are so rich in cytoskeletal proteins that the focal contacts represent a significant percentage of the total platelet protein. Further, unlike the situation in cultured cells, in which focal contacts dissociate when cells are removed from the matrix for analysis, platelets in suspension can be readily solubilized and the focal contact-like structures isolated by centrifugation (Fig. 4).

In the past, it has been shown that one of the events induced in platelets as a consequence of ligand binding to the GP IIb-IIIa complex in platelets is that several proteins become phosphorylated on tyrosine residues (Ferrell and Martin, 1988; Golden et al, 1990). In recent studies, we have now shown that most of the proteins phosphorylated on tyrosine residues are components of the focal contact-like structures, suggesting that activation of protein tyrosine kinase(s) may be one mechanism by which the ligand-induced cytoskeletal reorganizations are induced. Another ligand-induced change detected recently is the activation of calpain. Thus, incubation of platelets with GP IIb-IIIa antibodies that inhibit fibrinogen binding prior to activation of platelets with physiological agonists inhibited the activation of calpain that normally occurs in aggregating platelets (Fox et al, 1993). Moreover, thrombasthenic platelets that lack the GP IIb-IIIa complex (George et al, 1984) did not undergo activation of calpain when stirred with physiological agonists (Fox et al, 1993). The major proteins cleaved by calpain in aggregating platelets are actin-binding protein, talin, and spectrin, proteins known to be components of the platelet cytoskeleton. Thus, the ligand-induced activation of calpain appears to be another mechanism by which the ligand-induced cytoskeletal changes are manifested in platelets.

If tyrosine kinases and calpain are enzymes activated as consequence of ligand binding to the GP IIb-IIIa complex, one might expect these proteins to be associated with the focal contact-like structures. Evidence from a number of laboratories has now indicated that pp 60$^{c\text{-}src}$, the major tyrosine kinase present in platelets (Golden et al, 1986), is associated with the cytoskeleton of aggregating platelets (Grondin et al, 1991; Zhang et al, 1992; Fox et al, manuscript submitted). In addition, other potential regulatory proteins, including protein kinase C, an enzyme known to be concentrated in focal-contacts of adherent cultured cells (Burridge et al, 1988), and phosphoinositide-3-kinase (Zhang et al, 1992), have also been detected in association with the GP IIb-IIIa-rich cytoskeletal structures that form when platelets aggregate, making these regulatory proteins additional candidates for enzymes that may induce the integrin-mediated cytoskeletal changes in aggregating platelets. As yet, calpain has not been detected in the focal contact like structures, perhaps because it is dissociated from these structures in the EGTA-containing lysis buffers typically used to isolate platelet cytoskeletons.

ROLE OF THE GP IIb-IIIa-CYTOSKELETON INTERACTION IN REGULATING THE ADHESIVE PROPERTIES OF THE GP IIb-IIIa COMPLEX

As in focal contacts of adherent cells, it appears possible that the association of GP IIb-IIIa with the cytoskeleton in platelets may regulate the adhesive properties of the integrin. In cultured cells, the evidence for this function of the cytoskeleton came from experiments in which cells were incubated with cytochalasins, the stress fibers disrupted, and adhesive interactions dissociated (Burridge et al, 1988). Recently, we used a similar approach to test the hypothesis that the cytoskeleton regulates the adhesive properties of the GP IIb-IIIa complex in platelets. The adhesive function of the GP IIb-IIIa complex was assayed by the ability of the glycoprotein complex to bind the monoclonal antibody PAC-1. This antibody binds to the GP IIb-IIIa complex, apparently at the same site as fibrinogen (Shattil et al, 1985). Like fibrinogen, this antibody binds only to the GP IIb-IIIa complex on activated platelets. We found that preincubation of platelets with cytochalasins inhibited the subsequent agonist-induced binding of PAC-1. Inhibition occurred in a dose-dependent manner and was specific in that the agonist-induced secretion of granule contents was not inhibited. Morphological examination of the platelets and their isolated cytoskeletons revealed that cytochalasins depolymerized the cytoplasmic actin filaments in platelets. In contrast, biochemical analysis revealed that they did not dissociate GP IIb-IIIa from the membrane skeletal components, indicating that the formation of the focal contact-like structures shown in the right hand side of Fig. 6 is a critical step in stabilizing the GP IIb-IIIa-in-fibrinogen interactions in activated platelets.

It also appears possible that the cytoskeleton could regulate the initial activation of the GP IIb-IIIa complex, i.e., step 1 in Fig. 6. However, this is a more difficult question to address since cytochalasins do not dissociate the interaction between the GP IIb-IIIa complex and the membrane skeleton proteins. One can envisage several ways in which the association of integrin with the membrane skeleton could regulate integrin function. For example, by localizing signaling enzymes, the cytoskeleton may allow the generation of regulatory molecules at the appropriate site. One candidate molecule is phosphatidic acid, which is generated in activated platelets and has recently been shown to be able to render purified GP IIb-IIIa complex competent to bind fibrinogen (Smyth et al, 1991). Another possibility is that platelet activation causes modifications of components of the membrane skeleton that could in turn induce a conformational change in the cytoskeleton-associated integrin. The observation that several components of the membrane skeleton became rapidly phosphorylated on tyrosine residues upon platelet activation (Fox et al, manuscript submitted) indicates that alterations to the membrane skeleton do occur with a time course that could allow them to play a role in inducing the adhesive function of the GP IIb-IIIa complex. Clearly, future experiments will be required in order to test the hypothesis that the membrane skeleton is important in regulating the functional activity of the GP IIb-IIIa complex in platelets and to identify mechanisms involved.

As a first step in directly determining the functional importance of the association of the GP IIb-IIIa complex with the membrane skeleton, it will be of interest to elucidate the molecular details of the interaction between the GP IIb-IIIa complex and the membrane skeleton. This information could potentially be utilized to devise means for dissociating the interaction and thus assessing its functional importance. One approach that has been used to gain insight into the cytoskeletal protein(s) that serves to link the cytoskeleton to the cytoplasmic domain of other integrins has been to determine whether any of the cytoskeletal proteins that are present in focal contacts can bind to the cytoplasmic domain of integrins in *in vitro* binding assays. This approach has identified two cytoskeletal proteins that are able to associate with integrins. One is talin, a protein that was shown to bind to the fibronectin receptor (α_5 β_1) in an assay using equilibrium gel filtration (Horwitz et al, 1986). Because talin is known to bind to vinculin, which in turn can bind to α-actinin, which can bind to actin, a model containing this chain of proteins was proposed (Burridge et al, 1988).

Recent studies suggest a second potential candidate for mediating the actin-integrin interaction (Otey et al, 1990). When cell lysates were passed over an affinity column containing the cytoplasmic domain of the β_1-integrin subunit, several proteins bound. One of them was α-actinin (Otey et al, 1990). An assay in which the synthetic peptide was coated onto microtiter wells confirmed that α-actinin was able to bind to the integrin's cytoplasmic domain. The affinity of this interaction was approximately 1.6×10^{-8} M, which was much higher than that of the binding of integrin to talin, and apparently involved the second of four homologous repeats in the rod-shaped portion of the α-actinin molecule (Otey et al, 1991). Further studies using purified chicken gizzard integrin (a β_1-containing integrin) and purified platelet GP IIb-IIIa complex ($\alpha_{IIb}\beta_3$) showed that α-actinin was able to bind to the intact integrins. Interestingly, the affinity of the interaction with the intact integrins was about 100-fold lower than that with the isolated β_1 cytoplasmic domain, suggesting that conformational restraints may exist in the intact complexes (Otey et al, 1990).

Yet another mechanism by which integrins might associate with the cytoskeleton is through direct interaction with actin. Some evidence for this suggestion comes from work on the platelet. Painter and coworkers (Painter et al, 1985) isolated membranes from unstimulated platelets and showed that approximately 10% of the GP IIb-IIIa was associated with actin filaments, apparently in the absence of other detectable proteins. Further studies will be needed to determine which of the potential interactions occurs in intact cells. It is conceivable that more than one interaction might occur and that this could provide a means for regulating the functional activity of integrins.

SUMMARY

The platelet cytoskeleton contains two actin filament-based components. One is the cytoplasmic actin filaments that fill the cytoplasm and mediate contractile events. The other is the membrane skeleton that coats the plasma membrane and regulates properties of the membrane such as its contours and stability and the lateral distribution of membrane glycoproteins. Recent work reviewed in this article indicates that the GP IIb-IIIa complex can associate with the membrane skeleton. Upon platelet activation, GP IIb-IIIa becomes competent to bind its adhesive ligand, fibrinogen. This induces a reorganization of the cytoskeleton such that the membrane skeletal proteins with which GP IIb-IIIa is associated become associated with underlying cytoplasmic filaments. As in focal contacts of cultured cells, this ligand-induced association of GP IIb-IIIa with cytoplasmic actin filaments regulates the ability of GP IIb-IIIa to bind adhesive ligand. Intracellular enzymes that are activated as a consequence of ligand binding to the GP IIb-IIIa complex include tyrosine kinase(s) and calpain, making these potential candidates for enzymes inducing the two-way signaling across the membrane. Additional candidates include phosphoinositide 3-kinase and protein kinase C, other enzymes that have been detected in focal contacts of aggregating platelets. Future studies identifying interactions between the GP IIb-IIIa complex and membrane skeletal proteins should help to further elucidate the significance of the GP IIb-IIIa in cytoskeleton interaction in regulating integrin-mediated transmembrane signaling in platelets.

REFERENCES

Banga, H.S., Simons, E.R., Brass, L.F., and Rittenhouse, S.E., 1986, Activation of phospholipases A and C in human platelets exposed to epinephrine: role of glycoproteins IIb/IIIa and dual role of epinephrine, *Proc. Natl. Acad. Sci. USA.* 83:9197.

Bernard, J., and Soulier, J.P., 1948, Sur une nouvelle variete de dystrophie thrombocytaire hemorragipare congenitale, *Sem. Hop. Paris.* 24:3217.

Boyles, J.K., Fox, J.E.B., and Berndt, M.C., 1987, The distribution of GP Ib and the stability of the plasma membrane are dependent upon an intact membrane skeleton, *Thromb. Haemost.* 58:225 (abstract).

Burridge, K., Fath, K., Kelly, T., Nuckolls, G., and Turner, C., 1988, Focal adhesions: transmembrane junctions between the extracellular matrix and the cytoskeleton, *Annu. Rev. Cell Biol.* 4:487.

Carlsson, L., Markey, F., Blikstad, I., Persson, T., and Lindberg, U., 1979, Reorganization of actin in platelets stimulated by thrombin as measured by the DNase I inhibition assay, *Proc. Natl. Acad. Sci. USA.* 76:6376.

Casella, J.F., Flanagan, M.D., and Lin, S., 1981, Cytochalasin D inhibits actin polymerization and induces depolymerization of actin filaments formed during platelet shape change, *Nature.* 293:302.

Ferrell, J.E., Jr., and Martin, G.S., 1988, Platelet tyrosine-specific protein phosphorylation is regulated by thrombin, *Mol. Cell Biol.* 8:3603.

Fox, J.E.B., 1985a, Linkage of a membrane skeleton to integral membrane glycoproteins in human platelets. Identification of one of the glycoproteins as glycoprotein Ib, *J. Clin. Invest.* 76:1673.

Fox, J.E.B., 1985b, Identification of actin-binding protein as the protein linking the membrane skeleton to glycoproteins on platelet plasma membranes, *J. Biol. Chem.* 260:11970.

Fox, J.E.B., 1987, The platelet cytoskeleton, *in:* "Thrombosis and Haemostasis," M. Verstraete, J. Vermylen, R. Lignen, and J. Arnout, eds., Leuven University Press, Belgium, (pp. 175-225).

Fox, J.E.B., Austin, C.D., Boyles, J.K., and Steffen, P.K., 1990, Role of the membrane skeleton in preventing the shedding of procoagulant-rich microvesicles from the platelet plasma membrane, *J. Cell Biol.* 111:483.

Fox, J.E.B., Austin, C.D., Reynolds, C.C., and Steffen, P.K., 1991, Evidence that agonist-induced activation of calpain causes the shedding of procoagulant-containing microvesicles from the membrane of aggregating platelets, *J. Biol. Chem.* 266:13289.

Fox, J.E.B., Boyles, J.K., Berndt, M.C., Steffen, P.K., and Anderson, L.K., 1988, Identification of a membrane skeleton in platelets, *J. Cell Biol.* 106:1525.

Fox, J.E.B., Lipfert, L., Clark, E.A., Reynolds, C.C., Austin, C.D., and Brugge, J.S., Association of the protein tyrosine kinase $pp60^{c-src}$ and phosphotyrosine-containing proteins with the platelet membrane skeleton, manuscript submitted.

Fox, J.E.B., and Phillips, D.R., 1982, Role of phosphorylation in mediating the association of myosin with the cytoskeletal structures of human platelets, *J. Biol. Chem.* 257:4120.

Fox, J.E.B., Taylor, R.G., Taffarel, M., Boyles, J.K., and Goll, D.E., 1993, Evidence that activation of platelet calpain is induced as a consequence of binding of adhesive ligand to the integrin, Glycoprotein IIb-IIIa. *J. Cell. Biol.* 120:1501-1507.

Fox, J.E.B., Reynolds, C.C., Morrow, J.S., and Phillips, D.R., 1987, Spectrin is associated with membrane-bound actin filaments in platelets and is hydrolyzed by the Ca^{2+}-dependent protease during platelet activation, *Blood.* 69:537.

George, J.N., Nurden, A.T., and Phillips, D.R., 1984, Molecular defects in interactions of platelets with the vessel wall, *N. Engl. J. Med.* 311:1084.

Golden, A., Brugge, J.S., and Shattil, S.J., 1990, Role of platelet membrane glycoprotein IIb-IIIa in agonist-induced tyrosine phosphorylation of platelet proteins, *J. Cell Biol.* 111:3117.

Golden, A., Nemeth, S.P., and Brugge, J.S.,1986, Blood platelets express high levels of the $pp60^{c-src}$-specific tyrosine kinase activity, *Proc. Natl. Acad. Sci. USA.* 83:852.

Grondin, P., Plantavid, M., Sultan, C., Breton, M., Mauco, G., and Chap, H., 1991, Interaction of $pp60^{c-src}$, phospholipase C, inositol-lipid, and diacylglycerol kinases with the cytoskeletons of thrombin-stimulated platelets, *J. Biol. Chem.* 266:15705.

Hartwig, J.H., and DeSisto, M.,1991, The cytoskeleton of the resting human blood platelet: structure of the membrane skeleton and its attachment to actin filaments, *J. Cell Biol.* 112:407.

Horwitz, A., Duggan, K., Buck, C., Beckerle, M.C., and Burridge, K., 1986, Interaction of plasma membrane fibronectin receptor with talin - a transmembrane linkage, *Nature.* 320:531.

Isenberg, W.M., McEver, R.P., Phillips, D.R., Shuman, M.A., and Bainton, D.F., 1987, The platelet fibrinogen receptor: an immunogold-surface replica study of agonist-induced ligand binding and receptor clustering, *J. Cell Biol.* 104:1655.

Jennings, L.K., Fox, J.E.B., Edwards, H.H., and Phillips, D.R., 1981, Changes in the cytoskeletal structure of human platelets following thrombin activation, *J. Biol. Chem.* 256:6927.

Otey, C.A., Parr, T., Blanchard, A.D., Critchley, D., and Burridge, K., 1991, Mapping of interactive sites on α-actinin and the ß-1 integrin cytoplasmic domain, *J. Cell Biol.* 115:166 (abstract).

Otey, C.A., Pavalko, F.M., and Burridge, K., 1990, An interaction between α-actinin and the ß$_1$ integrin subunit *in vitro*, *J. Cell Biol.* 111:721.

Painter, R.G., and Ginsberg, M.H., 1984, Centripetal myosin redistribution in thrombin stimulated platelets: relationship to platelet factor 4 secretion, *J. Exp. Cell Res.* 155:198.

Painter, R.G., Prodouz, K.N., and Gaarde, W., 1985, Isolation of a subpopulation of glycoprotein IIb-IIIa from platelet membranes that is bound to membrane actin, *J. Cell Biol.* 100:652.

Phillips, D.R., Charo, I.F., Parise, L.V., and Fitzgerald, L.A.,1988, The platelet membrane glycoprotein IIb-IIIa complex, *Blood.* 71:831.

Phillips, D.R., Jennings, L.K., and Edwards, H.H.,1980, Identification of membrane proteins mediating the interaction of human platelets, *J. Cell Biol.* 86:77.

Shattil, S.J., Hoxie, J.A., Cunningham, M., and Brass, L.F., 1985, Changes in the platelet membrane glycoprotein IIb-IIIa complex during platelet activation, *J. Biol. Chem.* 260:11107.

Smyth, S.S., Hillery, C.A., and Parise, L.V., 1991, Phosphatidic and lysophosphatidic acid modulate the fibrinogen binding activity of purified platelet glycoprotein IIb-IIIa, *Blood.* 78:278a (abstract).

Zhang, J., Fry M.J., Waterfield, M.D., Jaken S., Liao L., Fox, J.E.B., and Rittenhouse, S.E., 1992. Activated phosphoinositide 3-kinase associates with membrane skeleton in thrombin exposed platelets. *J. Biol. Chem.* 267, 4686.

CYTOSKELETAL INTERACTIONS OF Rap1b IN PLATELETS

Gilbert C. White, II, Neville Crawford*, and Thomas H. Fischer,

Center for Thrombosis and Hemostasis
University of North Carolina
Chapel Hill, NC 27599-7035

*Department of Biochemistry and Cell Biology
Royal College of Surgeons of England
London, United Kingdom

INTRODUCTION

Low molecular weight GTP binding proteins (G proteins) are membrane-associated proteins which reversibly bind guanine nucleotides and regulate cellular processes, such as growth and differentiation (Evans et al., 1991; Macara, 1991). Members of this superfamily of proteins show considerable sequence homology and share structural features, including an effector domain which interacts with GTPase activating proteins or GAPs and post-translational modification at the carboxy terminus by polyisoprenyl groups, either farnesyl or geranyl-geranyl (Maltese, 1990; Gibbs, 1991). To date, more than 50 low molecular weight G proteins in four subfamilies have been reported. The prototype for this group of proteins is $p21^{ras}$, the 21 kDa protein product of the ras protooncogene. At least seven distinct G proteins are present in platelets (Bhullar & Haslam, 1988; Ohmori et al., 1988; Polakis et al., 1989; Polakis et al., 1989; Farrell et al., 1990; Nemoto et al., 1992) (Table I).

Although the precise function of these proteins remains unsure, current observations suggest increasingly diverse activities. Ras itself is oncogenic and mutant forms of ras lead to cell transformation. This and the similarity of the low molecular weight G proteins to the heterotrimeric G proteins led to the original hypothesis that ras and the other low molecular weight G proteins are involved in cell signalling. This hypothesis has been strengthened by recent findings that ras-GAP is associated with phospholipase C (Torti & Lapetina, 1992) and with the insulin receptor (Pronk et al., 1992). However, among the low molecular weight G proteins, ras is the only protein that is oncogenic and the role of the other low molecular weight G proteins in signalling remains to be established. Some low molecular weight G proteins, including members of the rab subfamily, are localized to intracellular membranes in cells and have been implicated in vesicular trafficking (Balch, 1990). More recently, $p21^{rac}$ has been shown to promote the assembly of a protein complex that leads to activation of NADPH oxidase in neutrophils

Mechanisms of Platelet Activation and Control, Edited by
K.S. Authi et al., Plenum Press, New York, 1993

(Abo et al., 1991; Knaus et al., 1991). Taken together, these observations point to a wide spectrum of activities by low molecular G proteins. In this Chapter, work supporting a potentially new and unique function for rap1b in platelets will be presented. We will show that rap1b is attached to the plasma membrane in resting platelets and becomes associated with the activation-dependent cytoskeleton during platelet aggregation. The possible interaction of rap1b with membrane glycoproteins IIb/IIIa will be explored and the effect of phosphorylation of rap1b on its association with the cytoskeleton will be examined.

CELLULAR LOCALIZATION OF Rap1b

The low molecular weight G proteins are membrane-bound in intact cells. Some, including ras, are localized to the plasma membrane while others, especially members of the rab subfamily, are localized to intracellular membranes in the endoplasmic reticulum

Table 1	Low molecular weight G proteins in platelets	
G Protein	Isoprenoid	Features
rap1b (smg-p21)	geranyl-geranyl	cAMP-kinase substrate
rap2		
rap2b		
G25K (Gp)	geranyl-geranyl	
rac1 (p25)	geranyl-geranyl	
ral (p28, Gn)	geranyl-geranyl	
rhoA		botulinum C3 substrate

and Golgi apparatus. Attachment to the membrane is through an isoprenyl modification that is added posttranslationally by thioester linkage to the C-terminal cysteine residue (Maltese, 1990; Gibbs, 1991). Rap1b, like other members of the low molecular weight G protein family, is membrane bound in the intact cell through a geranyl-geranyl group (Kawata et al., 1990). In order to determine whether rap1b was attached to the plasma membrane or intracellular membranes, resting whole platelet were lysed and separated into plasma and intracellular membranes as described by Menashi, et al (1981). The resulting membrane fractions were analyzed for the presence of rap1b using M90, a murine monoclonal antibody which recognizes rap1b and other ras-like G proteins (Lacal & Aronson, 1986). Figure 1 shows that most of the rap1b is present in the plasma membrane fraction and none is present in the intracellular membrane fraction. Interestingly, a small amount of rap1b is nearly always found in the cytoplasmic fraction after cell lysis (data not shown). This soluble rap1b is associated with other protein(s) and migrates with an apparent molecular mass of 500 kDa on Sephacryl S300 (Fischer & White, unpublished observations). The identity of the protein(s) with which this soluble form of rap1b interacts remains uncertain.

When platelets are activated by a strong agonist like thrombin, there are remarkable changes in the cellular localization of rap1b (Fisher et al.,1990). In resting platelets, rap1b is associated with the plasma membrane but fractionates with the cytoplasmic fraction after solubilization in Triton X-100 (Figure 2, lane 8, panel A). In contrast, after thrombin stimulation, all of the immunodetectable rap1b is found in a 10,000 g fraction that contains the cell cytoskeleton (Figure 2, lane 4, panel B), an activation-dependent structure that contains various contractile proteins such as actin, myosin, actin binding protein, and a-actinin, as well as more recently reported components such as phosphoinositide-3-kinase (Zhang et al., 1992), tyrosine kinases such as pp60[src] (Horvath et al., 1992), membrane glycoproteins GPIIb/GPIIIa (Phillips et al., 1980) and PECAM-1 (Newman et al., 1992). Incorporation of rap1b into the cytoskeleton occurred in two phases. Initial incorporation

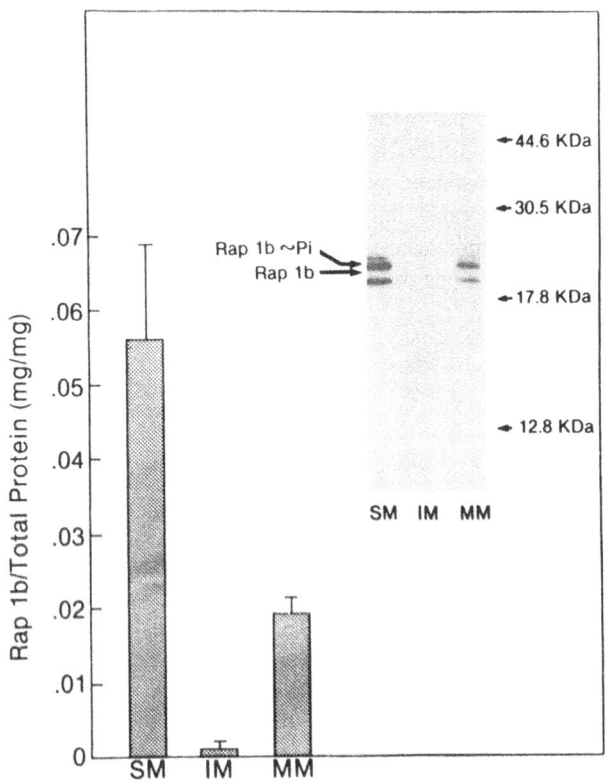

Figure 1   Rap1b distribution in plasma membranes (SM), intracellular membranes (IM), and mixed membranes (MM). Rap1b was quantified from Western blots using recombinant rap1b as a standard. INSET: Western blot of each fraction developed with M90.

occurred within seconds and resulted in association of approximately 20% of the total cellular rap1b with the cytoskeleton (Figure 3). The second phase was much slower and led to incorporation of the remaining rap1b by ten minutes. Furthermore, as shown in Figure 3, association of rap1b with the cytoskeleton was aggregation-dependent. In the absence of external calcium, aggregation was inhibited and only the initial phase of rap1b association with the cytoskeleton occurred. While thrombin stimulation resulted in the efficient incorporation of rap1b into the cytoskeleton, other agonists were much less effective (Figure 4), in part because the cytoskeleton was less formed with the other agonists, but even when normalized for cytoskeletal weight, the other agonists were less effective.

The aggregation-dependent incorporation of rap1b into the cytoskeleton of thrombin activated platelets is similar to the incorporation of membrane glycoproteins IIb and IIIa (Phillips et al., 1980). Because of this and since GPIIb/IIIa have been reported to be associated with a low molecular weight G protein (Torti et al., 1991), we undertook a series of studies to determine if rap1b was the low molecular weight G protein associated with GPIIb/IIIa and if the incorporation of rap1b into the cytoskeleton was mediated by GPIIb/IIIa. When resting platelets were lysed in non-ionic detergents and fractionated on Sephacryl S300, the rap1b chromatographed with a mobility distinct from that of GPIIb/IIIa. Gel overlay with [^{32}P]GTP showed multiple bands, all coeluting with immunoreactive rap1b. There were no GTP binding proteins associated with GPIIb/IIIa. In similar studies, we were also unable to immunoprecipitate rap1b from platelet lysates with antibodies to GPIIb/IIIa and vice versa. Thus, under the experimental conditions used, we are unable to demonstrate an interaction of rap1b with GPIIb/IIIa in resting platelets. Specifically, we were able to show that rap1b does not appear to be the 22 kDa GTP binding protein identified by Torti, et al. (1991). To test the role of GPIIb/IIIa in the incorporation of rap1b into the cytoskeleton, platelets from patients with Glanzmann's thrombasthenia, which lack GPIIb/IIIa, were examined for incorporation of rap1b into the cytoskeleton. When thrombasthenic platelets were activated in the presence or absence of calcium, rap1b was incorporated into the cytoskeleton in a manner similar to control platelets. Thus, rap1b incorporation into the cytoskeleton was independent of GPIIb/IIIa. Our results do not exclude the possibility that GPIIb/IIIa incorporation into the cytoskeleton is dependent on rap1b.

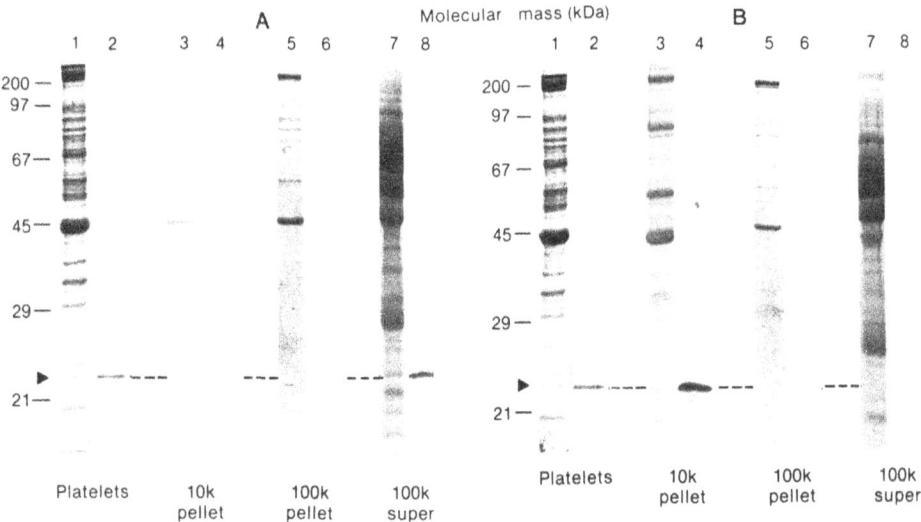

Figure 2   Cytoskeletal assocation of rap1b. Control (panel A) or thrombin-activated (panel B) platelets were lysed in 1% Triton X-100 and separated into a 10,000 g fraction containing the cell cytoskeleton, a 100,000 g pellet containing the membrane skeleton, and a 100,000g supernatant containing soluble proteins. Odd numbered lanes are Coomassie Blue stained gels. Even numbered lanes are Western blots with M90.

EFFECT OF cAMP ON Rap1b ASSOCIATION WITH THE CYTOSKELETON

Rap1b is a substrate for cAMP-dependent protein kinase in either intact cells or membrane fractions. Under optimal conditions, approximately one mole of phosphate per

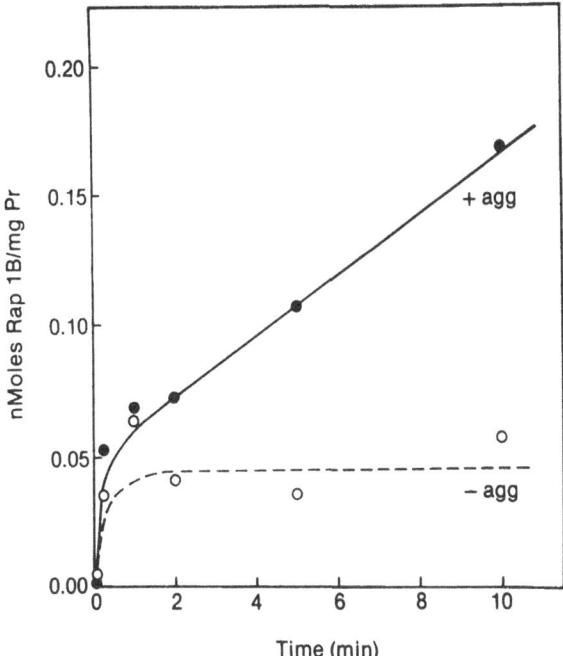

Figure 3   Time course of rap1b assocation with the cytoskeleton after thrombin stimulation in the presence (●) or absence (○) of calcium.

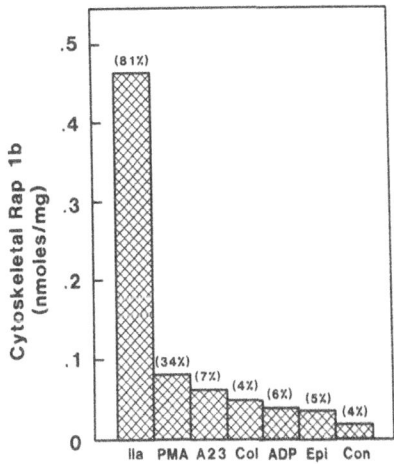

Figure 4   Effect of different agonists on rap1b association with the cytoskeleton. Washed platelets were activated in the presence of calcium with 1 unit/ml thrombin (IIa), 10 μM phorbol myristate acetate (PMA), 3 μM ionophore A23187 (A23), 2 μg/ml collagen (Col), 1 mM ADP (ADP), or 1 mM epinephrine (Epi). Control platelets (Con) were treated with buffer. Serotonin release is indicated in parentheses above each bar.

mole of protein is incorporated into rap1b by the catalytic subunit of cAMP-dependent protein kinase. The site of phosphorylation is serine179, very near the carboxy terminal cysteine 181 residue which is the site of isoprenyl modification and the site of membrane attachment (Siess et al., 1990; Hata et al., 1991; Fischer et al., 1991). Since cytoskeleton

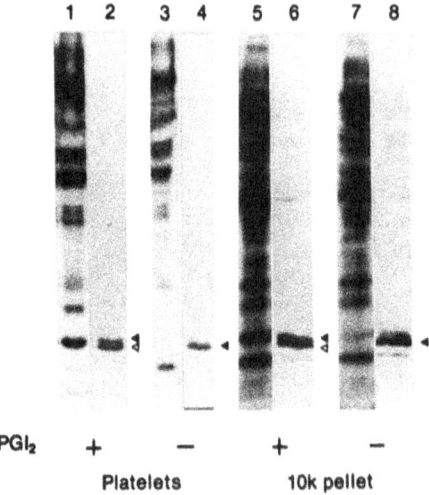

Figure 5 Control (-) or PGI₂-treated (+) [³²P]ATP-loaded platelets were activated with thrombin at 1 unit/ml and cytoskeleton fractions isolated by centrifugation at 10,000 g after lysis in Triton X-100. Lanes 1-4, whole platelets. Lanes 5-8, cytoskeletons. Odd lanes are radioautograms showing [³²P]phosphoproteins. Even lanes are Western blots with M90. The closed arrows point to phosphorylated rap1b. The open arrows point to unphosphorylated rap1b.

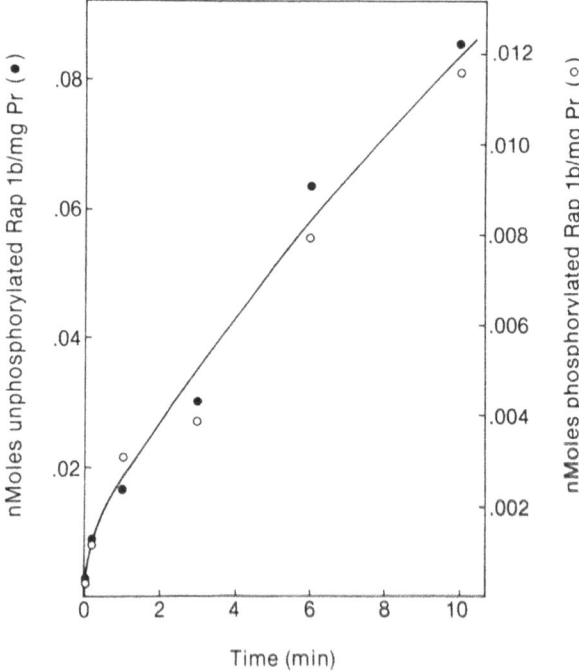

Figure 6 Comparison of the rates of association of phosphorylated (●) and unphosphorylated (○) rap1b with the cytoskeleton of thrombin-activated platelets.

assembly occurs with platelet activation and rap1b is a component of the cytoskeleton, an important question was whether phosphorylation of rap1b by cAMP influenced cytoskeleton incorporation of rap1b. We speculated that phosphorylation of rap1b might prevent its incorporation into the cytoskeleton and this in turn might inhibit cytoskeleton formation and platelet activation. To test this hypothesis, washed platelets loaded with [^{32}P]ATP were treated with 1 μM prostacyclin (PGI$_2$), a concentration sufficient to phosphorylate at least some rap1b but not high enough to inhibit platelet activation, then activated with thrombin. Cytoskeleton fractions were then isolated and analyzed for the incorporation of phosphorylated rap1b. In intact PGI$_2$ treated platelets, two immunoreactive rap1b bands were observed (Figure 5, lane 2). The upper band comigrated with a radiolabeled band on radioautogram (lane 1) and corresponded to phosphorylated rap1b while the lower band comigrated with rap1b in unlabeled platelets (lane 4) and corresponded to unlabeled rap1b. In cytoskeletons from activated platelets (lane 6), both phosphorylated and unphosphorylated rap1b were present. Figure 6 compares the rate of incorporation of phosphorylated and unphosphorylated rap1b, showing that both forms of rap1b are incorporated into the cytoskeleton at the same rate. Thus, phosphorylation of rap1b by cAMP-dependent protein kinase does not appear to inhibit incorporation of rap1b into the activation-dependent cytoskeleton.

SUMMARY

We have presented evidence that rap1b, a 22 kDa low molecular weight GTP binding protein, becomes associated with the cytoskeleton in thrombin-activated platelets. The initial incorporation is very rapid and occurs as fast as we can measure it. Thus, some rap1b is associated with the cytoskeleton as fast as it is formed. The remainder of the rap1b is incorporated more slowly. This biphasic incorporation of rap1b is similar to the incorporation of GPIIb/IIIa into the cytoskeleton, but no interaction between GPIIb/IIIa and rap1b could be demonstrated. Phosphorylation of rap1b by cAMP-dependent protein kinase did not inhibit its association with the cytoskeleton. We conclude that rap1b is one of an increasing number of proteins that associate with the cytoskeleton during cell activation.

The function of rap1b in the cytoskeleton is unclear at this time. However, it is possible to speculate on potential roles. There is growing evidence that low molecular weight G proteins participate in the formation of multi-molecular aggregates. For example, p21^{rac} promotes the assembly of a membrane-associated complex composed of NADPH oxidase, p47, and p67 and this complex is important for activation of NADPH oxidase in neutrophils. Similarly, in yeast, BUD1, a homolog of rap1, forms a complex with BUD5 (a homolog of GDI), BEMI, CDC24, and CDC42 (a homolog of G25K). This multi-protein aggregate may be important in cytoskeletal structure in yeast. In platelets, rap1b, which is membrane associated, may promote the assembly of a complex of proteins during cell activation and may localize this complex to the plasma membrane. Although the components of the complex are unknown, phosphorylation of rap1b does not inhibit its association with the cytoskeleton. Phosphorylation may, however, inhibit some subsequent function of rap1b or the complex. We are currently attempting to identify proteins in the cytoskeleton which complex with rap1b and the effect of phosphorylation on the assembly and function of the complex.

REFERENCES

Abo, A., Pick, E., Hall, A., Totty, N., Teahan, C. and Segal, A.W. 1991, Activation of NADPH oxidase involved the small GTP-binding protein p21rac, *Nature* 353:668.

Balch, W.E. 1990, Small GTP binding proteins in vesicular transport, *Trends Biochem. Soc.* 15:473.

Bhullar, R.P. and Haslam, R.J. 1988, Gn-proteins are distinct from ras p21 and other known low molecular mass GTP-binding proteins in platelets, *FEBS Letts.* 237:168.

Evans, T., Hart, M.-J., and Cerione, R.A. 1991, The ras superfamilies: Regulatory proteins and post-translational modifications, *Curr. Opin. Cell Biol.* 3:185.

Farrell, F.X., Ohmstede, C.A., Reep, B.R. and Lapetina, E.G. 1990, cDNA sequence of a new ras-related gene (rap2b) isolated from human platelets with sequence homology to rap2, *Nucl. Acids Res.* 18:4281.

Fisher, T.H., Gatling, M.N., Lacal, J.C. and White II, G.C. 1990, Rap 1b, a cAMP-dependent protein kinase substrate, associates with the platelet cytoskeleton. *J. Biol. Chem.* 265: 19405.

Fischer, T.H., Collins, J.H. and White, II, G.C. 1991, The localization of the cAMP-dependent protein kinase phosphorylation site in the platelet ras protein, rap1b, *FEBS Letts.* 282:173.

Gibbs, J.B., Ras C-terminal processing enzymes, New drug targets, *Cell* 65:1.

Hata, U., Kaibuchi, K., Kawamura, S., Hiroyoshi, M., Shirataki, H. and Takai, Y. 1991, Enhancement of the actions of smg p21 GDP/GTP exchange protein by the protein kinase A-catalyzed phosphorylation of smg p21, *J. Biol. Chem.* 266:6571.

Horvath, A.R., Muszbek, L. and Kellie, S. 1992, Translocation of pp60c-src to the cytoskeleton during platelet aggregation, *EMBO J.* 11:855.

Kawata, M., Farnsworth, C.C., Yoshida, Y., Gelb, M.H., Glomset, J.A. and Takai, Y. 1990, Posttranslationally processed structure of the human platelet protein smg p21B: Evidence for geranylgeranylation and carboxyl methylation of the C-terminal cysteine, *Proc. Natl. Acad. Sci.* 87:8960.

Knaus, V.G., Heyworth, P.G., Evans, T., Curnutte, J.T. and Bokoch, G. 1991, Regulation of phagocyte oxygen radical production by the GTP-binding protein rac-2, *Science* 254:1512.

Lacal, J.-C. and Aronson, S.A. 1986, Ras p21 deletion mutants and monoclonal antibodies as tools for localization of regions relevant to p21 function, *Proc. Natl. Acad. Sci.* 83:5400.

Macara, I.G. 1991, The ras superfamily of molecular switches, *Cell Signalling* 3:179.

Maltese, W.A. 1990, Posttranslational modification of proteins by isoprenoids in mammalian cells, *FASEB J.* 4:3319.

Menashi, S., Weintroub, H. and Crawford, N.G. 1981, Characterization of human platelet surface and intracellular membranes isolated by free flow electrophoresis, *J. Biol. Chem.* 256:4095.

Nemoto, Y., Namba, T., Teru-uchi, T., Ushikubi, F. and Narumiya, S. 1992, A rho gene product in human blood platelets. I. Identification of the platelet substrate for botulinum C3 ADP-ribosylation as rhoA protein, *J. Biol. Chem.* 267:20916.

Newman, P.J., Hillery, C.A., Albrecht, R., Parise, L.V., Berndt, M.C., Mazurov, A.V., Dunlop, L.C., Zhang, J. and Rittenhouse, S.E. 1992, Activation dependent changes in human platelet PECAM-1: Phosphorylation, cytoskeleton association and surface membrane redistribution, *J. Cell Biol.* 119:239.

Ohmori, T., Kikuchi, A., Yamamoto, K., Kawata, M., Kondo, J. and Takai, Y. 1988, Identification of a platelet Mr 22,000 GTP-binding proteins as the novel smg p21 gene product having the same effect domain as the ras gene product, *Biochem. Biophys. Res. Comm.* 157:670.

Phillips, D.R., Jennings, L.K. and Edwards, H.H. 1980, Identification of membrane proteins mediating the interaction of human platelets, *J. Biol. Chem.* 86:77.

Polakis, P.G., Snyderman, R. and Evans, T. 1989, Characterization of G25K, a GTP-binding protein containing a novel putative nucleotide binding domain, *Biochem. Biophys. Res. Comm.* 160:25.

Polakis, P.G., Weber, R.F., Nevins, B., Didsbury, J.R., Evans, T. and Snyderman, R. 1989, Identification of the ral and rac 1 gene products, low molecular mass GTP-binding proteins from human platelets, *J. Biol. Chem.* 264:16383.

Pronk, G.J., Medema, R.H., Burgering, B.M.Th., Clark, R., McCormick, F. and Bos, J.L. 1992, Interaction between the p21ras GTPase activating protein and the insulin receptor, *J. Biol. Chem.* 267:24058.

Siess, W., Winegar, D. and Lapetina, E.G. 1990, Rap1b is phosphorylated by protein kinase A in intact human platelets, *Biochem. Biophys. Res. Comm.* 170:944.

Torti, M. and Lapetina, E.G. 1992, Role of rap1b and p21ras GTPase-activating protein in the regulation of phospholipase C-γ1 in human platelets, *Proc. Natl. Acad. Sci.* 89:7796.

Torti, M., Sinigaglia, F., Ramaschi G. and Balduini, C. 1991, Platelet glycoprotein IIb-IIIa is associated with 21-kDa GTP-binding protein, *Biochim. Biophys. Acta* 1070:20.

Zhang, J., Fry, M.J., Waterfield, M.D., Jaken, S., Liao, L., Fox, J.E.B. and Rittenhouse, S.E. 1992, Activation phosphoinositide 3-kinase associates with the membrane skeleton in thrombin-exposed platelets, *J. Biol. Chem.* 267:4686.

MECHANISMS INVOLVED IN PLATELET PROCOAGULANT RESPONSE

Edouard M. Bevers, Paul Comfurius, Robert F.A. Zwaal

Department of Biochemistry
Cardiovascular Research Institute Maastricht
University of Limburg, P.O.Box 616
6200 MD Maastricht, The Netherlands

INTRODUCTION

Blood platelets are essential to the normal hemostatic process. Vessel wall injury produces several platelet agonists which through specific membrane receptors elicit a variety of cellular responses. Shape change, activation of binding sites for fibrinogen and other adhesive molecules, and secretion of intracellular granule contents ensure the rapid formation of large platelet aggregates, which prevent further loss of blood from the injured vessel. In addition, an important platelet response is the surface exposure of specific phospholipids, providing a catalytic surface for the assembly of enzyme complexes of the coagulation cascade. This platelet procoagulant response leads to a dramatic increase in the rate of thrombin formation, which allows rapid formation of an insoluble meshwork of fibrin, required to consolidate the primary haemostatic plug. On the other hand, the same catalytic surface is also instrumental in negative feedback control of the coagulation cascade by activated protein C. Platelets are the primary source of procoagulant lipid surfaces, but other cells such as erythrocytes or endothelial cells may, sometimes, under pathological conditions- also become procoagulant.

Although our knowledge about the more familiar platelet responses such as shape change, adhesion, aggregation and secretion has been increasingly growing due to numerous studies over the past decades, understanding of the mechanisms involved in the platelet procoagulant response is rather poor. This review presents current ideas about these mechanisms, which involve shedding of microvesicles as well as the contribution of a membrane protein, which transports specific phospholipids from the outer to the inner leaflet of the plasma membrane.

PHOSPHOLIPID-DEPENDENT COAGULATION REACTIONS

Negatively charged phospholipid surfaces provide the site of assembly of enzyme complexes of two consecutive reactions of the coagulation cascade (Mann et al., 1990). In the so called tenase complex, factor X is activated by a complex of factors IXa and VIIIa, whereas the prothrombinase reaction involves conversion of prothrombin to thrombin by the enzyme complex composed of factors Xa and Va. Binding of factors IX, X and prothrombin is accomplished through a calcium-mediated bridging between the gamma-carboxyglutamic acid residues of these proteins and the negatively charged polar headgroups of the phospholipid molecules. Binding to a lipid surface causes an increase in local concentration and favourable juxtaposition of the coagulation factors, which results in an increased rate of both reactions. The effect of phospholipids is kinetically reflected by the sharp decrease in Km of the two substrates, factor X and prothrombin, from far above to far below their respective plasma concentrations, allowing both reactions to proceed at close to Vmax conditions (Rosing et al., 1980; Van Dieijen et al., 1981). Since the two reactions occur in sequence, the presence of a catalytic phospholipid surface can produce an increase in the rate of thrombin formation by several orders of magnitude.

The major negatively-charged phospholipid responsible for the catalytic properties of a lipid surface is phosphatidylserine, although other anionic phospholipids might contribute as well (Rosing et al., 1988). Remarkably, membranes containing phosphatidylserine retain full procoagulant activity, irrespective of the actual surface charge of the membrane, suggesting that the binding of coagulation factors is not merely electrostatic.

Finally, it should be mentioned that phosphatidylserine also provides a major catalytic function in the extrinsic coagulation pathway, as part of the tissue thromboplastin complex. Since platelets are devoid of tissue factor, this activity will not be discussed here.

PLATELET PROCOAGULANT ACTIVITY

In platelets, the two major anionic phospholipids are phosphatidylserine and phosphatidylinositol. Procoagulant activity of the latter phospholipid is weak in comparison to phosphatidylserine. Moreover, in view of its rapid metabolic turnover during stimulus-response coupling, this lipid seems to be less suitable as procoagulant lipid. Therefore, phosphatidylserine might be considered as the exclusive molecule responsible for procoagulant activity in platelets. This does, however, not exclude a contribution of other platelet lipids in modulating the catalytic properties of the plasma membrane.

In the quiescent platelet, phosphatidylserine is virtual exclusively located in the cytoplasmic leaflet of the plasma membrane (Schick, 1976; Chap et al., 1977; Zwaal, 1978). The surface of unstimulated platelets is therefore not particularly suitable for the assembly of either the factor X- and prothrombin activating complexes. Upon platelet activation, the normal asymmetric distribution of the various phospholipids is lost (fig. 1) (Bevers et al., 1983; Thiagarajan and Tait, 1991) resulting in concomitant increase in procoagulant activity (Comfurius et al., 1985; Rosing et al., 1985b) and formation of factor Va (Tracy et al., 1979; Kane and Majerus, 1982; Sims et al., 1989) and VIIIa (Ahmad et al., 1989; Gilbert et al., 1991) binding sites, the extent of which depends on the platelet activation procedure. Weak agonists such as ADP, epinephrine and PAF hardly affect the procoagulant properties of the platelet surface (Comfurius et al., 1985). Although being one of the most potent agonists in evoking aggregation and secretion, thrombin causes only a moderate increase in procoagulant activity. More pronounced is the response to collagen, which produces a 5-8 fold increase in procoagulant activity. The combined action of collagen and thrombin, the complement membrane attack complex C5b-9 and calcium ionophore elicit the highest platelet procoagulant response (Comfurius et al., 1985; Rosing et al., 1985b; Sims et al., 1989; Ahmad et al., 1989), increasing tenase and prothrombinase activity some 20-fold or more. In addition, treatment of platelets with local anaesthetics dibucaine or tetracaine (Verhallen et al., 1987; Fox et al., 1990a) or sulfhydryl oxidizing agents (diamide or pyridyldithioethanolamine) (Bevers et al., 1989; Comfurius et al., 1990) also causes an increase in procoagulant activity. In the absence of extracellular calcium, or rather in the presence of calcium-chelators, these activators are unable to provoke procoagulant activity, strongly suggesting the need for increased intracellular calcium levels in this response. Mild stirring enhances generation of a procoagulant surface upon stimulation with collagen and thrombin, but not with ionophore or complement.

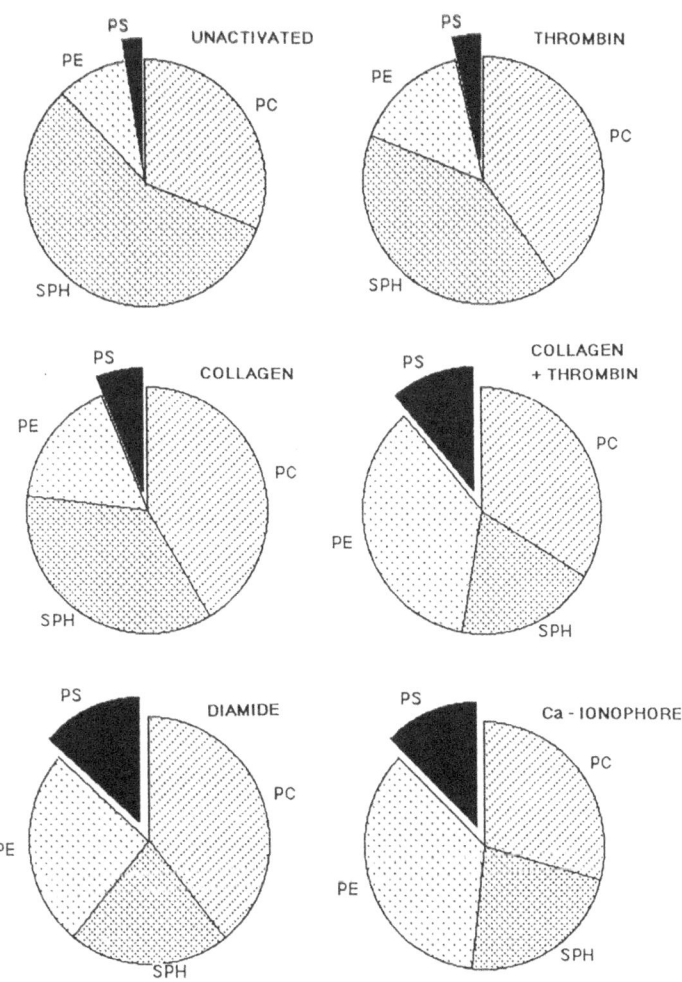

Fig.1   Phospholipid composition of the outer monolayer of platelets activated by different agonists as determined by susceptibility to sphingomyelinase and phospholipase A$_2$ (adapted from Bevers et al., 1983).

The procoagulant response is not strictly related to shape change, aggregation or secretion. This is supported by a number of observations: (i) a significant difference in procoagulant activity is observed between thrombin and collagen plus thrombin, whereas secretion and aggregation occur to the same extent; (ii) with the exception of ionophore stimulated platelets, the time course of generation of a procoagulant surface does not parallel that of platelet secretion and aggregation: at 37°C platelet release and aggregation are complete within 2 min, while full expression of a procoagulant membrane surface usually requires 5 min or more; (iii) local anaesthetics as well as sulphhydryl reactive compounds inhibit aggregation and release but evoke a procoagulant platelet surface; (iv) finally, storage pool deficient- and thrombasthenic platelets are indistinguishable from normal platelets in their ability to become procoagulant (Bevers et al.,1986).

MECHANISMS INVOLVED IN EXPRESSION OF PROCOAGULANT ACTIVITY

A difference in phospholipid composition between both leaflets of the bilayer seems to be a general feature of the plasma membrane of eukaryotic cells (Op den Kamp, 1979). Choline-containing lipids, sphingomyelin and phosphatidylcholine are confined to the outer leaflet, while the aminophospholipids, in particular phosphatidylserine, are preferentially located in the cytoplasmic side of the membrane. This explains why unperturbed cells such as platelets, but also red blood cells, endothelial cells and leukocytes do not provide a phospholipid surface suitable for the binding of coagulation factors. Under normal conditions, the asymmetric distibution of phospholipids is maintained during the lifetime of the cells, which helps to prevent unwanted coagulation. Two distinct mechanisms have been proposed to be responsible for maintenance of phospholipid asymmetry (for a review, see Schroit and Zwaal, 1991): (i) selective interactions between aminophospholipids and proteins belonging to the cytoskeleton of the cell (Haest, 1982), (ii) the presence of an ATP- and sulfhydryl-dependent transport system, referred to as aminophospholipid translocase, which specifically moves aminophospholipids from outer to inner leaflet of the plasma membrane (Seigneuret and Devaux, 1984).

Phospholipid analysis of the outer leaflet of the platelet membrane before and after activation has revealed that not only phosphatidylserine loses its asymmetric distribution, but that randomization concerns all phospholipid classes (fig 1.) (Bevers et al, 1983). The progressive loss of lipid asymmetry upon activation, with corresponding increase in surface exposed phosphatidylserine, is closely associated with the cell's ability to stimulate tenase and prothrombinase activity. It will be obvious that procoagulant activity will be expressed upon cell lysis, thereby exposing cytoplasmic oriented phosphatidylserine. However, platelet damage (lysis), which to a limited extent always accompanies the activation process, can be excluded to be responsible, since no gross differences in the amount of cell lysis is observed under various activation conditions.

As part of the platelet activation process, membrane fusion events take place, for instance during the release reaction but also as a result of formation of platelet derived microparticles (vide infra). These fusion events provide a means for rapid transbilayer movements (flip-flop) of phospholipids (Sims et al., 1989; Baldwin et al., 1990; Schroit and Zwaal, 1991), which may be facilitated by structural rearrangements of the membrane skeleton leading to decreased fixation of phosphatidylserine to the cytoplasmic side of the membrane (Comfurius et al., 1985; Fox et al., 1990a and b; Fox et al., 1991). The extent to which in particular phosphatidylserine asymmetry is lost, is also determined by the degree of inhibition of the aminophospholipid translocase, as explained below.

PLATELET MICROVESICLE FORMATION

In 1967 Wolf described the presence of platelet derived material ('platelet dust') with clot promoting activity in human plasma. Crawford (1972) demonstrated the presence of contractile proteins in isolated microparticles, and Sandberg and coworkers (1982;1985) have suggested that platelet microparticles were derived from intracellular membranes, externalized during secretion.

Recently it became evident that platelet microparticles originate from the plasma membrane as they were shown to contain membrane glycoproteins Ib, IIb and IIIa as well as membrane skeletal proteins (Sims et al., 1988, 1989; Fox et al., 1991). Microparticles are shed from the plasma membrane during platelet activation and electronmicroscopy reveals these particles as unilamellar vesicles with an average diameter of about 0.2 μm (fig 2). Their time course of appearance is slower than that of platelet aggregation and secretion, but closely parallels the time course of generation of procoagulant activity. A close relationship is also observed between the extent of microvesicle formation and the extent of procoagulant activity caused by various platelet activation procedures, i.e. extensive following ionophore-, collagen plus thrombin-, or complement C5b-9 activation, moderate after collagen activation, and low after stimulation by thrombin, ADP, or adrenaline. Also with local anaesthetics such as dibucaine and sulphhydryl reactive compounds like diamide and pyridyldithioethylamine, appearance of microvesicles coincides with generation of procoagulant activity. Usually, some 25-30% of the procoagulant activity and factor Va binding sites are associated with the microvesicles while the remainder is present on the remnant cells (fig 3). An exception is observed with platelets activated by complement pore proteins C5b-9, where most of the factor Va binding sites are associated with the microvesicles (Sims et al., 1989).

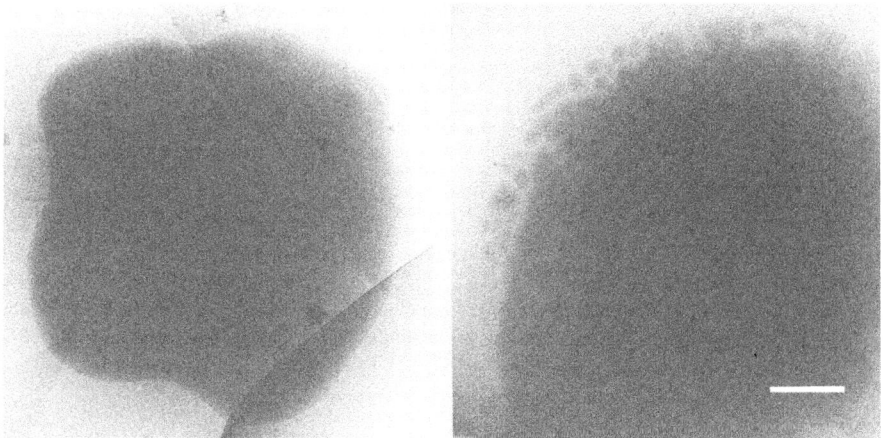

Fig.2   Cryo-electron micrograph of a quiescent platelet (left) and a platelet 15 sec after activation by ionophore A23187 in the presence of extracellular calcium. Cryo-electron microscopy allows the study of platelet morphology by direct virtification from suspension without the use of chemical fixation (Frederik et al., 1991). Bar represents 1 μm (*Courtesy Mr M.C.A. Stuart.*)

Shedding of microvesicles appears to be closely associated with surface exposure of phosphatidylserine. Since blebbing and shedding of microvesicles from the platelet surface must entail eversion and fusion of apposing segments of plasma membrane, this process was suggested to lead to transient formation of non-bilayer lipid phases at the point where the plasma membrane fuses to form the budding vesicle (Sims et al., 1989; Comfurius et al., 1990; Schroit and Zwaal, 1991). This would lead to a concomitant localized collapse in membrane phospholipid asymmetry and the appearance of phosphatidylserine in the cell's outer leaflet, including that of the microvesicle itself.

The shedding of vesicular material is certainly not unique to platelets, but has been shown for a variety of eukaryotic cells (Allan and Michell, 1975; Van de Water et al., 1985; Hamilton et al.,1990; Beaudoin and Grondin, 1991). For erythrocytes, endothelial cells and tumor cells the shedding phenomenon is also accompanied by loss of phospholipid asymmetry and surface exposure of procoagulant phosphatidylserine (Chandra et al., 1987; Comfurius et al., 1990; Hamilton et al., 1990). Although the cellular mechanisms responsible for microvesicle production are largely unknown, one common feature seems to be the absolute requirement of extracellular Ca^{2+}. Fox and coworkers (1990b,1991) found a close relationship between platelet microvesicle production and cytoskeletal breakdown caused by the endogenous SH-dependent protease, calpain, and suggest that

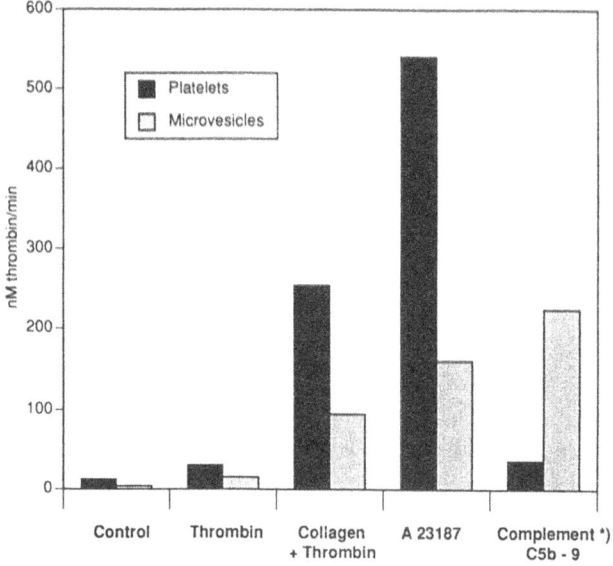

Fig.3   Distribution of expression of prothrombinase activity between platelets and platelet-derived microparticles, as a function of the platelet agonist. * Since microvesicles from complement-activated platelets cannot be satisfactorily separated from remnant platelets by centrifugation, this distribution was calculated from exposure of factor Va binding sites as visualized by fluorescence-gated flow cytometry using a fluorescently labeled anti-Factor Va antibody (Sims et al., 1989)

shedding may result from a dissociation of the platelet membrane skeleton from its membrane attachment sites. However, conflicting results exist about the role of calpain, since others have found that inhibition of cytoskeletal breakdown by calpain inhibitors did not lead to appreciable decrease in microvesicle production nor expression of procoagulant activity (Comfurius et al., 1990; Wiedmer et al., 1990). In addition, platelets of a patient with Scott syndrome (vide infra) have been reported to exhibit normal hydrolysis of cytoskeletal components when treated with calcium ionophore, while expression of procoagulant activity is severely impaired, indicating that calpain is not exclusively responsible (Comfurius et al., 1985; Rosing et al., 1985a; Bevers et al., 1992). Further evidence to exclude a direct role of calpain is provided by the observation that sulfhydryl reactive compounds, which completely block calpain activity, are capable of inducing procoagulant activity (Bevers et al., 1989; Comfurius et al., 1990) and corresponding microvesicle release (unpublished observations). Finally, formation of microvesicles and corresponding procoagulant activity in red blood cells upon treatment with calcium ionophore is not affected by calpain inhibitors (unpublished). Together, these data make a direct involvement of calpain in the mechanism of microvesicle formation and exposure of procoagulant activity unlikely. However, this does not rule out a putative regulatory role of the cytoskeleton, the structure of which can also be modulated independent of calpain by changes in cytoplasmic Ca^{2+} levels.

SCOTT SYNDROME

In 1979 Weiss et al described a patient with a rare bleeding disorder, presently referred to as Scott syndrome, which they defined as a deficiency in the expression of 'platelet factor 3', indicating the platelet's inability to form a procoagulant surface. In a variety of subsequent studies, it was observed that upon activation, these platelets generated a reduced number of binding sites for factor Va (Sims et al., 1989) and VIIIa (Ahmad et al., 1989), showed impaired development of procoagulant surface for tenase- and prothrombinase complexes in conjunction with a decreased surface exposure of phosphatidylserine (Rosing et al., 1985a), and were markedly deficient in microvesicle formation (Sims et al., 1989). These findings strongly support the view that in normal platelets the formation of a procoagulant surface is coupled to shedding of microvesicles.

The close association between microvesicle formation and development of procoagulant activity is further substantiated by the fact that the aberrant phenomena in Scott syndrome are not restricted to the patient's platelets, but also extend to the erythrocytes (Bevers et al., 1992). While in normal red blood cells calcium ionophore induces shedding of microvesicles, accompanied by exposure of phosphatidylserine and expression of membrane procoagulant activity, in red blood cells from a patient with Scott syndrome these events are strongly impaired. Moreover, the aberrant response to ionophore was also found with resealed erythrocyte ghosts, suggesting a defect in the membrane or membrane-associated cytoskeleton. Although no obvious abnormality in composition of cytoskeletal proteins neither in their degradation by calpain was apparent, the remarkable ability of this patient's red blood cells and resealed ghosts to retain a biconcave shape upon Ca^{2+} influx, without appreciable echinocyte formation, may point at an aberration in Ca^{2+}-regulated rearrangements of the membrane skeleton. The fact that the defect is common to both platelets and erythrocytes suggests a mutation in an early stem cell. In this respect it is of interest to mention that preliminary experiments with T lymphocytes obtained from this patient also showed reduced formation of microvesicles upon treatment with ionophore.

REGULATION OF EXPRESSION OF PROCOAGULANT ACTIVITY: INVOLVEMENT OF AMINOPHOSPHOLIPID TRANSLOCASE

The presence in the plasma membrane of an active pump system for aminophospholipids may have important consequences with respect to the regulation of membrane lipid asymmetry. This putative protein, referred to as aminophospholipid translocase was first proposed for red blood cells by Seigneuret and Devaux (1984) based on the observation that introduction of spin-labeled aminophospholipid in the outer leaflet of the red cell membrane is followed by rapid transfer to the inner leaflet. Transport was demonstrated to be dependent on hydrolysable ATP, to be inhibited by sulfhydryl-reactive compounds and to be selective only for the L-isomer of aminophospholipids. Moreover, transport is inhibited by increased levels of intracellular Ca^{2+} (Bitbol et al., 1987; Comfurius et al., 1990). At present, the identity of the protein is still obscure: likely candidates include the Mg^{2+}-dependent -ATPase (Morrot et al., 1990) and an integral membrane protein of 31 kDa, possibly the membrane integral part of the rhesus- factor (Schroit et al., 1990). Evidence for the presence of an aminophospholipid translocase has meanwhile been obtained for a variety of other cells, including platelets (Schroit and Zwaal, 1991).

Platelet activation in the presence of extracellular Ca^{2+}, not only induces microvesicle formation, but also leads to inactivation of the aminophospholipid translocase (Bitbol et al., 1987; Comfurius et al., 1990). Subsequent lowering of intracellular Ca^{2+} restores translocase activity, provided that irreversible inactivation of translocase by intracellular calpain is prevented. Under these conditions, surface exposed phosphatidyl-serine is pumped back to the inner leaflet of the bilayer, with corresponding loss of procoagulant activity. This process appears to be ATP- and sulfhydryl-dependent, and

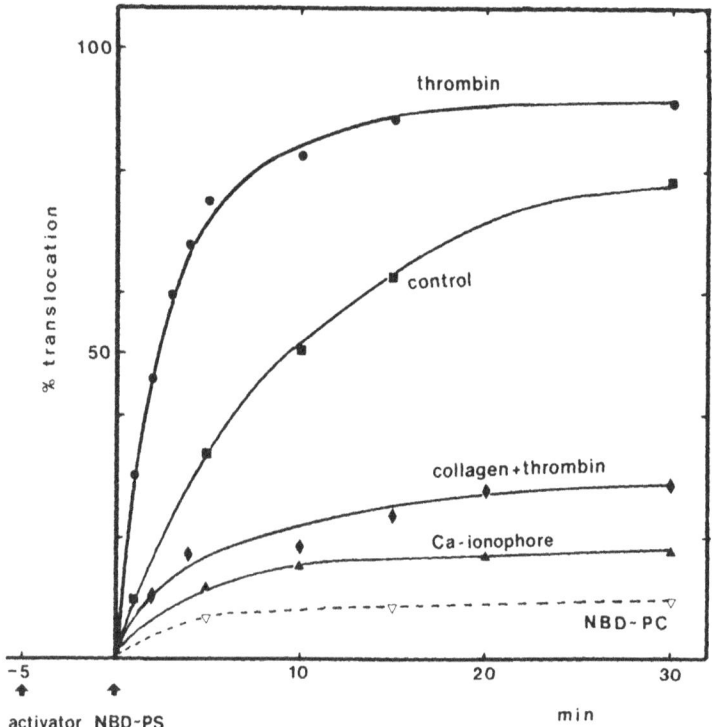

Fig. 4 Aminophospholipid translocase activity in platelets after various activation procedures. Inward transport of fluorescent (7-nitrobenz-2-oxa-1,3-diazol-4-yl) (NBD)-labeled lipid analogues was monitored using the 'back exchange' procedure developed for red blood cells (Connor and Schroit, 1988). ■ , unstimulated platelets; ● , thrombin-activated platelets; ◆ , platelets activated by collagen plus thrombin; ▲ , ionophore-activated platelets (adapted from Tilly et al., 1990).

occurs only in the remnant cells, not in the platelet-derived vesicles, possibly because of a lack of ATP. Very similar phenomena can be observed with red blood cells (Connor and Schroit, 1988; Comfurius et al., 1990); moreover, even erythrocyte ghosts, resealed in the presence of Mg^{2+}-ATP, still retain full aminophospholipid translocase activity (Seigneuret and Devaux, 1984; Connor et al., 1990), indicating that the transporter is restricted to the plasma membrane (including the membrane skeleton). Although evidence is accumulating that the aminophospholipid translocase plays an important role in the maintenance of membrane lipid asymmetry, its mere inhibition by sulfhydryl-oxidizing agents or ATP depletion in red cells and platelets in the absence of extracellular Ca^{2+} does not lead to an appreciable loss of lipid asymmetry (Schroit and Zwaal, 1991, and references therein).

Inward transport of aminophospholipids is not inhibited to the same extent by various platelet stimulators (fig. 4). Whereas Ca^{2+}-ionophore produces virtual complete inhibition, the combined action of collagen plus thrombin results in strongly impaired, but still measurable residual translocase activity (Tilly et al., 1990). In contrast, the complement pore proteins C5b-9 do not produce appreciable loss of translocase activity (unpublished). As mentioned above (cf. fig 3), platelet activation by the complement membrane attack complex causes extensive shedding of procoagulant microvesicles, but the respective remnant cells remain virtually non-coagulant (Sims et al., 1989). Apparently, membrane fusion events only lead to scrambling of membrane phospholipids when aminophospholipid translocase is inactive, which prevents aminophospholipids from being pumped back to the inner leaflet.

Interestingly, stimulation of platelets by thrombin is associated with a 4-5 fold increase in translocase activity compared to the quiescent platelet (fig 4) (Tilly et al., 1990). Whether this increased activity is due to an intracellular regulatory process or whether it merely reflects additional copies of aminophospholipid translocase, from the incorporation of granula membranes in the plasma membrane upon secretion remains to be elucidated. As suggested by Tilly et al. (1990), aminophospholipid translocase might play an important role in the maintenance of phospholipid asymmetry in secretory cells, where the secretory event would otherwise lead to scrambling of membrane phospholipids due to disturbance of the normal bilayer structure at the site of fusion of granula membrane with the plasma membrane.

Another example in which shedding of microvesicles does not lead to phosphatidylserine exposure on the remnant cells is observed with sickled red blood cells (Franck et al., 1985). Inward transport of phosphatidylserine in these cells is comparable to normal red blood cells and becomes slightly reduced under anoxic conditions (Blumenfeld et al., 1991) under which these cells assume their sickled shape with the formation of relatively large membrane protrusions. Repeated sickling of reversibly sickled cells by cycles of deoxygenation and reoxygenation causes membrane microvesicles to pinch off from the protrusions with concomitant increase in procoagulant activity. This increase in procoagulant activity is, however, virtually exclusively present on the microvesicles, presumably because scrambling of phosphatidylserine in remnant cells is instantaneously corrected by a sufficiently active translocase. Microvesicles released from sickle cells have been identified in the plasma of patients with sickle cell disease. Of special importance, the thrombotic episodes accompanying sickle cell crisis may be secondary to the procoagulant activity of these microvesicles.

CONCLUDING REMARKS

The platelet procoagulant response is the result of a process in which membrane phospholipid asymmetry is lost, resulting in a gradual increase in surface exposed phosphatidylserine which enables an efficient assembly of tenase and prothrombinase enzyme complexes. The significance of this process for hemostasis and thrombosis seems obvious, if only because of a patient with a partial impairment of this phenomenon has a moderately severe bleeding disorder (Weiss et al., 1979). Although surface exposed phosphatidylserine provides binding sites for tenase and prothrombinase, it also promotes the assembly of a complex composed of activated protein C and protein S, causing the proteolytic inactivation of factors Va and VIIIa (Esmon, 1992). The ability of platelets and platelet-derived microvesicles to support protein C-catalyzed inactivation of the prothrombinase cofactor Va appears to be comparable to their capacity to stimulate prothrombinase activity (Tans et al., 1991).

Exposure of procoagulant surfaces is not restricted to platelets or (pathological) red blood cells. Evidence has been presented that the surface of endothelial cells, which is normally thrombo-resistant, can exhibit procoagulant properties associated with microvesicle formation, when challenged with agonists which increase cytosolic Ca^{2+}, including ionophore, histamine, thrombin and complement proteins C5b-9 (Hamilton et al., 1990). In addition, increased prothrombinase activity has been observed in virus-infected endothelial cells (Van Dam-Mieras et al., 1987; Visser et al., 1988). Finally, also tumor cells have been demonstrated to expose procoagulant phosphatidylserine (Connor et al., 1989) and to release procoagulant microvesicles (Van de Water et al., 1985). The physiological relevance of release of microvesicles, if any, in the environment remains to be elucidated, the more so as this seems to be counterproductive in localizing the hemostatic process to the site of injury. It appears that surface exposure of phosphatidylserine in blood cells significantly affects their recognition by the reticulo-endothelial system (Schroit et al., 1985), which could be instrumental in removal of microvesicles and circulating activated platelets.

Many questions about the cellular mechanisms that regulate and control the procoagulant response remain to be answered. The tight coupling observed between microvesicle formation and development of procoagulant activity has led to the hypothesis that lipid scrambling is caused by the shedding process. However, a reversed cause and effect relationship cannot be excluded, i.e. blebbing and shedding are the result of a change in phospholipid asymmetry. Alternatively, one should consider the possibility that both processes are in fact epiphenomena, governed by a rise in cytosolic Ca^{2+}. Although cytoskeletal degradation by activated calpain doesn't seem to be required to evoke a procoagulant response, it remains to be investigated whether calcium-induced uncoupling of membrane-cytoskeletal interactions forms an essential part of the mechanism. Finally, future research should also focus on the role of the aminophospholipid translocase in the procoagulant response: the identity of this putative protein and the cellular processes that regulate its function have to be elucidated. Answers to these questions may contribute to a better understanding of the procoagulant response and may provide a means to manipulate this process in vivo which could be beneficial to regulate normal hemostasis.

ACKNOWLEDGEMENTS

A significant part of this review is based on previously published studies that were supported by NATO travel grants 0746/88 and 910436 and Program grant 900-526-093 from the Dutch Foundation for Medical and Health Research (Medigon). We thank Mr. M.C.A. Stuart (Electron-microscopy Unit, University of Limburg) for kindly providing the electron-micrograph of platelet-derived microvesicles.

REFERENCES

Ahmad, S.S., Rawala-Sheikh, R., Ashby, B. and Walsh, P.N., 1989, Platelet receptor-mediated factor X activation by factor IXa. High-affinity IXa receptors induced by factor VIII are deficient on platelets in Scott syndrome, J. Clin. Invest. 84:824.

Allan, D. and Michell, R.H., 1975, Accumulation of 1,2 diacylglycerol in the plasma membrane may lead to echinocyte transformation of erythrocytes, Nature 258:348.

Baldwin, J.M., O'Reilly, R., Whitney, M. and Lucy, J.A., 1990, Surface exposure of phosphatidylserine is associated with the swelling and osmotically-induced fusion of human erythrocytes in the presence of Ca^{2+}, Biochim. Biophys. Acta 1028:14.

Beaudoin, A.R. and Grondin, G., 1991, Shedding of vesicular material from the cell surface of eukaryotic cells: different cellular phenomena, Biochim. Biophys. Acta 1071:203.

Bevers, E.M., Comfurius, P. and Zwaal, R.F.A., 1983, Changes in membrane phospholipid distribution during platelet activation, Biochim. Biophys. Acta 736:57.

Bevers, E.M., Comfurius, P., Nieuwenhuis, H.K., Levy-Toledano, S., Enouf, J., Belluci, S., Caen, J.P. and Zwaal, R.F.A., 1986, Platelet prothrombin converting activity in hereditary disorders of platelet function, Br. J. Haematol. 63:335.

Bevers, E.M., Tilly, R.H.J., Senden, J.M.G., Comfurius, P. and Zwaal, R.F.A., 1989, Exposure of endogenous phosphatidylserine at the outer surface of stimulated platelets is reversed by restoration of aminophospholipid translocase activity, *Biochemistry* 28:2382.

Bevers, E.M., Wiedmer, T., Comfurius, P., Shattil, S.J., Weiss, H.J., Zwaal, R.F.A. and Sims, P.J., 1992, Defective Ca^{2+}-induced microvesiculation and deficient expression of procoagulant activity in erythrocytes from a patient with a bleeding disorder: a study of the red blood cells of Scott syndrome, *Blood* 79:380.

Bitbol, M., Fellmann, P., Zachowski, A. and Devaux, P.F., 1987, Ion regulation of phosphatidylserine and phosphatidylethanolamine outside-inside translocation in human erythrocytes, *Biochim. Biophys Acta* 904:268.

Blumenfeld, N., Zachowski, A., Galacteros, F., Beuzard, Y. and Devaux, P.F., 1991, Transmembrane mobility of phospholipids in sickle erythrocytes: the effect of deoxygenation on diffusion and asymmetry, *Blood* 77:849.

Chandra, R., Joshi, P.C., Bajpai, V.K. and Gupta, C.H., 1987, Membrane phospholipid organization in calcium-loaded human erythrocytes, *Biochim. Biophys. Acta* 902:253.

Chap, H.J., Zwaal, R.F.A. and van Deenen, L.L.M., 1977, Action of highly purified phospholipases on blood platelets. Evidence for an asymmetric distribution of phospholipids in the surface membrane. *Biochim. Biophys. Acta* 467:146.

Comfurius, P., Bevers, E.M. and Zwaal, R.F.A., 1985, The involvement of cytoskeleton in the regulation of transbilayer movement of phospholipids in human blood platelets, *Biochim. Biophys. Acta* 815:143.

Comfurius, P., Senden, J.M.G., Tilly, R.H.J., Schroit, A.J., Bevers, E.M. and Zwaal, R.F.A,, 1990, Loss of membrane phospholipid asymmetry in platelets and red cells may be associated with calcium induced shedding of plasma membrane and inhibition of aminophospholipid translocase, *Biochim. Biophys. Acta* 1026:153.

Connor, J. and Schroit, A.J., 1988, Transbilayer movement of phosphatidylserine in erythrocytes; inhibition of transport and preferential labeling of a 31 kD protein by sulfhydryl reactive reagents, *Biochemistry* 27:848.

Connor, J., Bucana, C., Fidler, I.J. and Schroit, A.J., 1989, Differentiation-dependent expression of phosphatidylserine in mammalian plasma membranes: quantitative assessment of outer leaflet lipid by prothrombinase complex formation, *Proc. Natl. Acad. Sci. U.S.A.* 86:3184.

Connor, J., Gillum, K. and Schroit, A.J., 1990, Maintenance of lipid asymmetry in red blood cells and ghosts: effect of divalent cations and serum albumin on the transbilayer distribution of phosphatidylserine, *Biochim. Biophys. Acta* 1025:82.

Crawford, N., 1972, The presence of contractile proteins in platelet microparticles isolated from human and animal platelet-free plasma, *Br. J. Haematol.* 21:53.

Esmon, C.T., 1992, The protein-C anticoagulant pathway, *Arterioscleros. Thromb.* 12, 135.

Fox, J.E.B., Austin, C.D., Boyles, J.K. and Steffen, P.K., 1990a, Role of membrane skeleton in preventing the shedding of procoagulant-rich microvesicles from the plateletplasma membrane, *J. Cell Biol.* 111:483.

Fox, J.E.B., Reynolds, C.C. and Austin, C.D., 1990b, The role of calpain in stimulus-response coupling: evidence that calpain mediates agonist-induced expression of procoagulant activity in platelets, *Blood* 76:2510.

Fox, J.E.B., Austin, C.D., Reynolds, C.C. and Steffen, P.K., 1991, Evidence that agonist-induced activation of calpain causes shedding of procoagulant-containing microvesicles from the membrane of aggregating platelets, *J. Biol. Chem.* 266:13289.

Franck, P.F.H., Bevers, E.M., Lubin, B.H., Comfurius, P., Chiu, D.T.-Y., Op den Kamp, J.A.F., Zwaal, R.F.A., van Deenen, L.L.M. and Roelofsen, B., 1985, Uncoupling of the membrane skeleton from the lipid bilayer: the cause of accelerated phospholipid flip-flop leading to enhanced procoagulant activity of sickle cells, *J. Clin. Invest.* 75:183.

Frederik, P.M., Stuart, M.C.A., Bomans, P.H.H., Busing, W.M., Burger, K.N.J. and Verkley, A.J., 1991, Perspective and limitations of cryo-electron microscopy, *J. Microscopy* 161:253.

Gilbert, G.E., Sims, P.J., Wiedmer, T., Furie, B., Furie, B.C. and Shattil, S.J., 1991, Platelet-derived microparticles express high-affinity receptors for factor VIII, *J. Biol. Chem.* 266:17261.

Haest, C.W.M., 1982, Interactions between membrane skeleton proteins and the intrinsic domain of the erythrocyte membrane, *Biochim. Biophys. Acta* 694:331.

Hamilton, K.K., Hattori, R., Esmon, C.T. and Sims, P.J., 1990, Complement proteins C5b-9 induce vesiculation of the endothelial plasma membrane and expose catalytic surface for the assembly of the prothrombinase complex, *J. Biol. Chem.* 265:3809.

Kane, W.H. and Majerus, P.W., 1982, The interaction of human coagulation factor Va with platelets, *J. Biol. Chem.* 257:3963.

Mann, K.G., Nesheim, M.E., Church, W.R., Haley, P. and Krishnaswamy, S., 1990, Surface dependent reactions of the vitamin K-dependent enzyme complexes, *Blood* 76:1.

Morrot, G., Zachowski, A. and Devaux, P.F., 1990, Partial purification and characterization of the human erythrocyte Mg^{2+}-ATPase, *FEBS lett.* 266:29.

Op den Kamp, J.A.F., 1979, Lipid asymmetry in membranes, *Ann. Rev. Biochem.* 48:47.

Rosing, J., Tans, G., Govers-Riemslag, J.W.P., Zwaal, R.F.A. and Hemker, H.C., 1980, The role of phospholipids and factor Va in the prothrombinase complex, *J. Biol. Chem.* 255:274.

Rosing, J., Bevers, E.M., Comfurius, P., Hemker, H.C., van Dieijen, G., Weiss, H.J. and Zwaal, R.F.A., 1985a, Impaired factor X- and prothrombin activation associated with decreased phospholipid exposure in platelets from a patient with a bleeding disorder, *Blood* 65:1557.

Rosing, J., Van Rijn, J.L.M.L., Bevers, E.M., Van Dieijen, G., Comfurius, P. and Zwaal, R.F.A., 1985b, The role of activated human platelets in prothrombin and factor X activation, *Blood* 65:319.

Rosing, J., Speijer, H. and Zwaal, R.F.A., 1988, Prothrombin activation on phospholipid membranes with positive electrostatic potential, *Biochemistry* 27:8.

Sandberg, H., Andersson, L.-O. and Höglund, S., 1982, Isolation and characterization of lipid-protein particles containing platelet factor 3 released from human platelets, *Biochem. J.* 203:303.

Sandberg, H., Bode, A.P., Dombrose, F.A., Hoechli, M. and Lentz, B.R., 1985, Expression of procoagulant activity in human platelets: release of membranous vesicles providing platelet factor 1 and platelet factor 3, *Thromb. Res.* 39:63.

Schick, P.K., Kurica, K.B. and Chacko, G.K., 1976, Location of phosphatidylethanolamine and phosphatidylserine in the human platelet plasma membrane, *J. Clin. Invest.* 57:1221.

Schroit, A.J. and Zwaal, R.F.A., 1991, Transbilayer movement of phospholipids in red cell and platelet membranes, *Biochim. Biophys. Acta* 1071:313.

Schroit, A.J., Bloy, C., Connor, J. and Cartrou, J.P., 1990, Involvement of Rh blood group polypeptides in the maintenance of aminophospholipid asymmetry, *Biochemistry* 29:10303.

Schroit, A.J., Madsen, J.W. and Tanaka, Y., 1985, *In vivo* recognition and clearance of red blood cells containing phosphatidylserine in their plasma membranes, *J. Biol. Chem.* 260:5131.

Seigneuret, M. and Devaux, P.F., 1984, ATP-dependent asymmetric distribution of spin-labeled phospholipids in the erythrocyte membrane: relation to shape changes, *Proc. Natl. Acad. Sci. U.S.A.* 81:3751.

Sims, P.J., Faioni, E.M., Wiedmer, T. and Shattil, S.J., 1988, Complement proteins C5b-9 cause release of membrane vesicles from the platelet surface that are enriched in the membrane receptor for coagulation factor Va and express prothrombinase activity, *J. Biol. Chem.* 263:18205.

Sims, P.J., Wiedmer, T., Esmon, C.T., Weiss, H.J. and Shattil, S.J., 1989, Assembly of the platelet prothrombinase complex is linked to vesiculation of the platelet plasma membrane, *J. Biol. Chem.* 264:17049.

Tans, G., Rosing, J., Thomassen, M.C.L.G.D., Heeb, M.J., Zwaal, R.F.A. and Griffin, J.H., 1991, Comparison of anticoagulant and procoagulant properties of stimulated platelets and platelet-derived microparticles, *Blood* 77:2641.

Thiagarajan, P. and Tait, J.F., 1991, Collagen-induced exposure of anionic phospholipid in platelets and platelet-derived microparticles, *J. Biol. Chem.* 266: 24302.

Tilly, R.H.J., Senden, J.M.G., Comfurius, P., Bevers, E.M. and Zwaal, R.F.A., 1990, Increased aminophospholipid translocase activity in platelets during secretion, *Biochim. Biophys. Acta* 1029:188.

Tracy, P.B., Peterson, J.M., Nesheim, M.E., McDuffie, F.C. and Mann, K.G., 1979, Interaction of coagulation factor V and factor Va with platelets, *J. Biol. Chem.* 254:10345.

Van Dam-Mieras, M.C.E., Bruggeman, C.A., Muller, A.D., Debie, W.H.M. and Zwaal, R.F.A., 1987, Induction of endothelial cell procoagulant activity by cytomegalovirus infection., *Thromb. Res.* 47:69.

Van de Water, L., Tracy, P.B., Aronson, D., Mann, K.G. and Dvorak, H.F., 1985, Tumor cell generation of thrombin via functional prothrombinase assembly, *Cancer Res.* 45:5521.

Van Dieijen, G., Tans, G., Rosing, J. and Hemker, H.C., 1981, The role of phospholipids and factor VIIIa in the activation of bovine factor X, *J. Biol. Chem.* 256:3433.

Verhallen, P.F.J., Bevers, E.M., Comfurius, P. and Zwaal, R.F.A., 1987, Correlation between calpain-mediated cytoskeletal degradation and expression of procoagulant activity. A role for the platelet membrane skeleton in the regulation of membrane lipid asymmetry? *Biochim. Biophys. Acta* 903:206.

Visser, M.R., Tracy, P.B., Vercellotti, G.M., Goodman, J.L., White, J.G. and Jacob, H.S., 1988, Enhanced thrombin generation and platelet binding on herpes simplex virus-infected endothelium, *Proc. Natl. Acad. Sci. U.S.A.* 85:8227.

Weiss, H.J., Vi, W.J., Lages, B.A. and Rogers, J. (1979), Isolated deficiency of platelet procoagulant activity, *Am. J. Med.* 67:206.

Wiedmer, T., Shattil, S.J., Cunningham, M. and Sims, P.J., 1990, Role of calcium and calpain in complement-induced vesiculation of the platelet plasma membrane and in the exposure of the platelet factor Va receptor, *Biochemistry* 29:623.

Wolf, P., 1967, The nature and significance of platelet products in human plasma, *Br. J. Haematol.* 13:269.

Zwaal, R.F.A., 1978, Membrane and lipid involvement in blood coagulation, *Biochim. Biophys. Acta* 515:163.

HISTAMINE AS AN INTRACELLULAR MESSENGER IN

HUMAN PLATELETS

J.M. Gerrard, S.P. Saxena, and A. McNicol

Manitoba Institute of Cell Biology
Department of Pediatrics
University of Manitoba
Winnipeg, Manitoba
R3E OV9
Canada

INTRODUCTION

The fusion of cells from mouse and man to produce functional hybrids, first shown by Henry Harris in 1965 (Harris and Watkins, 1965) and since duplicated using many different cell types, has emphasized the remarkable consistency and similarity in the internal constituents of cells. At the same time, an incredible diversity in the extracellular coat and in the responses of cells to extracellular messengers gives specific properties and behaviours to individual cell types. In this context, while there are hundreds of extracellular messenger compounds, to date only a few intracellular messengers have been unequivocally identified. These include cyclic AMP, cyclic GMP, calcium, inositol trisphosphate, diacylglycerol, and polyamines.

During the last several years we have provided evidence, in platelets, that histamine is an additional intracellular messenger (Saxena et al., 1989a). Although these studies were the first to demonstrate an intracellular messenger role for histamine, Kahlson (1960) and Brandes et al. (1987) had previously postulated such a role in relation to cell growth. Indeed the investigations in platelets were derived from and were facilitated by the synthesis of N,N-diethyl-2-[4-(phenylmethyl)phenoxy]ethanamine HCl (DPPE) by Brandes and Hermonat (1984) and the discovery that it is an antagonist for an intracellular histamine receptor (H_{IC}) which is distinct from traditional H_1, H_2 and H_3 receptors (Saxena et al., 1989a).

The present review will focus on the findings in platelets, but will mention evidence in relation to neutrophils and to growth which suggests that histamine may have a wider role as an intracellular messenger.

Before a substance can be established as an intracellular messenger, five essential criteria must be met. The criteria, defined below, are refined and extended from those outlined by Robinson and Sutherland in their initial work with cyclic AMP (Robinson et al., 1971) and have been discussed, in part, previously (Brandes et al., 1990).

Essential criteria for an intracellular messenger are:

1. Physiologic agonist(s) must stimulate the formation of the substance. This formation must occur at concentrations of the agonist which are at least as small as the lowest levels capable of producing a physiologic response. Further, the formation of the proposed messenger should precede, or at least not follow, the physiologic response.

2. Inhibitors which prevent synthesis of the substance must block the physiologic response.

3. The physiologic effect should be mimicked by exogenous addition of physiologically reasonable concentrations of the substance. In the case of a chemical which does not readily cross cell membranes, this may require the use of permeabilized cells, liposomes or a permeable analog.

4. Specific receptor antagonist(s) of the proposed messenger should block the physiologic response at concentrations similar to their ability to inhibit its binding.

5. The formation and the site of action must be shown to be intracellular.

Additional supporting criteria are:

6. The mechanism of the formation of the substance needs to be demonstrated in a broken cell preparation.

7. A mechanism for "turning off" the action of the putative intracellular messenger needs to be provided.

8. There must be a mechanism of action of the putative intracellular messenger.

HISTAMINE IN PLATELETS

Evidence for relation to each of these criteria in relation to a role for histamine acting as an intracellular messenger in human platelets will be provided in the order presented above.

1.   Agonist stimulation of histamine formation: Table 1 shows that several agonists stimulate the formation of histamine in human platelets. Collagen and phorbol myristate acetate (PMA) are the most potent stimulants of histamine production (Saxena et al., 1989a, 1990). A concentration of 1.8 μg/ml collagen causes maximum histamine production which precedes full aggregation (Saxena et al., 1990). The lowest dose of collagen (0.4 μg/ml) to induce aggregation was associated with a lesser, but still significant increase in platelet histamine. Studies with PMA show that the production of histamine by this agonist coincides with, rather than precedes, the aggregation response (Saxena et al., 1989a).

Agonists such as thrombin, platelet activating factor, and the thromboxane analogue EP171 (Saxena et al., 1989b; McNicol et al., 1991) cause the formation of intermediate amounts of histamine. In contrast, the calcium ionophore A23187 (Saxena et al., 1991b),

ADP (McNicol et al., 1991), and low concentrations of thapsigargin can also aggregate platelets without elevating intracellular histamine. These results have been interpreted to indicate that activation of platelets by agents which stimulate primarily protein kinase C are associated with increased histamine, whereas the stimulation of platelets by agents which act primarily to elevate intracellular calcium are not associated with an increase in intracellular histamine.

2.   Effect of inhibiting histamine formation: Histamine is formed from the amino acid L-histidine by the enzyme histidine decarboxylase (HDC). This enzyme can be inhibited reversibly by α-methylhistidine or irreversibly by α-fluromethylhistidine, resulting in antagonism of both histamine formation and platelet aggregation in response to PMA or collagen (Saxena et al., 1989a; Saxena et al., 1990).

Table 1   Production of Histamine by Human Platelets in Response to Agonists

	Histamine (pmoles per 10^9 platelets)
Control platelets	12.2 ± 0.9
Collagen (1.8 μg/ml)	34.9 ± 1.5
Phorbol myristate acetate (30 nM)	36.6 ± 0.7
Thrombin (1 U/ml)	28.6 ± 0.9
Platelet activating factor (1 μM)	28.9 ± 0.8
EP171 (thromboxane analog) (10 nM)	16.1 ± 1.7
ADP (10 μM)	10.1 ± 1.8
A23187 (0.5 μM)	7.6 ± 1.7

Data are from Saxena et al., 1989a, 1989b, 1991a, and McNicol et al., 1991.

3.   Effects of exogenous histamine: Curiously, unlike IP_3 (Israels et al., 1985, Authi et al., 1986), histamine alone is not sufficient to initiate the platelet aggregatory response, suggesting that it must interact with another intermediary. Consistent with this concept, the addition of histamine to permeabilized, but not intact, platelets reverses inhibition by HDC inhibitors of agonist-induced aggregation (Saxena et al., 1989a; Saxena et al., 1990). The precise nature of the complimentary factor is unknown as attempts to demonstrate synergistic effects for histamine with thromboxane A_2 analogues or calcium have, to date, been unsuccessful. Interaction of histamine with a phosphorylated protein is a possibility. Of interest, the dose response for histamine in promoting platelet aggregation is bell shaped with decreasing potency at concentrations below 0.1 μM and above 10 μM. The demonstration of a lesser effect above 10 μM correlates with previous reports of an inhibitory action at higher concentrations of histamine added extracellularly to platelets (Klysner et al., 1980; Houston et al., 1983).

4.   Effect of histamine antagonists: DPPE and a number of traditional antihistaminergic agents have been tested for their ability to both antagonize [^{3}H]-histamine binding in permeabilized human platelets and to block platelet aggregation in response to PMA (Saxena et al., 1989a). Although DPPE was initially found to be the most potent inhibitor of histamine binding, other H_1 antagonists including astemizole, promethazine, chlorcyclizine, and cyproheptidine have recently been found to be more potent. The orthoisomer of DPPE, phenyltoloxamine, a traditional H_1 antagonist, is less potent than DPPE. Pyrilamine (H_1), ranitidine and cimetidine (H_2 antagonists) are only active at extremely high concentrations. Although one report suggests that pyrilamine may, under certain conditions, inhibit human platelet responses at lower concentrations (Murayama et al., 1990), we have been unable to reproduce these findings. The results suggest that histamine's action to promote platelet aggregation occurs at the H_{IC} receptor identified by DPPE and that this site differs from traditional H_1, H_2 and H_3 receptors. The order of potency is similar for inhibition of platelet aggregation and histamine binding. The concurrence between these two processes is reasonable but not precise (Saxena et al., 1992). The results may be due to differences between the conditions used to study aggregation and [^{3}H]-histamine binding. For example, the longer incubation used in the binding studies could affect the inhibitory ability of DPPE. Platelets may sequester or metabolize DPPE over time since DPPE added 15 min. before an agonist is less effective at inhibiting aggregation than when it is added only 30 sec. before.

5.   Formed histamine is intracellular: Since it does not cross the platelet membrane readily, the aggregation promoting effect of histamine can be demonstrated only in saponin-permeabilized platelets (Saxena et al., 1989a, 1990), strongly suggesting that its site of action is intracellular. In addition, the platelet histamine synthesized in response to PMA and thrombin is not present primarily in the dense granules, since under conditions which released up to 90% of platelet serotonin (an amine present in these granules) approximately 85% of the histamine was retained by the platelet (Saxena et al., 1989b). When non-permeabilized platelets were stimulated with PMA in the presence of DPPE to block histamine binding, 89% of histamine formed was associated with the pellet, whereas under identical conditions permeabilized platelets liberated 75% of their histamine (Saxena et al., 1989b). This provides further evidence that the histamine formed following PMA stimulation is intracellular and can be released only when the platelets are permeabilized and the histamine binding to H_{IC} prevented.

6.   Studies of histamine formation in cell homogenates: Preliminary studies to demonstrate histidine decarboxylase (HDC) activity in human platelet homogenates are in progress. Others have demonstrated the presence of HDC in rabbit platelets (Schayer and Kobayashi, 1956; Tuomisto, 1970), but these differ from human platelets in that they contain significant amounts of histamine in their dense granules. Thus, the presence of HDC and the mechanism of its regulation in human platelets remains to be demonstrated. Studies in other cells have suggested that HDC may be regulated by calcium (Huszti and Magyar, 1987); phosphorylation-mediated regulation of HDC through the action of protein kinase C is also an interesting possibility. Indeed, studies in platelets provide strong evidence that histamine synthesis is regulated by a protein kinase C mediated phosphorylation (Saxena et al., 1991b).

Studies of Gill and co-workers (1987) showing that agonists can stimulate platelet uptake of histamine suggest than an influx of histamine from the extracellular milieu might be an alternate mode of regulation of platelet cytoplasmic histamine levels in a manner analogous to the regulation of intracellular calcium.

7.   A mechanism for limiting or "turning off" the histamine: Time course studies of histamine production in response to the platelet agonist PMA suggest that the peak histamine level occurs at 2 min. followed by a subsequent decrease. Since histamine is metabolized either by diamine oxidase or histamine-methyltransferase, studies are required to evaluate the presence of these enzymes and the effect of their inhibitors. Studies with the calcium ionophore A23187 suggest that a rise in cytoplasmic calcium is associated with decreased histamine synthesis and/or enhanced histamine metabolism (Saxena et al., 1991a,b). Co-addition of A23187 with PMA significantly decreased the PMA-induced rise in platelet histamine. Aminoguanidine, an inhibitor of diamine oxidase can, under some conditions, increase the amount of platelet histamine measured in response to ADP or to the combination of A23187 + PMA. The results provide some evidence that histamine metabolism may be augmented by a rise in intracellular calcium levels in platelets.

8.   Studies of an intracellular mechanism of histamine action: While our results support a role for histamine as an intracellular messenger mediating platelet aggregation, its mechanism of action is not yet understood. Some evidence points to more than one site of action, however. Ultrastructural studies using DPPE show that this agent can inhibit granule membrane fusion and pseudopod function as well as the cell-to-cell attachment which occurs during aggregation (McNicol et al., 1989). Addition of histamine to permeabilized platelets can reverse all three effects of DPPE. The results imply a role for intracellular histamine in the fusion of granule membranes important for secretion (Gerrard et al., 1989), in the cytoskeletal assembly required for pseudopod formation (Carroll et al., 1982), and in modulating the availability of the fibrinogen receptor needed for aggregation (Mustard et al., 1987). An action of histamine at these three steps could result from its influence on the production or effects of other messengers. The H_{1C} antagonist DPPE does not inhibit phospholipase C activation or changes in cytosolic calcium in response to the thromboxane A_2 analog EP171 (McNicol et al., 1991), nor does it inhibit protein kinase C stimulation by PMA (Brandes et al., 1988). However, DPPE does significantly inhibit the collagen-induced release of arachidonic acid from platelet phospholipids, an action currently believed to be mediated by phospholipase A_2 (McNicol et al., 1991). This inhibition of arachidonic acid release by DPPE results in decreased thromboxane A_2 production which can be reversed by the addition of histamine to permeabilized platelets (Saxena et al., 1990). The effect of DPPE to inhibit thromboxane production can explain part but not all of the effects of this inhibitor on platelets. For example, in the presence of aspirin, histamine significantly reverses the effects of DPPE to block PMA-induced aggregation, suggesting that the effect of histamine is not solely to promote thromboxane synthesis. Recent evidence suggests a relationship between tyrosine phosphorylation and the effect of histamine on platelet function. The intracellular histamine antagonist DPPE has been found to inhibit protein tyrosine phosphorylation, but not protein kinase C mediated protein phosphorylation (Rendu et al., 1991). It is not yet clear whether this is a direct effect of DPPE on a tyrosine kinase or whether the effect of DPPE is mediated by its effect to inhibit histamine binding.

Limits to current evidence supporting a role for intracellular histamine in platelet functions

There are certain puzzling aspects of the findings to date. 1) Histamine added alone to permeabilized platelets does not stimulate aggregation. This implies that histamine must act in concert with some other factor which is produced when cells are stimulated by collagen or PMA. The nature of this factor remains unknown. 2) Addition of histamine to permeabilized platelets pretreated with α-fluoromethylhistidine or α-methylhistidine does not completely correct the inhibition of PMA and collagen-induced platelet aggregation. It is possible that the failure of full correction by histamine reflects the imperfect model of the permeabilized cell, or some effect of these agents to inhibit either histamine binding or some non-histamine mediated event, as well as inhibiting histamine synthesis. 3) There is not a perfect correlation between antagonism of histamine binding to the platelet H_{IC} receptor and inhibition of platelet aggregation. The difference may well reflect some differences in the conditions of these two assays, or other effects of these agents in addition to their ability to inhibit histamine binding, but point to an area where more work is needed. Already DPPE, at concentrations somewhat higher than those needed to inhibit histamine binding, has been shown to inhibit platelet aggregation by blocking a non-histamine related process (McNicol et al., 1991). 4) The fact that a rise in intracellular calcium, a biochemical change usually associated with platelet activation, appears to decrease platelet histamine is puzzling and points to a new level of complexity. It also emphasizes the fact that platelet activation by agonists which have a major effect to increase cytosolic calcium (A23187, ADP, thapsigargin) can occur independent of changes in platelet histamine. 5) Clear evidence for human platelet synthesis of histamine from histidine is still needed.

IMPLICATIONS IN HUMAN DISEASE

The discovery that histamine is an intracellular messenger mediating platelet aggregation may have relevance to our understanding of bleeding and/or thrombotic disorders. Various antihistamines have been previously reported to inhibit platelet aggregation (O'Brien, 1961; Mitchell and Sharp, 1964; Herrmann and Frank, 1966; Horton et al., 1983), and clinicians sometimes advise patients with bleeding diatheses to avoid antihistamines. Previous studies were unable to show any relationship between antiplatelet effects and H_1 or H_2 antagonist potency. Antagonism of histamine binding at the platelet H_{IC} site may explain the previous observations, and better knowledge of the relative potency of antihistaminics at this site may enable improved usage of antihistamines in patients with bleeding disorders.

A high level of histamine has been reported in platelets of patients with peripheral vascular disease (Gill et al., 1988), possibly reflecting in vivo platelet activation or indicating that some individuals may have an abnormality of histamine metabolism predisposing them to peripheral vascular disease. It is also of interest that the histidine decarboxylase inhibitor α-hydrazohistidine has been shown to decrease lipid and protein infiltration into the vessel wall (Owens and Hollins, 1979), a process which accompanies atherosclerosis.

HISTAMINE AS AN INTRACELLULAR MESSENGER IN OTHER CELL SYSTEMS

The finding that histamine is an intracellular messenger in platelets raises the question of a similar role in other cell and tissue activation events.

Studies in Neutrophils

We have recently found that DPPE and other antagonists of intracellular histamine can inhibit neutrophil superoxide production at concentrations very similar to those which inhibit platelet aggregation (Genaske EM, Cham B, Gerrard JM, unpublished). The results suggest that intracellular histamine may have a role in neutrophil as well as platelet function.

Histamine and Cell Growth

Evidence gathered over the last 30 years implicates histamine intracellularly as acting to promote the growth of some tissues (Kahlson, 1960; Brandes et al., 1987; Kahlson et al., 1960, 1962; Haartmann et al., 1966; Kahlson and Rosengren, 1972; Bini et al., 1972; Scheinmann et al., 1979; Chandra and Ganguly, 1987; Oh et al., 1988; Kameswaren and West, 1962; Shepherd and Woodcock, 1968; Grahn and Rosengren, 1970; Bartholeyns and Bouclier, 1984; Watanabe et al., 1981b, 1982; Nolte et al., 1987; Endo, 1983; Reid et al., 1963; Russell and Snyder, 1968; Ishikawa et al., 1970; Garcia-Caballero et al., 1988). The data in this regard are conflicting. High histamine production in the fetus was initially interpreted as indicating such a role in the growth of fetal tissues (Kahlson, 1960). More recent evidence using mast cell deficient mice suggests that much of the histamine is of mast cell origin (Watanabe et al., 1981a), and studies of growth in these mice and in mice treated with the histidine decarboxylase inhibitor α-fluoromethylhistidine suggest that the high level of histamine in the fetus may not be critical to growth (Watanabe et al., 1981a; Maeyama et al., 1985).

More compelling is evidence which relates the growth of some mitogen stimulated cells and tumor tissues to increased production of histamine (Brandes et al., 1987; Bini et al., 1972; Scheinmann et al., 1979; Chandra and Ganguly, 1987; Oh et al., 1988; Kameswaren and West, 1962; Shepherd and Woodcock, 1968; Grahn and Rosengren, 1970; Bartholeyns and Bouclier, 1984; Watanabe et al., 1981a, 1982; Nolte et al., 1987; Endo, 1983; Kahlson et al., 1962). Nevertheless, the proposal that histamine has a role in mitogen-stimulated or tumor cell growth has been controversial because not all tumor cells show increased histamine forming capacity, and even in those that do, no clear correlation of histamine decarboxylase activity and growth rate has been found (Shepherd and Woodcock, 1968; Moore et al., 1978). Recently, however, Brandes et al. have shown that the intracellular histamine antagonist DPPE can decrease the growth of MCF-7 cells and that such inhibition can be partially reversed by the histamine precursor histidine. This appears to argue strongly for a role for intracellular histamine to mediate the growth of this cell line (Brandes et al., 1988).

The identification of an H_{IC} binding site in rat brain microsomes (Brandes et al., 1988) suggests histamine may have an important intracellular messenger function in the brain. The nature of its role is uncertain, although regulation of prolactin secretion is one possibility. Histamine is implicated as mediating prolactin secretion under both basal and stress-induced conditions, yet studies with antagonists do not fit clearly into either H_1 or H_2 patterns (Nctti et al., 1987; Hough, 1988).

Relationship to polyamines: The polyamines, putrescine, spermine, and spermidine also have been implicated as intracellular messengers (Koenig et al., 1988). Thus, it is pertinent to investigate any interaction between histamine and polyamines in platelet aggregation. In our studies, histamine, but not the polyamines, reversed the inhibition of PMA-induced platelet aggregation by α-methylhistidine (Saxena et al., 1989a). At present there is no evidence to suggest that polyamines act as intracellular messengers in platelet aggregation. Indeed, like histamine (Brandes and Bogdanovic, 1986; Houston et al., 1983), very high concentrations of spermine (Israels et al., 1986; Joseph et al., 1987), have been reported to inhibit aggregation of intact platelets.

As opposed to platelets, histamine and polyamines may have similar roles in cell proliferation. Many cells and tumors produce significant amounts of histamine and polyamines during periods of growth (Russell and Snyder, 1968; Janne, 1967; Scalabrino and Lorenzini, 1991). It is curious, and possibly relevant, that some cells and tissues appear to preferentially synthesize large amounts of polyamines over histamine and vice versa (Reid et al., 1963; Ishikawa et al., 1970). Indeed, Kahlson proposed that growing cells might require either histamine or polyamines (Kahlson and Rosengren, 1972). Similarities between histidine decarboxylase and ornithine decarboxylase, the rate limiting enzyme controlling polyamine synthesis, provide a further parallel linking these two potential messenger systems.

Histamine and polyamines conceivably could share similar mechanisms of action. Consistent with this, both have been reported to be bound covalently to proteins (Cohen et al., 1985; Beninati et al., 1988). However, polyamines appear to play a role in modulating early calcium fluxes (Fan and Koenig, 1988; Koenig et al., 1989), whereas this does not seem to be the case for histamine in platelets. Thus, present evidence is most consistent with different, though perhaps complimentary, intracellular actions of histamine and polyamines.

SUMMARY

The results of investigations in platelets provide evidence for an intracellular messenger role for histamine. Studies of neutrophils and of cellular proliferation suggest that there may be a wider role for histamine as an intracellular messenger modulating activation processes in cells.

ACKNOWLEDGEMENTS

We thank L.J. Brandes, F.S. LaBella, A.B. Becker, and K.J. Simons for helpful discussions. Our investigations were supported by grant MA7396 from the Medical Research Council of Canada and by the National Cancer Institute of Canada. S.P.S. is the recipient of a scholarship from the Heart and Stroke Foundation of Canada and J.M.G. is the recipient of a Professorship from the Children's Hospital of Winnipeg Research Foundation.

REFERENCES

Authi, K.S. Evenden B.J., and Crawford, N., 1986, Metabolic and functional consequences of introducing inositol 1, 4, 5-trisphosphate into saponin-permeabilized human platelets, *Biochem. J.* 233:709.

Bartholeyns, J., and Bouclier, M., 1984, Involvement of histamine in growth of mouse and rat tumors: antitumoral properties of monofluoromethyl histidine, an enzyme-activated irreversible inhibitor of histidine decarboxylase, *Cancer Res.* 44:639.

Beninati, S., Piacentini, M., Cocuzzi, E.T., Autuori, F., and Folk, J.E., 1988, Covalent incorporation of polyamines as δ-glutamyl derivatives into CHO cell protein, *Biochim. Biophys. Acta.* 952:325.

Bini, A., Frontini, E., Nicolin, A., and Olivani, P., 1972, Histamine in DAB induced hepatoma and effects of tritoqualine and compound 48/80 on the incidence and development of the tumor, *Arch. Int. Pharmacodyn. There. Suppl.* 196:291.

Brandes, L.J., and Bogdanovic, R.P., 1986, New evidence that the antiestrogen binding site is a novel growth promoting histamine receptor (?H₃) which mediates the antiestrogenic and antiproliferative effects of tamoxifen, *Biochem. Biophys. Res. Commun.* 134:601.

Brandes, L.J., Bogdanovic, R.P., Cawker, M.C. and LaBella, F.S., 1987, Histamine and growth: interaction of antiestrogen binding site ligands with a novel histamine site that may be associated with calcium channels, *Cancer Res.* 47:4025.

Brandes, L.J. and Hermonat, M.W., 1984, A diphenylmethane derivative specific for the antiestrogen binding site in rat liver microsomes, *Biochem. Biophys. Res. Commun.* 123:724.

Brandes, L.J., Gerrard, J.M., Bogdanovic, P., Lint, D.W., Reid, R.E. and LaBella, F.S., 1988, Correlation of the antiproliferative action and diphenylmethane derivative antiestrogen binding site ligands with antagonism of histamine binding but not of protein kinase C mediated phosphorylation, *Cancer Res.* 48:3954.

Brandes, L.J., LaBella, F.S., Glavin, G.V., Paraskevas, F., Saxena, S.P., McNicol, A., and Gerrard, J.M., 1990, Histamine as an intracellular messenger, *Biochem. Pharm.* 40:1677.

Brandes, L.J., MacDonald, L.M. and Bogdanovich, R.P., 1985, Evidence that the antiestrogen binding site is a histamine or histamine-like receptor, *Biochem. Biophys. Res. Commun.* 126:905.

Carroll, R.C., Butler, R.G., Morris, P.A., and Gerrard, J.M., 1982, Separable assembly of platelet pseudopodal and contractile cytoskeletons, *Cell* 30:385.

Chandra, R., and Ganguly, A.K., 1987, Diamineoxidase activity and tissue histamine content of human skin, breast, and rectal carcinoma, *Cancer Let.* 34:207.

Cohen, I., Lim, C.T., Kahn, D.R., Glaser, T., Gerrard, J.M. and White, J.G., 1985, Disulfide-linked and transglutaminase-catalyzed protein assemblies in platelets, *Blood* 66:143.

Endo, Y., 1983, Induction of histidine decarboxylase in mouse tissues by mitogens in vivo, *Biochem. Pharm.* 32:3835.

Fan, C.C., and Koenig, H., 1988, The role of polyamines in beta-adrenergic stimulation of calcium influx and membrane transport in rat heart, *J. Mol. Cell Cardiol.* 20:789.

Garcia-Caballero, M., Neugebauer, E., Campos, R., Nunez de Castro, I., and Vara-Thorbeck, C., 1988, Increased histidine decarboxylase (HDC) activity in human colorectal cancer: results of a study of ten patients, *Agents and Actions* 23:357.

Gerrard, J.M., Beattie, L.L., Park, J-S., Israels, S.J., McNicol, A., Lint, D., Cragoe, E.J. Jr., 1989, A role for protein kinase C in the membrane fusion necessary for platelet granule secretion, *Blood* 74:2405.

Gill, D.S., Barradas, M.A., Fonseca, V.A., Gracey, L., and Dandona, P., 1988, Increased histamine content in leukocytes and platelets of patients with peripheral vascular disease, *Am. J. Clin. Pathol.* 89:622.

Gill, D.S., Barradas, M.A., Mikhailidis, D.P., and Dandora, P., 1987, Histamine uptake by human platelets, *Clin. Chim. Acta* 168:177.

Grahn, B., and Rosengren, E., 1970, Retardation of protein synthesis in rat tumors on inhibiting histamine formation, *Experientia* 26:125.

Haartmann, U Von., Kahlson, G., and Steinhardt, C., 1966, Histamine formation in germinating seeds, *Life Sci.* 5:1.

Harris, H., and Watkins, J.G., 1965, Hybrid cells derived from mouse and man: artificial heterokaryons of mammalian cells from different species, *Nature* 205:640.

Herrmann, R.G., and Frank, J.D., 1966, Effect of adenosine derivatives and antihistaminics on platelet aggregation, *Proc. Soc. Exp. Biol. Med.* 123:654.

Horton, M.A., Amos, R.J. and Jones, R.J., 1983, The effect of histamine H_2 receptor antagonists on platelet aggregation in man, *Scand. J. Haematol.* 31:15.

Hough, L.B., 1988, Cellular localization and possible functions for brain histamine: recent progress, *Prog. Neurobiol.* 30:469.

Houston, D.S., Gerrard, J.M., McCrea, J., Glover, S., and Butler, A.M., 1983, The influence of amines on various platelet responses, *Biochem. Biophys. Acta* 734:267.

Huszti, Z., and Magyar, K., 1987, Stimulation of hypothalamic histidine decarboxylase by calcium-calmodulin and protein kinase (cAMP-dependant) inhibitor, *Agents and Actions* 20:233.

Ishikawa, E., Toki, A., Moriyama, T., Matsioka, Y., Aikawa, T., and Suda, M., 1970, A study on the induction of histidine decarboxylase in tumor-bearing rat, *J. Biochem.* 68:347.

Israels, S.J., Gerrard, J.M. and Robinson, P., 1986, Differential effects of spermine of aggregation, inositol phosphate formation and protein phosphorylation in human platelets in response to thrombin, arachidonic acid and lysophosphatidic acid, *Biochim. Biophys. Acta* 883:247.

Israels, S.J., Robinson, P., Docherty, J.C. and Gerrard, J.M., 1985, Activation of permeabilized platelets by inositol-trisphosphate, *Thromb. Res.* 40:499.

Janne, J., 1967, Studies on the biosynthetic pathway of polyamines in rat liver, *Acta Physiol. Scand. Suppl.* 300:1.

Joseph, S., Krishnamarthi, S., and Kakkar, V.V., 1987, Effect of the polyamine spermine on agonist-induced human platelet activation - specific inhibition of "aggregation-independent" events induced by thrombin, but not by collagen, thromboxane mimetic, phorbol ester or calcium ionophore, *Thromb. Haemost.* 57:191.

Kahlson, G., 1960, A place for histamine in normal physiology, *Lancet* i:67.

Kahlson, G., Nilsson, K., Rosengren, E., and Zederfeldt, B., 1960, Wound healing as dependent on rate of histamine formation, *Lancet* ii:230.

Kahlson, G., and Rosengren, E., 1972, Histamine entering physiology, *Experientia* 28:993.

Kameswaren, L., and West, G.B., 1962, Studies concerned with the formation and inactivation of histamine, *Int. Arch. Allergy Appl. Immun.* 21:347.

Klysner, R., Geisler, A., Hansen, K.W., Skov, P.S., and Norn, S., 1980, Histamine H_2 receptor-mediated cyclic AMP formation in human platelets, *Acta Pharmacol. et Toxicol.* 47:1.

Koenig, H., Fan, C.C., Goldstone, A.D., Lu, C.Y., and Trout, J.J., 1989, Polyamines mediate androgenic stimulation of calcium fluxes and membrane transport in rat heart myocytes, *Circ. Res.* 64:415.

Koenig, H., and Goldstone, A.D., 1988, Polyamines are intracellular messengers in the beta-adrenergic regulation of Ca^{2+} fluxes, *Biochem. Biophys. Res. Commun.* 153:1179.

Kahlson, G., Rosengren, E., and Steinhardt, C., 1962, Activation of histidine carboxylase in tumor cells in mice, *Nature* 194:380.

Maeyama, K., Ohno, A., Taguchi, Y., Watanabe, T., and Wada, H., 1985, Effect of α-fluoromethylhistidine on increase in histidine decarboxylase activity of maternal mouse kidney observed during late pregnancy and evidence for its non-mast cell origin by using estrogen and W/W mice, *Japan J. Pharmacol.* 39:145.

McNicol, A., Saxena, S.P., Brandes, L.J. and Gerrard, J.M., 1989, A role for intracellular histamine in the ultrastructural changes induced in platelets by phorbol esters, *Arteriosclerosis* 9:684.

McNicol, A., Saxena, S.P., Becker, A.B., Brandes, L.J., and Gerrard, J.M., 1991, Further studies on the effects of the intracellular histamine antagonist DPPE on platelet function, *Platelets* 2:215.

Mitchell, J.R.A., and Sharp, A.A., 1964, Platelet clumping in vitro, *Br, J. Haematol.* 10:78.

Moore, T.C., Koppelmann, L.E., and Lemmi, C.A.E., 1978, Decreases in histamine forming enzyme activity of non-metastasizing fibrosarcomas in hamsters with progressive tumor growth, *Ann. Surg.* 188:175.

Murayama, T., Kajiyama, Y., Nomcera, Y., 1990, Histamine- stimulated and GTP-binding proteins - mediated phospholipase A2 activation in rabbit platelets, *J. Biol. Chem.* 265:4290.

Mustard, J.M., Kinlough-Rathbone, R.L., and Packham, M.A., 1987, Platelet activation: an overview, *Agents Actions [Suppl]* 21:23.

Netti, C., Guidobono, F., Sibilia, V., Villa, I., Cazzamalli, E., and Pecile, A., 1987, Involvement of brain histamine in basal and stress-induced release of prolactin in the rat, *Agents Actions* 20:236.

Nolte, H., Skovs, P.S. and Loft, H., 1987, Stimulation of histamine synthesis from tumor cells by concanavalin A and A23187, *Agents Actions* 20:291.

Norn, S., Skov, P.S., Hansen, K.W., Klysner, R., and Geisler, A., 1982, Possible regulatory role of histamine in human platelet function examined by thrombin-induced serotonin release, *Acta Pharmacol. et Toxicol.* 51:233.

O'Brien, J.R., 1961, The adhesiveness of platelets and its prevention, *J. Clin. Pathol. (London)* 14:140.

Oh, C., Suzuki, S., Nakishima, I., Yamashita, K., and Nakano, K., 1988, Histamine synthesis by non-mast cells through mitogen-dependent induction of histidine decarboxylase, *Immunology* 65:143.

Owens, G.A., and Hollins, T.M., 1979, Relationship between inhibition of aortic histamine formation, aortic albumin permeability, and atherosclerosis, *Atherosclerosis* 34:365.

Reid, J.D., Riley, J.F., and Shepherd, D.M., 1963, Histological and enzymatic changes in the livers of rats fed the hepatic carcinogen diethylnitrosamine, *Biochem. Pharm.* 12:1151.

Rendu, F., McNicol, A., Saleun, S., and Gerrard, J.M., 1991, Histamine formed in stimulated human platelets plays a role in tyrosine kinase activation, *Thrombos. Haemostas.* 65:729.

Robinson, G.A., Butcher, R.A., and Sutherland, E.W., 1971, "Cyclic AMP." Academic Press, New York.

Russell, O., and Snyder, S.H., 1968, Amine synthesis in rapidly growing tissues: ornithine decarboxylase activity in regenerating rat liver, chick embryo and various tumors, *Proc. Natl. Acad. Sci. U.S.A.* 60:1420.

Saxena, S.P., Brandes, L.J., Becker, A.B., Simons, K.J., LaBella, F.S. and Gerrard, J.M., 1989a, Histamine is an intracellular messenger mediating platelet aggregation, *Science* 243:1596.

Saxena, S.P., McNicol, A., Brandes, L.J., Becker, A.B. and Gerrard, J.M., 1989b, Histamine formed in stimulated human platelets is cytoplasmic, *Biochem. Biophys. Res. Commun.* 164:164.

Saxena, S.P., McNicol, A., Brandes, L.J., Becker, A.B. and Gerrard, J.M., 1990, A role for intracellular histamine in collagen-induced platelet aggregation, *Blood* 75:407.

Saxena, S.P., Robertson, C., Becker, A.B. and Gerrard, J.M., 1991a, Synthesis of intracellular histamine is associated with activation of protein kinase C but not with mobilization of Ca^{2+}, *Biochem. J.* 273:405.

Saxena, S.P., Becker, A.B., and Gerrard, J.M., 1991b, Synthesis of intracellular histamine in activated human platelets is down-regulated by a rise in cytosolic calcium, *Thrombos. Hemostas.* 65:668.

Saxena, S.P., Yhap, M. and Gerrard, J.M., 1992, Characterization of a novel intracellular histamine receptor (H_{IC}) in human platelets *FASEB J.* 6:3826.

Scalabrino, G., Lorenzini, E.C., 1991, Polyamines and mammalian hormones part II: paracrine signals and intracellular regulators, *Mol. Cell Endocrinol.* 77:37.

Schayer, R.W., and Kobayashi, Y., 1956, Histidine decarboxylase and histamine binding in rabbit platelets, *Proc. Soc. Exp. Biol. Med.* 92:653.

Scheinmann, P., Lebel, B., Lynch, N.R., Salomon, J.C., Paupe, J.R., and Burtin, C., 1979, Histamine levels in blood and other tissues of male and female C3H mice II: Mice carrying a 3-methyl-cholanthrene-induced tumor, *Agents Actions* 9:95.

Shepherd, D.M., and Woodcock, B.G., 1968, Histamine formation in normal, regenerating and malignant liver tissues of the rat, *Biochem. Pharm.* 17:23.

Tuomisto, J., 1970, Histamine synthesis and its inhibition by rabbit platelets in vitro, *Ann. Med. Exp. Fenn.* 48:164.

Watanabe, T., Kitamura, Y., Maeyama, K., Go, S., Yamatodani, A., and Wada, H., 1981, Absence of increase of histidine decarboxylase activity in mast dell deficient W/W mouse embryos before parturition, *Proc. Natl. Acad. Sci. U.S.A.* 78:4209.

Watanabe, T., Taguchi, Y., Sasaki, K., Tsuyama, K., and Kitamura, Y., 1981, Increase in histidine decarboxylase activity in mouse skin after application of the tumor promotor tetradecanoylphorbol acetate, *Biochem. Biophys. Res. Commun.* 100:427.

PLATELET ACTIVATION VIA BINDING OF MONOCLONAL ANTIBODIES TO THE Fcγ RECEPTOR II

J. Michael Wilkinson,[1] Edward J. Hornby[2] and Kalwant S. Authi[1]

[1]Department of Biochemistry and Cell Biology
Hunterian Institute, Royal College of Surgeons
35-43 Lincoln's Inn Fields
London, WC2A 3PN U.K.
[2]Department of Peripheral Pharmacology
Glaxo Group Research
Ware, Hertfordshire, U.K.

INTRODUCTION

During the last ten years a number of mouse monoclonal antibodies (mAbs) have been shown to cause platelet aggregation accompanied by a full activation response. Most commonly these mAbs recognise either the CD9 antigen or GPIIb-IIIa, two of the most abundant platelet surface glycoproteins, but more recently mAbs to other, less plentiful, surface proteins, including GPIV (CD36), CD69 and β_2 microglobulin, have been shown to give similar responses (for list see Table 1). As more data have accumulated a pattern of activation has emerged. In general, aggregation occurs following a concentration-dependant lag phase, at mAb concentrations of between 1 and 20 μg/ml. This aggregation is accompanied by an activation response which includes the metabolism of inositol phospholipids, the mobilisation of Ca^{2+}, thromboxane synthesis and the secretion of the contents of both dense and α-granules. Characteristically the mAbs involved belong to the IgG_1 subclass and are inhibited in their action by their F(ab')$_2$ or Fab fragments, thus implicating the Fc receptor in the activation process. The involvement of the FcγRII, the only Fcγ receptor expressed on platelets, was first demonstrated by Worthington et al. (1990) by blocking the activation caused by anti-CD9 mAbs with a mAb, IV.3, to the FcγRII. Since that time it has become clear that this is the major route of activation.

We describe here the activation caused by PM6/248, a mAb which recognises a conformation-dependant epitope on the GPIIb-IIIa molecule and which provokes a strong aggregation response in platelet-rich plasma (PRP) or gel-filtered platelets (GFP). We then compare this activation with that produced by mAbs to other platelet surface epitopes in order, as far as possible, to define a comprehensive view of platelet activation caused by the binding of mAbs.

Mechanisms of Platelet Activation and Control, Edited by
K.S. Authi *et al.*, Plenum Press, New York, 1993

Table 1. mAbs which have been shown to cause platelet activation.

mAb	Antigen	Subclass	Inhibition		Reference
			Fab	IV.3	
ALB6	CD9	IgG$_1$	Fab	ND	Boucheix et al. (1983)
B1.12	CD9	IgG$_1$	ND	ND	Thiagarajan et al. (1983)
FMC56	CD9	IgG$_1$	ND	ND	Gorman et al. (1985)
TP82	CD9	IgG$_1$	†F(ab')$_2$	ND	Higashihara et al. (1985)
AG-1	CD9	IgG$_1$	Fab	ND	Miller et al. (1986)
PMA2	CD9	IgG$_1$	ND	ND	Hato et al. (1988)
SYB-1	CD9	IgG$_1$	F(ab')$_2$	Yes	Carroll et al. (1990b)
p24/mAb7	CD9	IgG$_1$	ND	ND	Jennings et al. (1990)
50H.19	CD9	IgG$_{2a}$	F(ab')$_2$	Yes	Griffith et al. (1991)
6C9	GPIIb-IIIa	IgG$_1$	†F(ab')$_2$	ND	Modderman et al. (1988)
PL2-49	GPIIb-IIIa	IgG$_1$	F(ab')$_2$	ND	Morel et al. (1989)
P256	GPIIb-IIIa	IgG$_1$	†F(ab')$_2$	Yes	Bachelot et al. (1990)
PM6/248	GPIIb-IIIa	IgG$_1$	F(ab')$_2$	Yes	Hornby et al. (1991)
UR1	GPIIb-IIIa	IgG$_1$	F(ab')$_2$	Yes	Anderson et al. (1991)
Raj-1	GPIIb-IIIa	ND	F(ab')$_2$	Yes	Horsewood et al. (1991)
OKM5	GPIV, CD36	IgG$_1$	F(ab')$_2$	ND	Ockenhouse et al. (1989)
LeoA1	p67	IgG$_1$	ND	Yes	Scott et al. (1989)
F11	? CD69	IgG$_1$	Fab	ND	Kornecki et al. (1990)
Anti-Leu-3	CD69	IgG$_1$	Fab	ND	Testi et al. (1990)
B2.62.2	β_2m	IgG$_1$	F(ab')$_2$	Yes	Rubinstein et al. (1991)
JS-1	p155	ND	F(ab')$_2$	Yes	Horsewood et al. (1991)
Jun-1	?	ND	F(ab')$_2$	Yes	Horsewood et al. (1991)

ND - Not Determined; † - causes aggregation; IV.3 - anti FcγRII antibody.

PLATELET ACTIVATION BY PM6/248

During experiments to prepare mAbs to human platelet membrane proteins, seven mAbs were isolated which were shown, by crossed immunoelectrophoresis, to recognise the GPIIb-IIIa complex. The effect of all these mAbs on platelet aggregation was tested by the addition of increasing amounts to PRP followed, after an interval, by the addition of 10^{-5}M ADP. One of these mAbs, PM6/248, which is an IgG$_1$, gave a strong aggregation response, while the other mAbs had no such effect nor did they inhibit aggregation caused by subsequent addition of ADP. The characterisation of PM6/248, together with that of an IgG$_{2a}$ anti-CD9 mAb, MM2/57, has been described in detail by Hornby et al. (1991).

The initial observations on the aggregation response caused by PM6/248 are shown in Fig. 1. Addition to PRP of concentrations of PM6/248 of the order of 2 μg/ml caused a shape change followed by aggregation after a lag of approximately 8 min. No further aggregation was caused by addition of ADP. As increasing concentrations of PM6/248 were added the lag phase preceding the shape change was reduced progressively to about 3 min at a mAb concentration of 20 μg/ml. Despite the persistence of platelet shape change at all concentrations used, aggregation was inhibited at mAb concentrations of 10μg/ml or greater.

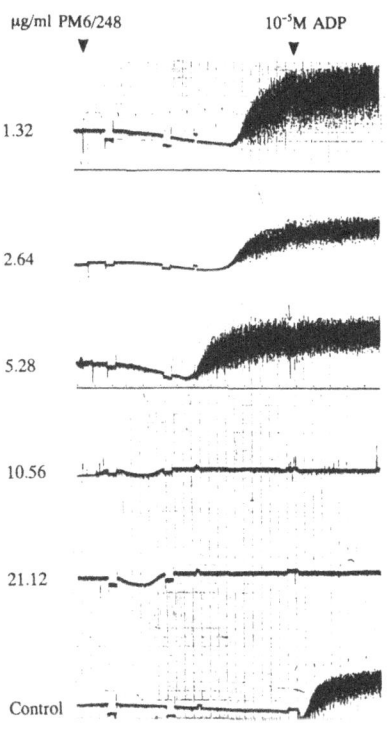

Figure 1 Effect on the aggregation response of the addition of increasing amounts of PM6/248 to PRP.

Subsequent experiments to clarify the details of this phenomenon were carried out using GFP suspensions containing 500 μg/ml of added fibrinogen. As shown in Fig. 2a, a maximum aggregation response was obtained at 8 μg/ml of PM6/248 with no aggregation occurring at concentrations in excess of 30 μg/ml. In order to investigate the platelet activation response to PM6/248, the release of 5HT from platelets was measured as were the synthesis of TxB$_2$ and the mobilisation of Ca^{2+} within the cytosol. Secretion of ^{14}C-5HT was concentration-related, being maximal at between 8 and 12 μg/ml of PM6/248 (Fig. 2b). TxB$_2$ synthesis also increased with increasing amounts of PM6/248, reaching a maximum at 16 μg/ml (Fig. 2c). However in neither case was secretion or biosynthesis inhibited by higher concentrations of mAb. The increase in cytosolic Ca^{2+} was studied using cells labelled with Fura 2 (Fig. 3) and was found to be concentration-related between 0.5 and 10 μg/ml of antibody, again following a lag phase, with no inhibition at higher

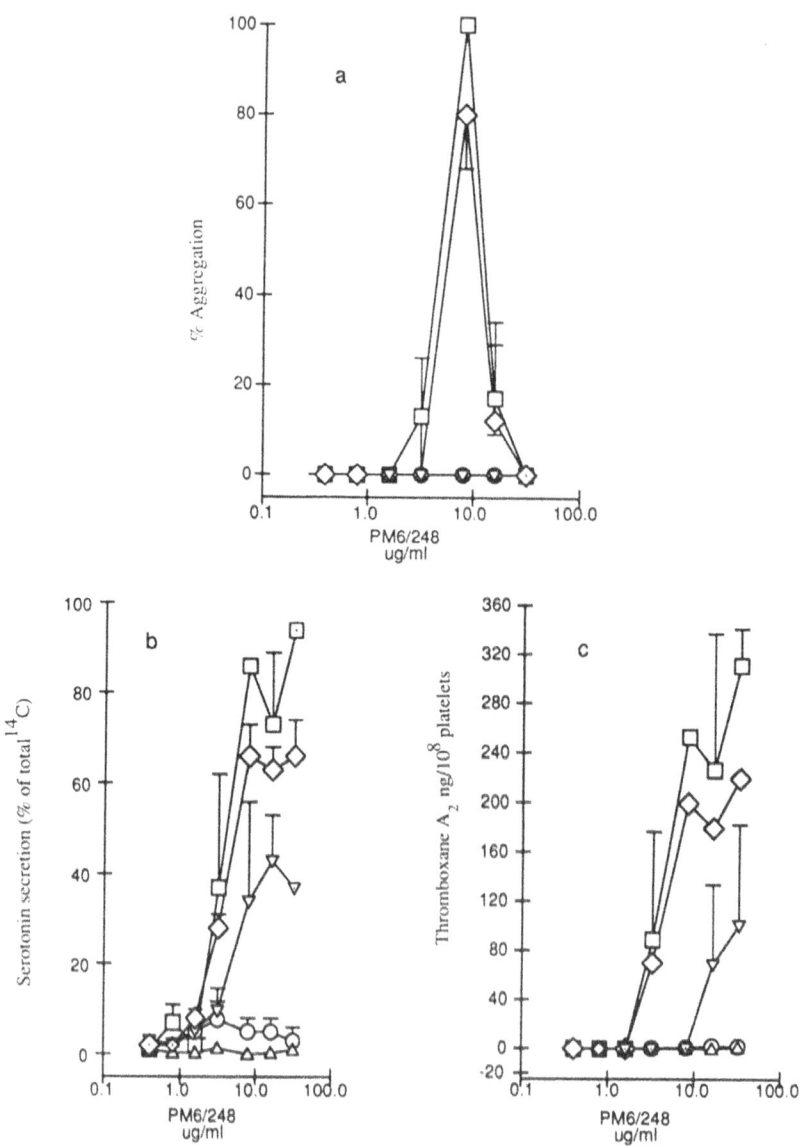

Figure 2. Effect of inhibitors on GFP responses to increasing concentrations of PM6/248. □ PM6/248 alone; ◇ GRGDS (0.1 mM); ▽ GRGDS (0.3 mM); ○ GR32191 (a TxA$_2$ antagonist) (1 μM); △ PGI$_2$ (50 nM); (*a*) Aggregation; (*b*) Serotonin secretion; (*c*) TxB$_2$ biosynthesis. (Reproduced from Br. J. Haematol. **79**, 277 (1991), with permission).

concentrations. A typical experiment is shown in Fig 3A using 10 μg/ml of PM6/248.

The mechanisms of these functional responses were studied with the use of a variety of inhibitors, including PGI$_2$, the TxA$_2$-receptor agonist, GR32191, the cyclooxygenase inhibitors, aspirin and indomethacin, and the fibrinogen antagonist, GRGDS. PGI$_2$, at 50nM, totally inhibited aggregation, secretion and TxB$_2$ synthesis (Fig 2), indicating that PM6/248 does not posses agglutinating activity. Similar inhibition was produced by

GR32191 (Fig 2), and aspirin and indomethacin inhibited the mobilisation of Ca^{2+}, demonstrating that TxA_2 and its receptor have an important role in all these responses. Measurement of the increase in cytosolic Ca^{2+} levels in the presence of EGTA (Fig. 3B) shows that Ca^{2+} is mobilised by influx from the extracellular medium as well as from intracellular stores. Aggregation was completely inhibited by 0.3 mM GRGDS but secretion and TxB_2 synthesis were only partially inhibited (Fig.2).

The mechanism by which signal transduction and aggregation are stimulated by binding of PM6/248 to the platelet surface was explored by using F(ab')$_2$ fragments of PM6/248 and also of the anti-FcγRII mAB, IV.3. PM6/248 F(ab')$_2$ did not cause platelet activation at any concentration tested and preincubation with either PM6/248 F(ab')$_2$ or IV.3 F(ab')$_2$ inhibited all the responses generated by subsequent treatment with PM6/248. These results suggest that the binding of PM6/248 Fc to the FcγRII is the mechanism by which mAb binding causes platelet activation. As activation is not inhibited by incubation with an irrelevant IgG$_1$ mAb, it is likely that a conformational change in the mAb Fc, caused by antigen binding, is necessary for binding to the FcγRII.

Preincubation of platelets with PM6/248 F(ab')$_2$ at 20 μg/ml was able to inhibit aggregation, but not secretion, induced by collagen (30 μg/ml) or the thromboxane A_2 mimetic, U46619 (10μM), and also aggregation induced by ADP (10 μM). These experiments, together with the inhibition of aggregation produced by high concentrations of PM6/248, suggest that PM6/248 binds to an epitope close to the fibrinogen-binding site of GPIIb-IIIa and is capable of inhibiting binding, perhaps by stearic hindrance.

ACTIVATION CAUSED BY OTHER ANTIBODIES

As noted above, mAbs to the CD9 antigen and GPIIb-IIIa, together with those to a

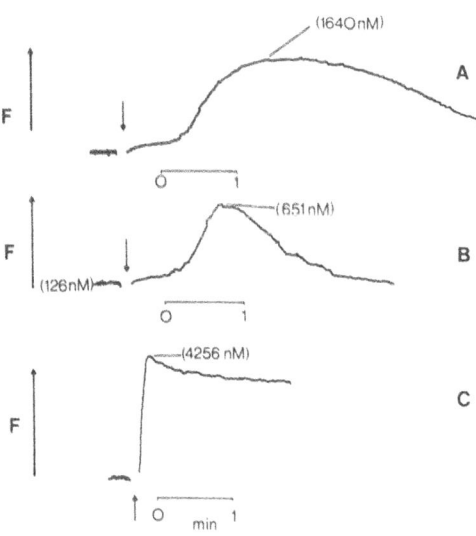

Figure 3. Increase of cytosolic Ca^{2+} levels by PM6/248. Typical traces measuring the increase in fluorescence of Fura 2 are shown which have been repeated on at least four different platelet preparations. Small arrows indicate the point of addition of either 10 μg/ml PM6/248 (traces A & B) or 3 μM ionomycin (trace C). (A) 1 mM Ca^{2+}; (B) 1 mM EGTA. Figures in brackets represent calculated levels of cytosolic Ca^{2+} (in nM). Bar represents time interval of 1 min. (Reproduced from Br. J. Haematol. **79**, 277 (1991), with permission).

number of other glycoproteins on the platelet surface, have been shown to give activation responses similar to PM6/248 (for references see Table 1). With one exception (mAb 50H.19) they are all of the IgG_1 subclass and, in all cases tested, activation is inhibited by mAb IV.3. With three exceptions (TP82, 6C9 and P256) aggregation is also inhibited by either Fab or $F(ab')_2$ fragments. There seems little doubt that aggregation and activation stimulated by each of these mAbs are mediated by binding of the mAb Fc region to the FcγRII following initial binding of the mAb to its antigen on the platelet surface, as was first described by Worthington et al. (1990) and as has been shown to be the case for PM6/248.

Although $F(ab')_2$ fragments of 6C9 (Modderman et al., 1988) and P256 (Bachelot et al., 1990) cause platelet "aggregation" this is not accompanied by the other activation responses (e.g. dense granule secretion, Ca^{2+} elevation etc.) typical of the whole mAbs. It is possible, as suggested by Rubinstein et al. (1991), that these $F(ab')_2$ fragments are capable of causing the exposure of the fibrinogen binding site of GPIIb-IIIa without stimulating further activation, in a manner similar to certain mAbs to GPIIb (Gulino et al., 1990) and to GPIIIa (Kouns and Jennings, 1991). If this is so there must be at least three epitopes on GPIIb-IIIa where binding of mAbs is able to cause cross-linking to the FcγRII. One, represented by PM6/248, which is close to the fibrinogen-binding site and inhibits fibrinogen binding, a second, represented by 6C9 and P256, where binding also activates the fibrinogen binding site and a third, represented by PL2-49 and UR1, which causes neither of these effects.

The recent completion of the CD9 sequence, derived from cDNA cloning (Boucheix et al., 1991, Lanza et al., 1991) suggests that the CD9 antigen has four transmembrane and two short extracellular domains. The finding that many, if not most, IgG_1 mAbs to CD9 cause activation may reflect the fact that this relatively small glycoprotein has only a limited number of epitopes available on the cell surface but that these are oriented in such a way that cross-linking of CD9 and the FcγRII can take place.

A further platelet response to anti-CD9 mAbs has been demonstrated by Carroll et al. (1990a), who showed that some mAbs of the IgG_{2a}, IgG_{2b} and IgG_3 subclass including MM2/57 are able to cause platelet lysis, and apparent aggregation, in PRP by C1q-complement mediated lysis. This does not, however, seem to be the case for mAb 50H.19, as all experiments with this mAb were carried out in GFP (Griffith et al., 1991).

ACTIVATION VIA FcγRII

The structure and biology of Fcγ receptors have been reviewed recently by van de Winkel and Anderson (1991). They report that the specificity of FcγRII for mouse IgG is $1 > 2b > > > 2a,3$ which is in good agreement with the finding that the vast majority of anti-platelet mAbs giving activation are IgG_1's. The reported polymorphism of FcγRII, where individuals may be either high or low responders, is also in accord with the polymorphism in the platelet response described by Testi et al. (1990) and Zuzel et al. (1991).

The mechanism of activation signalled through the Fc receptor has been investigated by Anderson and Anderson (1990) using the mAb IV.3. Following binding of this mAb to platelets, it was cross-linked using $F(ab')_2$ fragments of an anti-mouse IgG antibody. This cross-linking was shown to lead to aggregation and activation, following a lag phase, which was accompanied by inositol phospholipid and arachidonate metabolism, together with Ca^{2+} mobilisation, in a manner essentially indistinguishable from that demonstrated for activation by mAb binding. They speculated that the FcγRII may be coupled directly to a G protein. Similar experiments were reported by Worthington et al. (1990). It is thus clear that such activation results from the immobilisation and clustering of FcγRII, although the precise method of signal transduction remains to be elucidated.

The question of whether the cross-linking of antigen and FcγRII occurs between molecules on the same or adjacent platelets has been addressed by Anderson et al. (1991) and Horsewood et al. (1991). There is agreement between these groups that cross-linking between neighbouring platelets leads to activation, but while Horsewood et al. found evidence for intra-platelet activation, Anderson et al. did not, and this question thus remains open.

It has been recognised for some years that aggregated human IgG or immune complexes containing IgG are able to activate platelets and that this is mediated by the FcγRII (Rosenfeld et al., 1985). The mechanisms involving the FcγRII which have been elucidated by the use of mAbs to platelet surface antigens may be among the ways in which anti-platelet autoantibodies are able to initiate the destruction of platelets within the circulation.

REFERENCES

Anderson, G.P. and Anderson, C.L. 1990, Signal transduction by the platelet Fc receptor, *Blood* 76:1165.

Anderson, G.P., van de Winkel, J.G.J. and Anderson, C.L. 1991, Anti-GPIIb/IIIa (CD41) monoclonal antibody-induced platelet activation requires Fc receptor-dependent cell-cell interaction, *Br.J.Haematol.* 79:75.

Bachelot, C., Rendu, F., Boucheix, C., Hogg, N. and Levy-Toledano, S. 1990, Activation of platelets induced by mAb P256 specific for glycoprotein IIb-IIIa. Possible evidence for a role for IIb-IIIa in membrane signal transduction, *Eur.J.Biochem.* 190:177.

Boucheix, C., Soria, C., Mirshahi, M., Soria, J., Perrot, J.-Y., Fournier, N., Billard, M. and Rosenfeld, C. 1983, Characteristics of platelet aggregation induced by the monoclonal antibody ALB6 (acute lymphoblastic leukemia antigen p 24), *FEBS Lett.* 161:289.

Boucheix, C., Benoit, P., Frachet, P., Billard, M., Worthington, R. E., Gagnon, J. and Uzan, G. 1991, Molecular cloning of the CD9 antigen. A new family of cell surface proteins, *J.Biol.Chem.* 266:117.

Carroll, R.C., Rubinstein, E., Worthington, R.E. and Boucheix, C. 1990a, Extensive C1q-complement initiated lysis of human platelets by IgG subclass murine monoclonal antibodies to the CD9 antigen, *Thromb.Res.* 59:831.

Carroll, R.C., Worthington, R.E. and Boucheix, C. 1990b, Stimulus-response coupling in human platelets activated by monoclonal antibodies to the CD9 antigen, a 24 kDa surface-membrane glycoprotein, *Biochem.J.* 266:527.

Gorman, D.J., Castaldi, P.A., Zola, H. and Berndt, M.C. 1985, Preliminary functional characterization of a 24,000 dalton platelet surface protein involved in platelet activation, *Nouv.Rev.Fr.Hematol.* 27:255.

Griffith, L., Slupsky, J., Seehafer, J., Boshkov, L. and Shaw, A.R.E. 1991, Platelet activation by immobilized monoclonal antibody: Evidence for a CD9 proximal signal, *Blood* 78:1753.

Gulino, D., Ryckewaert, J.-J., Andrieux, A., Rabiet, M.-J. and Marguerie, G. 1990, Identification of a monoclonal antibody against platelet GPIIb that interacts with a calcium-binding site and induces aggregation, *J.Biol.Chem.* 265:9575.

Hato, T., Ikeda, K., Yasukawa, M., Watanabe, A. and Kobayashi, Y. 1988, Exposure of platelet fibrinogen receptors by a monoclonal antibody to CD9 antigen, *Blood* 72:224.

Higashihara, M., Maeda, H., Shibata, Y., Kume, S. and Ohashi, T. 1985, A monoclonal anti-human platelet antibody: a new platelet aggregating substance, *Blood* 65:382.

Hornby, E.J., Brown, S., Wilkinson, J.M., Mattock, C. and Authi, K.S. 1991, Activation of human platelets by exposure to a monoclonal antibody, PM6/248, to glycoprotein IIb-IIIa, *Br.J.Haematol.* 79:277.

Horsewood, P., Hayward, C.P.M., Warkentin, T.E. and Kelton, J.G. 1991, Investigation of the mechanisms of monoclonal antibody-induced platelet activation, *Blood* 78:1019.

Jennings, L.K., Fox, C.F., Kouns, W.C., McKay, C.P., Ballou, L.R. and Schultz, H.E. 1990, The activation of human platelets mediated by anti-human platelet p24/CD9 monoclonal antibodies, *J.Biol.Chem.* 265:3815.

Kornecki, E., Walkowiak, B., Naik, U.P. and Ehrlich, Y.H. 1990, Activation of human platelets by a stimulatory monoclonal antibody, *J.Biol.Chem.* 265:10042.

Kouns, W.C. and Jennings, L.K. 1991, Activation-independent exposure of the GPIIb-IIIa fibrinogen receptor, *Thromb.Res.* 63:343.

Lanza, F., Wolf, D., Fox, C. F., Kieffer, N., Seyer, J. M., Fried, V. A., Coughlin, S. R., Phillips, D. R. and Jennings, L. K. 1991, cDNA cloning and expression of platelet p24/CD9. Evidence for a new family of multiple membrane-spanning proteins, *J.Biol.Chem.* 266:10638.

Miller, J.L., Kupinski, J.M. and Hustad, K.O. 1986, Characterization of a platelet membrane protein of low molecular weight associated with platelet activation following binding by monoclonal antibody AG-1, *Blood* 68:743.

Modderman, P.W., Huisman, H.G., Van Mourik, J.A. and Von dem Borne, A.E.G.Kr. 1988, A monoclonal antibody to the human platelet glycoprotein IIb/IIIa complex induces platelet activation, *Thromb.Haemost.* 60:68.

Morel, M.-C., Lecompte, T., Champeix, P., Favier, R., Potevin, F., Samama, M., Salmon, C. and Kaplan, C. 1989, PL2-49, a monoclonal antibody against glycoprotein IIb which is a platelet activator, *Br.J.Haematol.* 71:57.

Ockenhouse, C.F., Magowan, C. and Chulay, J.D. 1989, Activation of monocytes and platelets by monoclonal antibodies or malaria-infected erythrocytes binding to the CD36 surface receptor in vitro, *J.Clin.Invest.* 84:468.

Rosenfeld, S.I., Looney, R.J., Leddy, J.P., Phipps, D.C., Abraham, G.N. and Anderson C.L. 1985, Human platelet Fc receptor for immunoglobulin G. Identification as a 40,000 molecular weight protein shared by monocytes. *J. Clin. Invest.* 76;2317.

Rubinstein, E., Boucheix, C., Urso, I. and Carroll, R.C. 1991, Fcgamma receptor-mediated interplatelet activation by a monoclonal antibody against β2 microglobulin, *J.Immunol.* 147:3040.

Scott, J.L., Dunn, S.M., Jin, B., Hillam, A.J., Walton, S., Berndt, M.C., Murray, A.W., Krissansen, G.W. and Burns, G.F. 1989, Characterization of a novel membrane glycoprotein involved in platelet activation, *J.Biol.Chem.* 264:13475.

Testi, R., Pulcinelli, F., Frati, L., Gazzaniga, P.P. and Santoni, A. 1990, CD69 is expressed on platelets and mediates platelet activation and aggregation, *J.Exp.Med.* 172:701.

Thiagarajan, P., Perusia, B., DeMarco, L., Wells, K. and Trichieri, G. 1983, Membrane proteins on human megakaryocytes and platelets identified by monoclonal antibodies, *Am.J.Hematol.* 14:255.

van de Winkel, J.G.J. and Anderson, C.L. 1991, Biology of human immunoglobulin G Fc receptors. *J. Leuk. Biol.* 49; 511.

Worthington, R.E., Carroll, R.C. and Boucheix, C. 1990, Platelet activation by CD9 monoclonal antibodies is mediated by the FcγII receptor, *Br.J.Haematol.* 74:216.

Zuzel, M., Walton M., Burns, G.F., Berndt, M.C. and Cawley, J.C. 1991, A monoclonal antibody to a 67 kD cell membrane glycoprotein directly induces persistent platelet aggregation independently of granule secretion. *Br. J. Haematol.* 79;466.

FUNCTIONAL RELATIONSHIP BETWEEN CYCLIC AMP-DEPENDENT

PROTEIN PHOSPHORYLATION AND PLATELET INHIBITION

Wolfgang Siess, Bernd Grünberg, Karin Luber

Institüt für Prophylaxe und Epidemiologie der
Kreislaufkrankheiten b.d. Universität München
Pettenkoferstr. 9, 8000 München 2

INTRODUCTION

Platelet agonists such as the prostaglandins prostacyclin, prostaglandin E_1 and prostaglandin D_2 which raise cyclic AMP levels, are the most potent inhibitors of platelet activation. Platelet shape change, secretion and aggregation induced by all physiological stimuli and platelet adhesion to subendothelium are inhibited by pretreatment of platelets with these prostaglandins (Siess, 1989). These substances bind to specific receptors on the platelet surface that couple to the guanine nucleotide-binding protein G_s (Fig.1). Activation of G_s stimulates adenylate cyclase and leads to an increase in intracellular cyclic AMP, the activation of protein kinase A and the subsequent phosphorylation of specific proteins.

TARGET SITES FOR INHIBITION OF PLATELET ACTIVATION BY cAMP

Cyclic AMP inhibits platelet activation at several steps. One of the most important steps is the inhibition of receptor-mediated phosphoinositide hydrolysis by phospholipase C activation (Fig.2). A possible mechanism could be the direct phosphorylation of phospholipase C by protein kinase A, as recently described for phospholipase C-γ1 in Jurkat T-cell lines treated with forskolin (Park et al, 1992). The inhibitory effects of cyclic AMP on protein kinase C activation, Ca^{2+}-mobilization, fibrinogen-receptor exposure, myosin light chain phosphorylation, actin polymerization and cytoskeletal assembly are believed to be consequences of the inhibition of receptor-mediated phospholipase C activation (Siess, 1989). Other studies have indicated that further target sites exist that are important for platelet inhibition by cyclic AMP. For example, it has been reported by several investigators that cyclic AMP inhibits secretion and aggregation induced by Ca^{2+}-ionophore at steps distal to calcium mobilisation (Siess, 1989). Also, modest increments in platelet cyclic AMP have been reported to abolish platelet-activating factor-induced aggregation and secretion, but to have little effect on phosphoinositide hydrolysis and elevation of cytosolic calcium induced by platelet-activating factor (Bushfield et al, 1985).

Mechanisms of Platelet Activation and Control, Edited by
K.S. Authi *et al.*, Plenum Press, New York, 1993

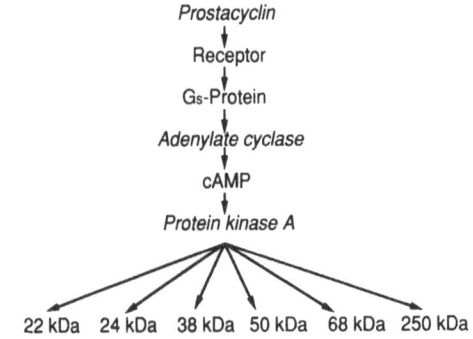

Figure 1   Signal transduction of prostacyclin in platelets.

Therefore, additional target sites for platelet inhibition by cyclic AMP must exist.

Many studies indicate that phosphorylation of the 20 and 47 kDa proteins are related closely to platelet functional responses: myosin light chain kinase-dependent phosphorylation of myosin light chain may trigger shape change, whereas protein kinase C (PKC)-dependent phosphorylation of the 47 kDa polypeptide may regulate aggregation and secretion. Therefore, the question emerged whether increasing platelet cyclic AMP could suppress aggregation by inhibiting PKC- or Ca^{2+}-dependent kinases involved in the activation of platelets.

Phorbol ester and Ca^{2+}-ionophores were used to probe specifically the PKC-dependent pathway and the Ca^{2+}-pathway of platelet activation, respectively (Siess and Lapetina, 1989). Phorbol esters, by intercalation into the membrane without mobilization

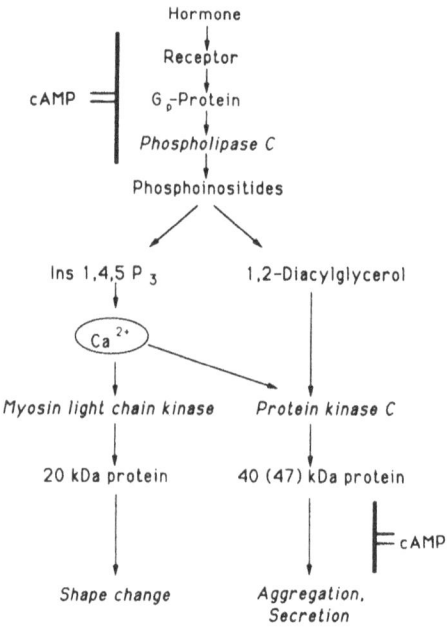

Figure 2   Target sites for inhibition of platelet activation by cyclic AMP.

of calcium, translocate PKC from the cytosol to the membrane and induce the PKC-dependent phosphorylation of various proteins, such as the 47 kDa polypeptide and the 20 kDa myosin light chain. After exposure of platelets to high concentrations (>50nM) of phorbol 12,13-dibutyrate, platelets do aggregate. A close temporal relationship between phorbol ester-induced 20 kDa myosin light chain phosphorylation and platelet aggregation was found, suggesting a role for PKC-dependent myosin phosphorylation in inducing platelet aggregation. Preincubation of platelets with PGI_2 completely suppressed aggregation but without significantly reducing the phosphorylation of 20 or 47 kDa proteins evoked by phorbol ester. It was concluded from these data that phorbol esters induce platelet aggregation through PKC-dependent phosphorylation of specific proteins, but that distal to this event are steps crucial for the regulation of platelet aggregation, which are target sites for cyclic AMP-dependent protein kinases. Such sites could be the glycoprotein IIb/IIIa complex (which represents the fibrinogen receptor) or, more likely, as yet unknown proteins that regulate the conformational change and the expression of the fibrinogen receptor upon platelet activation.

Ca^{2+}-ionophores activate platelets through Ca^{2+}-influx and Ca^{2+}-mobilization from intracellular stores. The increase in cytosolic Ca^{2+} leads to the activation of Ca^{2+}/calmodulin-dependent kinases, such as the myosin light chain kinase, which phosphorylates the 20 kDa myosin light chain, and to a Ca^{2+}-dependent activation of PKC, which phosphorylates a 47 kDa protein. It was found that Ca^{2+}-ionophore A23187, at $0.1\mu M$, induced shape change and a pronounced phosphorylation of the 20 kDa myosin light chain, but no or only a small phosphorylation of the 47 kDa protein. It was observed that, by increasing the concentration of A23187 from 0.2 to 1 μM, aggregation and phosphorylation of the 47 kDa protein increased progressively, whereas the stimulation of the 20 kDa myosin light chain phosphorylation remained unchanged. These results also support the idea that myosin light chain phosphorylation is related to shape change and a prerequisite for aggregation, whereas PKC-dependent phosphorylation of the 47 kDa protein is more involved in regulation of aggregation and secretion. Preincubation of platelets with PGI_2 neither inhibited Ca^{2+} mobilization, 20 kDa myosin light chain phosphorylation and shape change, nor reduced 47 kDa protein phosphorylation, but it inhibited platelet aggregation induced by Ca^{2+}-ionophore A23187. These results indicate that cyclic AMP can inhibit platelet aggregation at steps that are downstream of Ca^{2+}-mobilization and Ca^{2+}-dependent phosphorylation of 20 and 47 kDa proteins (Fig.2). Similar results have been obtained by using nitroprusside which activates guanylate cyclase in platelets. Elevation of cGMP in platelets inhibited both the PKC- and the Ca^{2+}-dependent pathways of platelet activation also at steps following protein phosphorylation (Doni et al, 1991).

PROTEINS PHOSPHORYLATED BY PROTEIN KINASE A IN PLATELETS

Inhibition of platelet activation by cyclic AMP is associated with the phosphorylation of proteins with molecular masses of 22, 24, 38, 50, 68 and 250 kDa. The 22, 24 and the 250 kDa proteins have been identified (Table 1).

We and others recently identified the 22 kDa protein as *rap*1B, a *ras*-like protein that belongs to the family of small GTP-binding proteins (Siess et al, 1990; Fischer et al, 1990; Hata et al, 1991). *Rap*1B is highly expressed in human platelets and is present in platelet membranes and cytosol (Lapetina et al, 1989). It was shown that phosphorylation of *rap*1B occurs at a C-terminal serine residue (serine-179) in close vicinity to cysteine-181 (Siess et al, 1990; Hata et al, 1991). Through a series of post-translational modifications that amino acid becomes C-terminal, carboxymethylated and geranyl-geranylated; these modifications and the positive charge of the polybasic C-terminal tail are essential for the

Table 1   Proteins phosphorylated by protein kinase A in human platelets

MW	Protein	Function
22 kDa	*rap* 1B	?
24 kDa	glycoprotein 1b (ß-chain)	Receptor for *von Willebrand factor* Binding site for *thrombin*
38 kDa	?	
50 kDa	?	
68 kDa	?	
250 kDa	Actin-binding protein	Links glycoprotein Ia and Ib to actin filaments

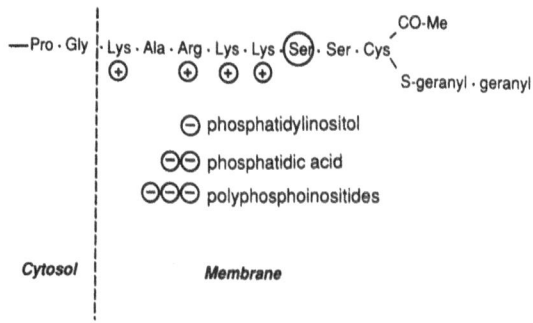

Figure 3   Hypothetical interactions of the C-terminal tail of *rap*
1B with membrane phospholipids. The phorphorylation site on Ser179 is circled.

membrane attachment of *rap*1B (Fig.3). *Rap*1B is present in a GDP-bound inactive or a GTP-bound active form that is interconvertible by GDP-GTP exchange and GTPase reactions. A stimulatory type of protein named GDP dissociation stimulator (GDS) has been described, that increases the GDP/GTP exchange thereby converting *rap*1B into the active form (Kaibuchi et al, 1991). GDS interacts specifically with the lipidated polybasic C-terminal tail of *rap*1B (Hiroyoshi et al, 1991; Shirataki et al, 1991). A recent study suggests that the insertion of a negatively charged phosphate group by protein kinase A into the polybasic C-terminal tail of *rap*1B facilitates the interaction of GDS with *rap*1B thereby converting it into the active GTP-bound form (Hata et al, 1991; Fig.4).

PHOSPHORYLATION OF *RAP*1B IS NOT INVOLVED IN PLATELET INHIBITION BY cAMP

In order to elucidate the function of *rap*1B in platelets, we explored the relationship between cyclic AMP-dependent protein phosphorylation and platelet inhibition. Time-course studies showed that cyclic AMP-dependent phosphorylation of all proteins except *rap*1B was maximal 1 min after exposure of platelets to high concentrations of PGI$_2$, iloprost or PGE$_1$; the time-course of *rap*1B phosphorylation was distinctly slow: maximal phosphorylation of *rap*1B occurred only after 45 min of incubation. Inhibition of thrombin-induced platelet activation was observed already after 30 s incubation with PGI$_2$ or iloprost;

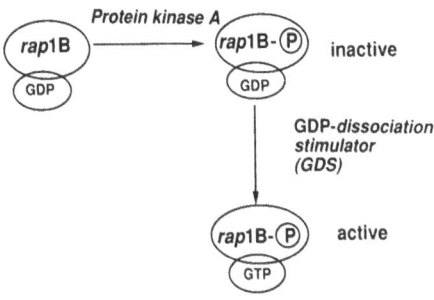

Figure 4   Facilitation of *rap*1B activation by protein kinase A-induced protein phosphorylation.

at this time phosphorylation of *rap*1B was only slightly increased. Platelets that had been preincubated and washed in the presence of PGE₁ and later resuspended in the absence of PGE₁ responded fully to activation by thrombin despite maximal phosphorylation of *rap*1B and P24, the ß-chain of glycoprotein Ib (Siess and Lapetina, 1990). Also, platelets which had been preincubated (for 2 hours) in platelet-rich plasma with a high concentration of iloprost (100 nM) to convert *rap*1B completely into its phosphorylated form and that were then resuspended in plasma in the absence of iloprost, showed normal aggregation and dense granule secretion upon stimulation by a wide variety of agonists (epinephrine, ADP, phorbol ester, vasopressin, collagen, thrombin-receptor activating peptide; Siess and Grünberg, 1993). Furthermore, addition of PGI₂ or iloprost to PGE₁-washed platelets prevented thrombin-induced platelet activation, but did not evoke further phosphorylation of *rap*1B or P24. These experiments indicate that the phosphorylation of *rap*1B and glycoprotein Ib is not related to platelet inhibition by cyclic AMP. It was found that the phosphorylation of the proteins P38 and P50 correlated better with the PGI₂-induced inhibition of platelet activation. The phosphorylation of these proteins is more likely to be involved in mediating cyclic AMP-dependent platelet inhibition. One project of our laboratories is to elucidate the structure of P38. The phosphorylation of P38 has been observed in platelets incubated with PGI₂, iloprost, or Sin-1.

*RAP*1B PHOSPHORYLATION IS A SENSITIVE MARKER FOR THE ACTION OF cAMP- AND cGMP-ELEVATING VASODILATORS

Possible explanations for the slow time-course of *rap*1B phosphorylation in platelets are a very active phosphatase specific for *rap*1B or a masking of the phosphorylation site that is located in the polybasic C-terminal tail of *rap*1B. Acidic membrane phospholipids such as phosphatidic acid or phosphatidylinositol might interact with the polybasic C-terminus of *rap*1B thereby hindering the accessibility for protein kinase A (Fig.3). Several substances that could influence the time-course of *rap*1B phosphorylation were studied. Mastoparan is an amphilic tetradecapeptide, accelerates GDP/GTP-exchange and is known to interact with the C-terminal tail of G_{ia} which shows homology to the C-terminus of *rap*1B. One would expect that if mastoparan interacts with *rap*1B, the C-terminal tail and the phosphorylation site of *rap*1B is exposed. Okadaic acid by inhibiting phosphatases could also enhance the phosphorylation of *rap*1B. It was found, however, that these substances were without effect on the time-course of *rap*1B phosphorylation.

Interestingly, the dephosphorylation of *rap*1B is also very slow: *rap*1B remains phosphorylated for at least 3 hours after removal of iloprost, PGI₂ or PGE₁ from platelets. A more rapid dephosphorylation of *rap*1B after removal of iloprost from platelets could neither be induced by epinephrine that inhibits adenylate cyclase and descreases cAMP levels nor by Ca^{2+}-ionophore that stimulates Ca^{2+}-dependent phosphatases. The very slow dephosphorylation of *rap*1B establishes a memory-like property of phosphorylated *rap*1B and makes the measurement of phosphorylated *rap*1B a useful and sensitive tool to follow the action of cAMP-elevating substances *in vitro* and *in vivo*. When platelets were exposed for 2 hours to low concentrations of iloprost that barely inhibit platelet aggregation, *rap*1B phosphorylation slowly appeared and continuously stayed high in contrast to the phosphorylation of the 50 kDa protein that was reversible.

Sin-1 is a NO-containing agent known to stimulate the soluble guanylate cyclase in platelets. We found that incubation of platelets with low concentrations of Sin-1 also evoked *rap* 1B phosphorylation. Although we also observed *rap*1B phosphorylation after incubating platelets with the stable and membrane permeable cGMP-analog 8-PCPT-cGMP, it has not been described that *rap*1B can be directly phosphorylated by a cGMP-dependent protein kinase. cAMP also inhibits cAMP phosphodiesterases (Bowen et al, 1991), and could - especially after prolonged incubation periods - lead to a protein kinase A-dependent *rap*1B-phosphorylation. Interestingly, very low concentrations of Sin-1 and iloprost synergized in inducing *rap*1B phosphorylation. *Rap*1B phosphorylation that can be measured by Western blotting techniques is therefore a useful biochemical marker for the analysis of the *in vivo* action of cAMP- and cGMP-elevating vasodilators.

ACKNOWLEDGEMENTS

The study was supported by a grant from the Deutsche Forschungsgemeinschaft (DFG Si 274/2).

REFERENCES

Bowen, R., and Haslam, R.J., 1991, Effects of nitrovasodilators on platelet cyclic nucleotide levels in rabbit blood; role for cyclic AMP in synergistic inhibition of platelet function by SIN-1 and prostaglandin E$_1$, *J. Cardiovasc. Pharmacol.* 17:424.

Bushfield, M., McNichol, A., and MacIntyre, D.E., 1985, Inhibition of platelet-activating-factor-induced human platelet activation by prostaglandin D2, *Biochem. J.* 237:267.

Doni, M.G., Deana, R., Padoin, E., Ruzzene, M., and Alexandre A., 1991, Platelet activation by diacylglycerol or ionomycin is inhibited by nitroprusside, *Biochim. Biophys. Acta* 1094:323.

Fischer, T.H., Gatling, M.N., Lacal, J.C. and White, G.C., 1990, Rap 1B, a cAMP-dependent protein kinase substrate, associates with the platelet cytoskeleton, *J. Biol. Chem.* 265:19405.

Hata, Y., Kaibuchi, K., Kawamura, S., Hiroyoshi, M., Shirataki, H., and Takai, Y., 1991, Enhancement of the actions of smg p21 GDP/GTP exchange protein by the protein kinase-A-catalyzed phosphorylation of smg p21, *J. Biol. Chem.* 266:6571.

Hiroyoshi, M., Kaibuchi, K., Kawamura, S., Hata, Y., and Takai, Y., 1991, Role in the C-terminal region of smg p21-like small GTP-binding protein, in membrane and smg p21 GDP/GTP exchange protein interactions, *J. Biol. Chem.* 266:2962.

Kaibuchi, K., Mizuno, T., Fujioka, H., Yamamoto, T., Kishi, K., Fukumoto, Y., Hori, Y., and Takai, Y., 1991, Molecular cloning of the cDNA for stimulatory GDP/GTP exchange protein for smg p21s (ras p21-like small GTP-binding proteins) and characterization of stimulatory GDP/GTP exchange protein, *Mol. Cell. Biol.* 11:2873.

Lapetina, E.G., Lacal, J.C., Reep, B.R., Molina y Vedia, L., 1989, A *ras*-related protein is phosphorylated and translocated by agonists that increase cAMP levels in human platelets, *Proc. Natl. Acad. USA* 86:3131.

Park, D.J., Min, H.K., and Rhee, S.G., 1992, Inhibition of CD3-linked phospholipase C by phorbol ester and by cAMP is associated with decreased phosphotyrosine and increased phosphoserine contents of PLC-γ1, *J. Biol. Chem.* 267:1496.

Shirataki, H., Kaibuchi, K., Hiroyoshi, M., Isomura, M., Araki, S., Sasaki, T., and Takai, Y., 1991, Inhibition of the action of the stimulatory GDP/GTP exchange protein for smg p21 by the geranylgeranylated synthetic peptides designed from its C-terminal region, *J. Biol. Chem.* 266:20672.

Siess, W., 1989, Molecular mechanisms of platelet activation, *Physiol. Rev.* 69:58.

Siess, W. and Grünberg, B., 1993. Phosphorylation of *rap* 1B by protein kinase A is not involved in inhibition by cyclic AMP. *Cell Signal.* 5: 209.

Siess, W., and Lapetina, E.G., 1989, Prostacyclin inhibits platelet aggregation induced by phorbol ester or Ca^{2+}-ionophore at steps distal to activation of protein kinase C and Ca^{2+}-dependent protein kinases, *Biochem. J.* 258:57.

Siess, W., and Lapetina, E.G., 1990, Functional relationship between cyclic AMP-dependent protein phosphorylation and platelet inhibition, *Biochem. J.* 271:815.

Siess, W., Winegar, D.A., and Lapetina, E.G., 1990, *Rap*1B is phosphorylated by protein kinase A in intact human platelets, *Biochem. Biophys. Res. Commun.* 170:944.

ROLE OF CYCLIC NUCLEOTIDE-DEPENDENT PROTEIN KINASES AND THEIR

COMMON SUBSTRATE VASP IN THE REGULATION OF HUMAN PLATELETS

Ulrich Walter, Martin Eigenthaler, Jörg Geiger and Matthias Reinhard

Medizinische Universitätsklinik
Klinische Forschergruppe
Josef-Schneider-Str. 2
97080 Würzburg, Germany

INTRODUCTION

Numerous hormones, eicosanoids, drugs and other vasoactive substances stimulate or inhibit the activation of platelets. Activators such as thrombin, thromboxane A_2, vasopressin, platelet activating factor, ADP, collagen and epinephrine cause platelet shape change, adhesion, aggregation and degranulation. Most platelet agonists activate phospholipase C (PLC), mobilize intracellular lipid-derived messengers, elevate cytosolic free Ca^{2+} and stimulate the activity of distinct protein kinases including myosin light chain kinase (MLCK), protein kinase C (PKC) and protein tyrosine kinases (Haslam, 1987; Siess, 1989; Rink and Sage, 1990; Halbrügge and Walter, 1993). In contrast, agents which elevate either platelet cAMP levels [e.g. PG-I_2, PG-E_1, adenosine] or cGMP levels [e.g. endothelium-derived relaxing factor (EDRF), nitroprusside and other nitric-oxide-generating agents] are powerful inhibitors of platelet activation (Siess, 1989; Walter, 1989; Walter et al., 1991). The two classes of cyclic nucleotide-elevating platelet inhibitors, PG-I_2 and EDRF, are produced by vascular endothelial cells, and these two inhibitors represent a very important, if not the most important physiological antithrombotic mechanism (Vane et al., 1990). In this chapter, some current concepts concerning the role of cAMP- and cGMP-dependent protein kinases (cAMP-PK, cGMP-PK) and their substrates including the vasodilator-stimulated phosphoprotein (VASP) in mediating the effects of cyclic nucleotide-elevating platelet inhibitors will be discussed. Results obtained in the last two years are emphasized. The reader is referred to the above mentioned reviews for a comprehensive summary of the older literature.

DIVERSITY OF CYCLIC NUCLEOTIDE REGULATION AND ACTION

As shown in Fig. 1, many hormones and agents regulate cellular cAMP levels by

Mechanisms of Platelet Activation and Control, Edited by
K.S. Authi *et al.*, Plenum Press, New York, 1993

stimulating or inhibiting the family of cell membrane-bound adenylyl cyclases via stimulatory or inhibitory GTP-binding proteins (Krupinsky, 1991). The cellular level of cGMP is regulated by both soluble and particulate enzymes, e.g. by the family of heterodimeric, heme-containing soluble guanylyl cyclases and by the family of cell-membrane spanning particulate guanylyl cyclases (Koesling et al., 1991). Soluble guanylyl cyclases are activated by endothelium-derived relaxing factor (EDRF), nitrovasodilators and other nitric oxide generating agents. Particulate guanylyl cyclases are activated by natriuretic peptides such as ANP and BNP or peptides such as E. coli heat-stable enterotoxin or the recently discovered guanylin (Koesling et al., 1991; Currie et al., 1992). However, the cellular level of cyclic nucleotides is not only regulated by synthetic enzymes (adenylyl and guanylyl cyclases) but also by degrading enzymes represented by the growing family of phosphodiesterases (PDEs) (Beavo, 1988; Beavo and Reifsnyder, 1990; Nicholson et al., 1991; Bentley and Beavo, 1992). These phospho-diesterases can also be modulated by distinct mechanisms.

Activation of cAMP-dependent protein kinases (cAMP-PK) is the most important mechanism of cAMP action, and multiple physiological events are regulated by cAMP-dependent protein phosphorylation (Taylor et al., 1990; Meinecke et al., 1990; McKnight, 1991). However, in olfactory cilial cells and cardiac pacemaker cells ion channels are directly regulated by cAMP without the involvement of protein kinases (Kaupp, 1991; DiFrancesco and Tortora, 1991). Under certain conditions, high levels of cellular cGMP or cAMP may also activate cAMP-PK or cGMP-PK, respectively (Walter, 1984; Jiang et al., 1992).

The diversity of cGMP action is even greater than that of cAMP (Walter, 1989; Lohmann et al., 1991; Butt et al., 1993). The direct regulation of cation channels by cGMP is well established for retinal rod outer segments and is likely to occur in other cell types as well (Kaupp, 1991). Other very important targets of cGMP are phosphodiesterases which may be stimulated (type II) or inhibited (type III) by cGMP (Beavo, 1988; Beavo and Reifsnyder, 1990; Nicholson et al., 1991; Bentley and Beavo, 1992). Therefore, cGMP can potentially enhance (via type III PDE) or decrease (via type II PDE) a cAMP response. Another important target of cGMP is the family of cGMP-PK. The soluble type I cGMP-PK (which appears to exist in at least two different splice variants) has been detected in high concentrations in vascular and non-vascular smooth muscle cells, fibroblasts, cerebellar Purkinje cells and platelets, whereas the exclusively particulate type II cGMP-PK has been detected so far only in brush border membranes (Lohmann et al., 1991; Butt et al., 1993). Considerable evidence suggests that activation of the type I cGMP-PK decreases cytosolic Ca^{2+} in smooth muscle cells (Felbel et al., 1988) and inhibits the preactivated L-type Ca^{2+} channel in mammalian cardiac myocytes (Méry et al., 1991). However, it should be emphasized that the multiple signal transduction pathways regulated by cAMP and cGMP often co-exist in the same cell, and all of these pathways may participate in the regulation of a given physiological response (Lohmann et al., 1991).

CYCLIC NUCLEOTIDE BINDING PROTEINS AND REGULATION OF CYCLIC NUCLEOTIDE LEVELS IN HUMAN PLATELETS

The most important and powerful stimulators of platelet adenylyl cyclase are prostaglandins (PG-I_2, PG-E_1, PG-E_2) and to a lesser degree adenosine and ß-adrenergic catecholamines, whereas α2-adrenergic agents, ADP, and thrombin are capable of inhibiting platelet adenylyl cyclase via inhibitory Gi-proteins (Aktories and Jakobs, 1985). At present, there is no evidence that human platelets contain a particulate, cell membrane-spanning guanylyl cyclase which is activated by ANP or other agents (Tremblay et al., 1988). In contrast, soluble, heme-containing guanylyl cyclases are clearly present

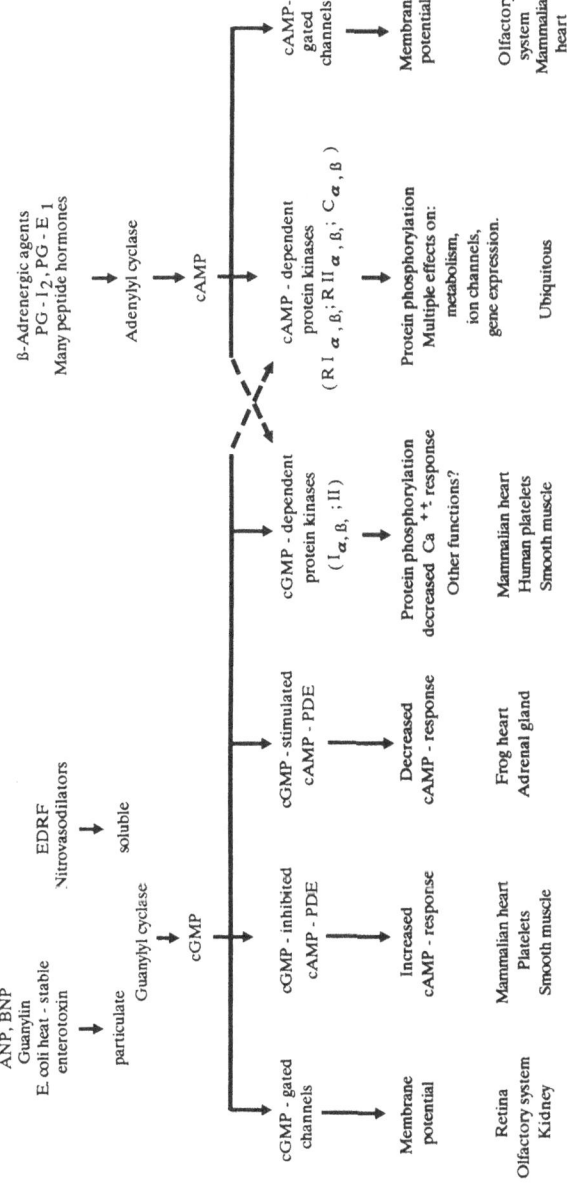

Figure 1   Regulation and action of cyclic nucleotides

in human platelets and activated by EDRF, nitrovasodilators and other NO-generating agents (Tremblay et al., 1988). It is of interest to note that the two most important endothelium-derived platelet inhibitors, PG-I$_2$ and EDRF, increase either cAMP or cGMP, respectively, in platelets as shown in Fig. 2 (Vane et al., 1990). With respect to cyclic nucleotide degradation, the major phosphodiesterases responsible for hydrolysis of cAMP and cGMP in human platelets are a cGI-PDE (type III PDE) and a cGMP-specific PDE (type V PDE), respectively (Beavo, 1988; Nicholson et al., 1991; Bentley and Beavo, 1992). The type III cGI-PDE which primarily hydrolyzes cAMP is inhibited by cGMP and activated by phosphorylation mediated by the cAMP-PK (Beavo, 1988). As shown in Fig. 2, elevation of cGMP in platelets inhibits the cGI-PDE resulting in decreased cAMP degradation. Therefore, cGMP is capable of increasing a cAMP response such as cAMP-PK-mediated protein phosphorylation. This pathway has been suggested as one mechanism for the synergism between cGMP- and cAMP-elevating agents with respect to platelet inhibition (Maurice and Haslam, 1990; Bowen and Haslam, 1991). In contrast, the cGI-PDE can also be phosphorylated and activated by the cAMP-PK, an effect which has been demonstrated in intact human platelets (Macphee et al., 1988). The cAMP-PK-mediated phosphorylation and activation of cGI-PDE would decrease cAMP levels and thus represent a mechanism for turning off a cAMP signal.

At present, there is very little evidence that human platelets contain ion channels directly regulated by either cGMP or cAMP.

In contrast, the presence of both cGMP-PK and cAMP-PK in human platelets is well established (Walter, 1989; Walter et al., 1991; Halbrügge and Walter, 1993). Therefore, it appears likely that the cyclic nucleotides cGMP and cAMP achieve their effects in human platelets primarily via cGMP-regulated PDE, cGMP-PK and cAMP-PK.

CYCLIC NUCLEOTIDE-DEPENDENT PROTEIN KINASES AND THEIR FUNCTIONAL ROLES IN HUMAN PLATELETS

Compared to other tissues and cell types, human platelets contain particularly high concentrations of both cGMP-PK and cAMP-PK (Waldmann et al., 1986; Eigenthaler et al., 1992). Several lines of biochemical and immunological evidence suggest that cGMP-PK type Iα and cAMP-PK types I and IIß represent the major forms of cyclic nucleotide-dependent protein kinases of human platelets (Halbrügge and Walter, 1993). Interestingly, the intracellular cAMP concentration in unstimulated platelets is similar to the concentration of cAMP-binding sites of the cAMP-PK whereas the concentration of cGMP under these conditions is 1-2 orders of magnitude lower than the concentration of cGMP binding capacity of the cGMP-PK (Table 1). These data suggest that small changes in platelet cAMP levels (e.g. even elevations less than 2-fold) would be sufficient to activate most of the cAMP-PK. In contrast, nitrovasodilator-caused platelet inhibition via activation of the cGMP-PK would be proportional to the intracellular cGMP concentration spanning across almost two orders of magnitude. Indeed, experimental evidence for this suggestion has been recently provided (Liebermann et al., 1991; Eigenthaler et al., 1992). The latter study was possible since cAMP-PK- and cGMP-PK-mediated phosphorylation of a 46/50 kDa protein called vasodilator-stimulated phosphoprotein (VASP; further details see below) could be quantitatively measured by an immunological method using intact human platelets (Halbrügge et al., 1990). PG-I$_2$ and other cAMP-elevating prostaglandins convert almost all of VASP from the dephospho- to phosphoform (Nolte et al., 1991a) whereas cGMP-elevating nitrovasodilators such as nitroprusside maximally convert up to 45-50% of VASP to its phosphoform as shown in Table 1 (see also Halbrügge et al., 1990; Eigenthaler et al., 1992).

It is interesting to speculate why human platelets contain such high concentrations of cAMP-PK and cGMP-PK. Second-messenger-regulated phosphorylation systems may have several functions including signal amplification, modulation of a maximal response, enhancement of sensitivity to allosteric effectors, integration of biological signals, and rate amplification. The rate of signal generation is directly proportional to the concentration of converter enzymes such as protein kinases, thus the high concentration of cAMP-PK and cGMP-PK in human platelets may indicate that the rapidity of reaching steady-state levels of protein phosphorylation in response to cyclic nucleotide-elevating platelet inhibitors is of particular importance. This seems very plausible since, in the flowing blood, platelets have to respond rapidly to endothelial-derived factors such as PG-I$_2$ and EDRF (which are both of very short-lived nature). In support of this hypothesis, rapid endothelial cell-dependent activation of platelet cAMP-PK and cGMP-PK mediated by PG-I$_2$ and EDRF, respectively, has been demonstrated (Nolte et al., 1991b).

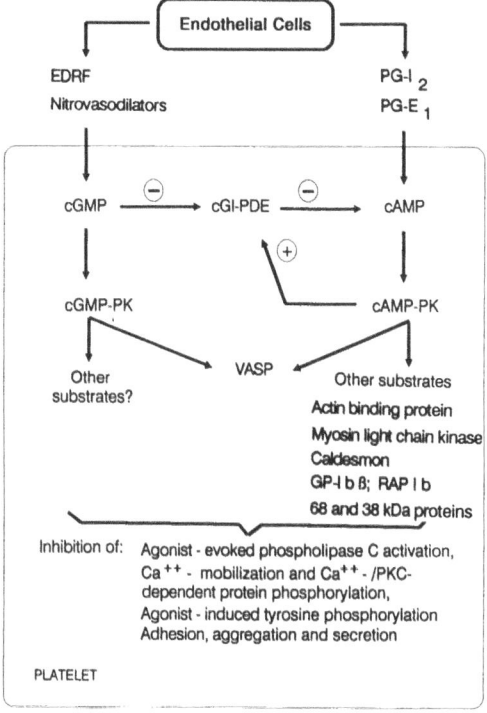

Figure 2   Regulation and action of cyclic nucleotides and cyclic nucleotide-dependent protein kinases in human platelets.

Considerable evidence suggests that the inhibition of human platelets caused by cAMP- and cGMP-elevating agents is mediated by cAMP-PK and cGMP-PK, respectively (Walter et al., 1991). For example, cell-membrane permeant and hydrolysis-resistant cyclic nucleotide analogs (which selectively activate either cAMP-PK or cGMP-PK and do not affect phosphodiesterases) inhibit platelet aggregation, and this inhibition correlates well with activation of either cGMP-PK or cAMP-PK in intact platelets (Sandberg et al., 1991; Geiger et al., 1992; Butt et al., 1992).

The molecular mechanisms of cyclic-nucleotide-mediated platelet inhibition have not been entirely elucidated. Multiple sites of cAMP/cGMP action resulting in inhibition of platelet activation have been considered and investigated (Walter et al., 1991; Halbrügge and Walter, 1993):

a)   inhibition/down-regulation of receptors for agonist such as thrombin, thromboxane etc.,

b) inhibition of PLC or PLC/G-protein/receptor coupling,
c) inhibition of agonist-induced calcium elevation, via activation of plasmalemmal/ intracellular Ca^{2+}-pumps or inhibition of plasmalemmal/intracellular Ca^{2+}-channels,
d) inhibition of MLCK,
e) inhibition of agonist-induced actin polymerization and reorganisation of the cytoskeleton, and
f) inhibition of fibrinogen binding.

Table 1   Intracellular concentration of cyclic nucleotides, cyclic nucleotide-dependent protein kinases and VASP in human platelets.

Substance	Intracellular Concentration		
	basal	after SNP treatment	after PG-I_2 treatment
cAMP	4.4 μM	4-5 μM	30 μM
cGMP	< 0.4 μM	3.8 μM	< 0.4 μM
cAMP-binding sites of cAMP-PK	6.2 μM	n.d.	n.d.
cGMP binding sites of cGMP-PK	14.6 μM	n.d.	n.d.
VASP	25 μM	n.d.	n.d.
% Dephospho-VASP	> 90 %	55 %	15 %
% Phospho-VASP	< 10 %	45 %	85 %

Analysis was performed with unstimulated platelets or cells incubated for 10 min with 100 μM sodium nitroprusside (SNP) or 5 μM prostacyclin (PG-I_2). SNP- or PG-I_2-treatment did not affect the concentration of protein kinases and VASP but the occupancy of kinase cyclic nucleotide binding sites (not shown) and therefore the state of VASP phosphorylation (n.d., when not determined). A binding capacity of 4 moles of cyclic nucleotides per mole of cAMP-PK or cGMP-PK is assumed. Other experimental details have been published (Eigenthaler et al., 1992).

Certainly, both cAMP and cGMP may have inhibitory effects at more than one site, and some of the suggested sites are functionally interrelated, i.e. alteration of Ca^{2+} level will have effects on MLCK, phospholipase A_2, proteases and the cytoskeleton.

There is considerable evidence that inhibition of agonist-induced PLC-activation in intact human platelets is a major effect of both cAMP-PK and cGMP-PK (Siess, 1989; Walter, 1989; Rink and Sage, 1990; Walter et al., 1991; Geiger et al., 1992). As shown in Fig. 3, activation of platelet cAMP-PK by PG-E_1 or by the membrane-permeant cAMP-analog 5,6-DCl-cBiMPS abolished the ADP-induced Ca^{2+}-mobilization from intracellular stores (best seen in the absence of extracellular Ca^{2+}, panel C) but did not inhibit the ADP-activated cation channel (measured as ADP-induced Mn^{2+} influx resulting in decreased Fura-2 fluorescence, panel A). Similar results were obtained when the platelet cGMP-PK was activated by nitrovasodilators and membrane-permeant cGMP analogs (Geiger et al., 1992). Activation of cAMP-PK or cGMP-PK in human platelets also inhibits the agonist-induced generation of inositol trisphosphate (IP_3) and 1,2-diacylglycerol (1,2-DG) and prevents the activation of both MLCK (measured by myosin light chain phosphorylation) and PKC (measured by pleckstrin phosphorylation) (Siess, 1989; Waldmann and Walter, 1989; Geiger et al., 1992; Butt et al., 1992). These data strongly suggest that cAMP-PK- and cGMP-PK-caused platelet inhibition is due to an inhibition at the level of IP_3- and 1,2-DG-generation, e.g. at the level of PLC or agonist receptor/G-protein/PLC-coupling. Direct phosphorylation of PLC-γ by the cAMP-PK has been observed in studies with fibroblasts (Olashaw et al., 1990) and with the human Jurkat T-cell line (Park et al., 1992). However, the molecular details of PLC inhibition by both cAMP-PK and cGMP-PK in platelets remain to be elucidated.

Activation of intracellular calcium pumps by both cAMP and cGMP (Grover and Khan, 1992; Johansson et al., 1992; Tao et al., 1992; Johansson and Haynes, 1992) and inhibition of MLCK by cAMP-PK-mediated protein phosphorylation (Sellers and Adelstein, 1987) may represent additional components of cAMP-and cGMP-induced platelet inhibition.

ROLE OF VASP AND OTHER SUBSTRATES OF CYCLIC NUCLEOTIDE-DEPENDENT PROTEIN KINASES IN HUMAN PLATELETS

Elucidation of the mechanism of cGMP/cAMP action in human platelets requires the identification of specific cGMP-PK and cAMP-PK substrates in studies with intact cells as well as characterization of the functional properties of these substrates.

A 46/50 kDa protein VASP recently purified from human platelets is the only established substrate for the cGMP-PK and an excellent substrate for the cAMP-PK as well (Waldmann et al., 1986; Waldmann et al., 1987; Halbrügge and Walter, 1989; Halbrügge et al., 1990; Eigenthaler et al., 1992).

In addition to VASP, several other proteins have been proven to be substrates of the platelet cAMP-PK (see also Fig.2). These proteins include the 240 kDa (240-260 kDa in different gel systems) actin-binding protein (also known as filamin) whose interaction with the glycoprotein Ib-IX complex (Andrews and Fox, 1991) or with G-proteins (Ueda et al., 1992) may be regulated by cAMP-PK-mediated phosphorylation.

Phosphorylation and activation of cGI-PDE type III (cGI-PDE) by cAMP-PK has been demonstrated both with purified proteins and with intact human platelets (Macphee et al., 1988). As discussed above, cAMP-PK-mediated phosphorylation of the cGI-PDE may represent a turn-off mechanism for the cAMP response. Phosphorylation of myosin light chain kinase (MLCK) by cAMP-PK in intact cells prevents the subsequent activation of MLCK by Ca^{2+}/calmodulin. As extensively discussed by others, cAMP-PK-mediated phosphorylation of MLCK may at least contribute to the inhibitory effects of cAMP-elevating agents on smooth muscle and platelet activation (Sellers and Adelstein, 1987). Phosphorylation of the ß-subunit of glycoprotein Ib by the cAMP-PK in-vitro and in intact cells may be responsible for inhibition of collagen-induced actin polymerization (Fox et al., 1987; Fox and Berndt, 1989). cAMP-PK phosphorylation of a small G-protein (smg p21, identified as Rap-1b) in intact platelets appears to affect the GDP/GTP exchange and intracellular localization of this protein (Kawata et al., 1989; Siess et al., 1990; Lapetina et al., 1989; Hata et al., 1991). Other proteins which are phosphorylated by cAMP-PK in intact platelets are caldesmon (Hettasch and Sellers, 1991) and two unidentified proteins with molecular masses of 68 and 38 kDa (Waldmann et al., 1987; Siess and Lapetina, 1990; Sandberg et al., 1991).

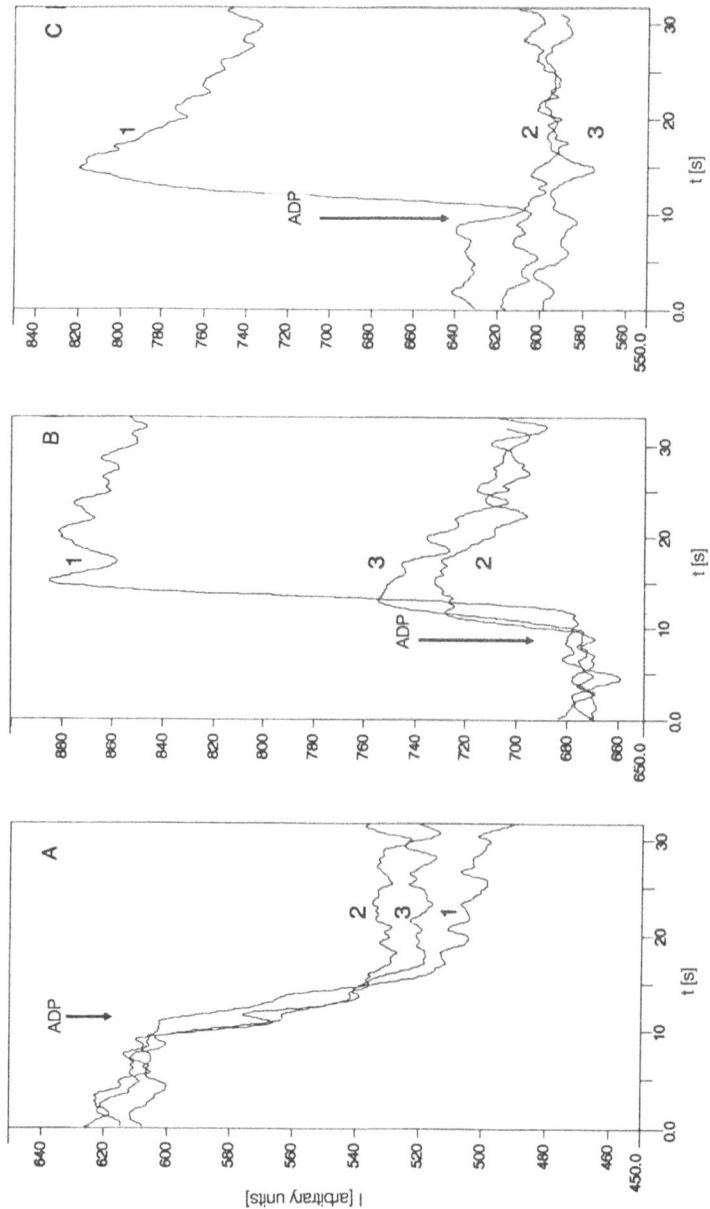

Figure 3   In Fura-2-loaded human platelets activation of platelet cAMP-PK inhibits the ADP-evoked calcium mobilisation from intracellular stores (B,C) but not the first phase of ADP-induced calcium influx (A). Calcium influx (measured as Mn^{2+} influx, A) was indicated by a reduction of Fura-2 fluorescence (Mn^{2+} quench). Calcium mobilisation in the presence (B) or absence (C) of extracellular Ca^{2+} was monitored as increased fluorescence. Intracellular calcium elevation in the absence of extracellular Ca^{2+} results from Ca^{2+} mobilisation from intracellular stores whereas the response in the presence of extracellular Ca^{2+} results from both Ca^{2+} influx and intracellular mobilisation. Platelet cAMP-PK was activated by a 10min preincubation with $5\mu M$ PG-E$_1$ (2) or 0.5mM 5,6-DCl-cBiMPS (3) before platelets were stimulated with $20\mu M$ ADP. Platelets preincubated with buffer alone (1) served as controls. Other experimental conditions have been published (Geiger et al. 1992).

Our group has extensively characterized and purified a 46/50 kDa protein designated as VASP since VASP is phosphorylated in intact human platelets in response to both cGMP- as well as cAMP-elevating platelet inhibitors whose effects are mediated by the cGMP-PK and cAMP-PK, respectively. Our current information concerning the properties and functions of VASP is summarized below.

a)   concentrations of VASP are present in human platelets (0.3 % of total platelet protein), low concentrations are detectable in most tissues and cell types examined (Halbrügge et al., 1992; Eigenthaler et al., 1992; Reinhard et al., 1992).

b)   VASP has an asymmetric, trimeric or tetrameric structure with at least one disulfide bond (Halbrügge et al., 1990; Reinhard and Walter, unpublished).

c)   Stoichiometric and reversible (upon stimulus removal) phosphorylation of VASP by both cAMP-PK and cGMP-PK can be demonstrated with intact human platelets, platelet extracts and purified proteins (Halbrügge and Walter, 1989; Halbrügge et al., 1990; Eigenthaler et al., 1992).

d)   VASP has at least two separate phosphorylation sites (on serine and threonine residues), and phosphorylation of one of these sites used by either cAMP-PK or cGMP-PK shifts the apparent molecular mass of VASP from 46 to 50 kDa on SDS-PAGE (Halbrügge and Walter, 1989; Eigenthaler et al., 1992; Butt et al., unpublished).

e)   VASP is a novel protein associated with actin filaments, focal contact areas (i.e. transmembrane junctions between microfilaments and the extracellular matrix) and cell-cell contacts (Reinhard et al., 1992).

f)   VASP phosphorylation in intact human platelets correlates very well with the platelet inhibition induced by both cAMP- and cGMP-elevating agents. VASP phosphorylation is therefore a useful marker for the action of nitrovasodilators, cAMP-elevating prostaglandins and for the interaction of endothelial cells and platelets (Waldmann et al., 1987; Halbrügge et al., 1990; Siess and Lapetina, 1990; Nolte et al., 1991a; Nolte et al., 1991b; Eigenthaler et al., 1992; Geiger et al., 1992).

A typical example of the intracellular localization of VASP in human platelets spread on glass is shown in Fig.4.

VASP phosphorylation correlates very well with platelet inhibition; however the precise functional role of VASP remains to be established. Hopefully, the properties and functions of VASP will be further elucidated by our ongoing cloning studies and determination of the VASP amino acid sequence. The subcellular localization of VASP in platelets and the association of VASP with actin filaments and focal contact areas suggest a possible role of VASP in regulating platelet cytoskeleton and adhesion. However, a role of VASP in regulating an enzymatic function cannot be ruled out. Recently, an association of pp60c-src, PLC, phosphoinositide 3-kinase and other enzymes involved in phosphatidylinositol metabolism with the cytoskeleton and membrane skeleton during platelet activation has been demonstrated (Grondin et al., 1991; Horvath et al., 1992; Zhang et al., 1992). Clearly, a possible interaction of VASP with some of these enzymes and regulation of their activities by phospho-/dephospho-VASP could be important in the mechanism of platelet inhibition. These different possibilities will have to be addressed in future investigations.

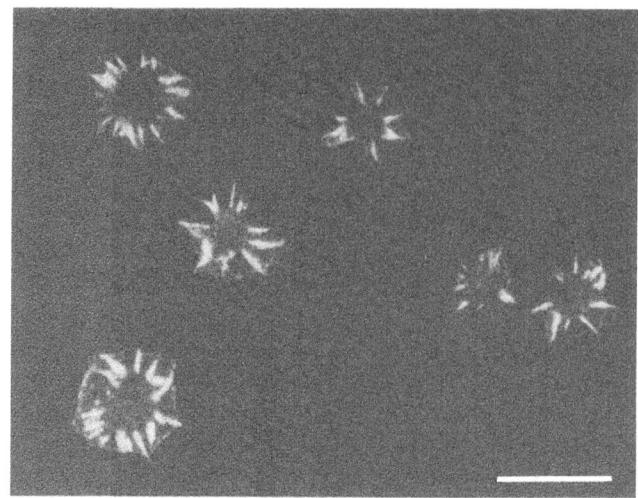

Figure 4   Subcellular distribution of VASP in human platelets spread on a glass surface. VASP distribution was studied by indirect immunofluorescence using a VASP antiserum and a TRITC-conjugated second antibody. Prominent VASP staining was found to be associated with the distal ends of microfilament bundles (sites of attachment to the underlying substratum, i.e. focal contacts) and with microfilaments outlining the periphery. The size of adhered and spread platelets is about 10 - 15 μM (bar: 10 μM). Other experimental details have been published (Reinhard et al., 1992).

SUMMARY

The activation of human platelets is inhibited by two intracellular pathways regulated by either cGMP- or cAMP-elevating agents. There is considerable evidence that the inhibitory effects of cGMP and cAMP are mediated by the cGMP-PK and cAMP-PK, respectively, in human platelets. The cGI-PDE is an additional target for cGMP, and the cGMP-mediated elevation of cAMP levels contributes to the well known synergism between cAMP- and cGMP-elevating platelet inhibitors. Stimulation of both cAMP-PK and cGMP-PK prevents the agonist-induced activation of MLCK and PKC and inhibits the agonist-induced calcium mobilization from intracellular stores without any major effect on the ADP-regulated cation channel. These studies suggest that the inhibition of an early event of platelet activation, e.g. activation of PLC, is an effect common to both cGMP-PK and cAMP-PK stimulation. A common substrate of both cGMP-PK and cAMP-PK, the 46/50 kDa protein VASP, has been recently identified as a novel microfilament- and focal contact-associated protein whose phosphorylation correlates very well with platelet inhibition. Future investigations will have to identify the precise molecular mechanism of cyclic nucleotide inhibition of Ca^{2+} discharge from intracellular stores and whether cGMP-PK- and cAMP-PK-mediated VASP phosphorylation is an important component of this effect of cyclic nucleotides in human platelets.

Acknowledgements

The research of our laboratory is supported by the Deutsche Forschungs-gemeinschaft (Ko 210/11-3; SFB 176, TP A11). The authors thank S. Ebert for the skillful preparation of the manuscript. The authors also thank their collaborators (in particular Drs. B.M. Jockusch (Bielefeld), J.A. Beavo (Seattle), B. Jastorff (Bremen) and S.O. Sage

(Cambridge) for important contributions to some of our original work reviewed here.

REFERENCES

Aktories, K., and Jakobs, K.H., 1985, Regulation of platelet cAMP formation, in "The platelets: physiology and pharmacology", G.L. Longenecker, ed., Academic Press, New York.

Andrews, R.K., and Fox, J.E.B., 1991, Interaction of purified actin-binding protein with the platelet membrane glycoprotein Ib-IX complex, *J. Biol. Chem.* 266:7144.

Beavo, J.A., 1988, Multiple isoenzymes of cyclic nucleotide phosphodiesterase, Adv. Second Messenger *Phosphoprotein Res.* 22:1.

Beavo, J.A., and Reifsnyder, D.H., 1990, Primary sequence of cyclic nucleotide phosphodiesterase isozymes and the design of selective inhibitors, *Trends Pharmacol. Sci.* 11:150.

Bentley, J.K., and Beavo, J.A., 1992, Regulation and function of cyclic nucleotides, *Curr. Opin. Cell. Biol.* 4:237.

Bowen, R., and Haslam, R.J., 1991, Effects of nitrovasodilators on platelet cyclic nucleotide levels in rabbit blood; role for cyclic AMP in synergistic inhibition of platelet function by sin-1 and prostaglandin E1, *J. Cardiovascul. Pharmacol.* 17:424.

Butt, E., Geiger, J., Jarchau, T., Lohmann, S.M., and Walter, U., 1993, cGMP-dependent protein kinase - gene, structure and function, *Neurochem. Res.*,18: 27.

Butt, E., Nolte, C., Schulz, S., Beltman, J., Beavo, J., Jastorff, B., and Walter, U., 1992, Analysis of the functional role of cGMP-dependent protein kinase in intact human platelets using a specific activator 8-pCPT-cGMP, *Biochem. Pharmacol.* 43: 2591.

Currie, M.G., Fok, K.F., Kato, J., Moore, R.J., Hamra, F.K., Duffin, K.L., and Smith, C.E., 1992, Guanylin: An endogenous activator of intestinal guanylate cyclase, *Proc. Natl. Acad. Sci. USA* 89:749.

DiFrancesco, D., and Tortora, P., 1991, Direct activation of cardiac pacemaker channels by intracellular cAMP. *Nature* 351:145.

Eigenthaler, M., Nolte, C., Halbrügge, M., and Walter, U., 1992, Concentration and regulation of cyclic nucleotides, cyclic nucleotide-dependent protein kinases and one of their major substrates in human platelets, *Eur. J. Biochem.* 205:471.

Felbel, J., Trockur, B., Ecker, T., Landgraf, W., and Hofmann, F., 1988, Regulation of cytosolic calcium by cAMP and cGMP in freshly isolated smooth muscle cells from bovine trachea, *J. Biol. Chem.* 263:16764.

Fox, J.E.B., and Berndt, M.C., 1989, Cyclic AMP-dependent phosphorylation of glycoprotein Ib inhibits collagen-induced polymerization of actin in platelets, *J. Biol. Chem.* 264:9520.

Fox, J.E.B., Reynolds, C.C., and Johnson, M.M., 1987, Identification of glycoprotein Ibß as one of the major proteins phosphorylated during exposure of intact platelets to agents that activate cyclic AMP-dependent protein kinase, *J. Biol. Chem.* 262:12627.

Geiger, J., Nolte, C., Butt, E., Sage, S. O., and Walter, U., 1992, Role of cGMP and cGMP-dependent protein kinase in nitrovasodilator inhibition of agonist-evoked calcium elevation in human platelets, *Proc. Natl. Acad. Sci. USA.* 89:1031.

Grondin, P., Plantavid, M., Sultan, C., Breton, M., Mauco, G., and Chap, H., 1991, Interaction of pp60c-src, phospholipase C, inositol-lipid, and diacylglycerol kinases with the cytoskeletons of thrombin-stimulated platelets, *J. Biol. Chem.*, 266:15705.

Grover, A.K., and Khan, I., 1992, Calcium pump isoforms: diversity, selectivity and plasticity, *Calcium* 13:9.

Halbrügge, M., and Walter, U., 1989, Purification of a vasodilator-regulated phosphoprotein from human platelets, *Eur. J. Biochem.* 185:41.

Halbrügge, M., Friedrich, C., Eigenthaler, M., Schanzenbächer, P., and Walter, U., 1990, Stoichiometric and reversible phosphorylation of a 46-kDa protein in human platelets in response to cGMP- and cAMP-elevating vasodilators, *J. Biol. Chem.*, 265:3088.

Halbrügge, M., and Walter, U., 1990, Analysis, purification and properties of a 50 000-dalton membrane-associated phosphoprotein from human platelets, *J. Chromatogr.* 521:335.

Halbrügge, M., Eigenthaler, M., Polke, C., and Walter, U., 1992, Protein phosphorylation regulated by cyclic nucleotide-dependent protein kinases in cell extracts and in intact human lymphocytes, *Cellular Signalling*, 4:189.

Halbrügge, M., and Walter, U., 1993, The regulation of platelet functions by protein kinases, in "Protein Kinases in Blood Cell Function", C.-K. Huang, R.I. Sha'afi, eds, CRC Press, USA.

Haslam, R.J., 1987, Signal transduction in platelet activation, in "Thrombosis and Haemostasis", M.

Verstraete, J. Vermylen, R. Lijnen, J. Arnout, eds., Leuven University Press, Leuven.

Hata, Y., Kaibuchi, K., Kawamura, S., Hiroyoshi, M., Shirataki, H., and Takai, Y., 1991, Enhancement of the actions of smg p21 GDP/GTP exchange protein by the protein kinase A-catalyzed phosphorylation of smg p21, *J. Biol. Chem.* 266:6571.

Hettasch, J.M., and Sellers, J.R., 1991, Caldesmon phosphorylation in intact human platelets by cAMP-dependent protein kinase and protein kinase C., *J. Biol. Chem.* 266:11876.

Horvath, A.R., Muszbek, L., and Kellie, S., 1992, Translocation of pp60c-src to the cytoskeleton during platelet aggregation, *EMBO J.* 11:855.

Jiang, H., Colbran, J.L., Francis, S.H., and Corbin, J.D., 1992, Direct evidence for cross-activation of cGMP-dependent protein kinase by cAMP in pig coronary arteries, *J. Biol. Chem.* 267:1015.

Johansson, J.S., and Haynes, D.H., 1992, Cyclic GMP increases the rate of the calcium extrusion pump in intact human platelets but has no direct effect on the dense tubular calcium accumulation system, *Biochim. Biophys. Acta.* 1105:40.

Johansson, J.S., Nied, L.E., and Haynes, D.H., 1992, Cyclic AMP stimulates Ca^{2+}-ATPase-mediated Ca^{2+} extrusion from human platelets, *Biochim. Biophys. Acta.* 1105:19.

Kaupp, U.B., 1991, The cyclic nucleotide-gated channels of vertebrate photoreceptors and olfactory epithelium, *Trends Neurosci.* 14:150.

Kawata, M., Kikuchi, A., Hoshijima, M., Yamamoto, K., Hashimoto, E., Yamamura, H. and Takai, Y., 1991, Phosphorylation of smg p21, a ras p21-like GTP-binding protein, by cAMP-dependent protein kinase in a cell-free system and in response to prostaglandin E1 in intact human platelets, *J. Biol. Chem.* 264:15688.

Koesling, D., Böhme, E., and Schultz, G., 1991, Guanylyl cyclases, a growing number of signal-transducing enzymes. *FASEB J.* 5:2785.

Krupinsky, J., 1991, The adenylyl cyclase family, *Mol. Cell. Biochem.* 104:73.

Lapetina, E.G., Lacal, J.C., Reep, B.R., and Molina y Vedia, L., 1989, A ras-related protein is phosphorylated and translocated by agonists that increase cAMP levels in human platelets, *Proc. Natl. Acad. Sci.* 86:3131.

Liebermann, E.H., O'Neill, S., and Mendelsohn, M.E., 1991, S-Nitrosocysteine inhibition of human platelet secretion is correlated with increases in platelet cGMP levels, *Circ. Res.* 68:1722..

Lohmann, S.M., Fischmeister, R., and Walter, U., 1991, Signal transduction by cGMP in heart, *Basic Research in Cardiology*, 86:503.

Macphee, C.H., Reifsnyder, D.H., Moore, T.A., Lerea, K.M., and Beavo, J.A., 1988, Phosphorylation results in activation of a cAMP phosphodiesterase in human platelets, *J. Biol. Chem.* 263:10353.

Maurice, D.H. and Haslam, R.J., 1990, Molecular basis of the synergistic inhibition of platelet function by nitrovasodilators and activators of adenylate cyclase: inhibition of cyclic AMP breakdown by cyclic GMP, *Mol. Pharmacol.* 37:671.

McKnight, G.S., 1991, Cyclic AMP second messenger systems, *Curr. Opin. Cell. Biol.* 3:213.

Meinecke, M., Büchler, W., Fischer, L., Lohmann, S.M., and Walter, U., 1990, cAMP-dependent protein kinase: subunit diversity and functional role in gene expression, in "Cellular and molecular biology of myelination", Jeserich, G., Althaus, H.H., and Waehnelt, T.V., eds., NATO ASI series H, Vol.43, Springer-Verlag, Heidelberg.

Méry, P.-F., Lohmann, S.M., Walter, U., and Fischmeister, R. 1991, Ca^{2+} current is regulated by cyclic GMP-dependent protein kinase in mammalian cardiac myocytes, *Proc. Nat. Acad. Sci. USA* 88:1197.

Nicholson, C.D., Challiss, R.A.J., and Shahid, M., 1991, Differential modulation of tissue function and therapeutic potential of selective inhibitors of cyclic nucleotide phosphodiesterase isoenzymes, *Trends Pharmacol. Sci.* 12:19.

Nolte, C., Eigenthaler, M., Schanzenbächer, P., and Walter, U., 1991a, Comparison of vasodilatory prostaglandins with respect to cAMP-mediated phosphorylation of a target substrate in intact human platelets. *Biochem. Pharmacol.* 42:253.

Nolte, C., Eigenthaler, M., Schanzenbächer, P., and Walter, U., 1991b, Endothelial cell-dependent phosphorylation of a platelet protein mediated by cAMP- and cGMP-elevating factors, *J. Biol. Chem.* 266:14808.

Olashaw, N.E., Rhee, S.G., and Pledger, W.J., 1990, Cyclic AMP agonists induce the phosphorylation of phospholipase C-γ and of a 76 kDa protein coprecipitated by anti-(phospholipase C-γ) monoclonal antibodies in BALB/ c-3T3 cells, *Biochem. J.* 272:297.

Park, D.J., Min, H.K., and Rhee, S.G., 1992, Inhibition of CD3-linked phospholipase C by phorbol ester and by cAMP is associated with decreased phosphotyrosine and increased phosphoserine contents of PLC-γ1, *J. Biol. Chem.* 267:1496.

Reinhard, M., Halbrügge, M., Scheer, U., Wiegand, C., Jockusch, B.M., and Walter, U., 1992, The 46/50 kDa phosphoprotein VASP purified from human platelets is a novel protein associated with actin filaments and focal contacts, *EMBO J.* 11:2063.

Rink, T., and Sage, S.O., 1990, Calcium signaling in human platelets, *Annu. Rev. Physiol.* 52:431.

Sandberg, M., Butt, E., Nolte, C., Fischer, L., Halbrügge, M., Beltman, J., Jahnsen, T., Genieser, H.-G., Jastorff, B., and Walter, U., 1991, Characterization of Sp-5, 6-dichloro-1-ß-D-ribofurano-sylbenzimidazole-3',5'-monophoshorothioate(Sp-5, 6-DCl-cBIMPS) as a potent and specific activator of cyclic-AMP-dependent protein kinase in cell extracts and intact cells, *Biochem. J.* 279:521.

Sellers, J.R., and Adelstein, R.S., 1987, Regulation of contractile activity in "The enzymes", third edition, P.D. Boyer, E.G. Krebs, eds., Vol. 17, Academic Press, New York.

Siess, W., 1989, Molecular mechanisms of platelet activation. *Physiol. Rev.* 69:58.

Siess, W., and Lapetina, E.G., 1990, Functional relationship between cyclic AMP-dependent protein phosphorylation and platelet inhibition, *Biochem. J.* 271:815.

Siess, W., Winegar, D.A., and Lapetina, E.G., 1990, Rap-1b is phosphorylated by protein kinase A in intact human platelets, *Biochem. Biophys. Res. Comm.*, 170:944.

Tao, J., Johansson, J.S. and Haynes, D.H., 1992, Stimulation of dense tubular Ca^{2+} uptake in human platelets by cAMP, *Biochim. Biophys. Acta* 1105:29.

Taylor, S.S., Buechler, J.A., and Yonemoto, W., 1990, cAMP-dependent protein kinase: framework for a diverse family of regulator enzymes. *Annu. Rev. Biochem.* 59:971.

Tremblay, J., Gerzer, R., and Hamet, P., 1988, Cyclic GMP in cell function. *Adv. Second Messenger and Phosphoprotein Research* 22:319.

Ueda, M., Oho, C., Takisawa, H., and Ogihara, S., 1992, Interaction of the low-molecular-mass, guanine-nucleotide-binding protein with the actin-binding protein and its modulation by the cAMP-dependent protein kinase in bovine platelets, *Eur. J. Biochem.* 203:347.

Vane, J.R., Äangard, E.E., and Botting, R.M., 1990, Regulatory functions of the vascular endothelium, *N. Engl. J. Med.* 323:27.

Waldmann, R., and Walter, U., 1989, Cyclic nucleotide-elevating vasodilators inhibit platelet aggregation at an early step of the activation cascade, *Eur. J. Pharmacol.* 159:317.

Waldmann, R., Bauer, S., Göbel, C., Hofmann, F., Jakobs, K. H., and Walter, U., 1986, Demonstration of cGMP-dependent protein kinase and cGMP-dependent phosphorylation in cell-free extracts of platelets, *Eur. J. Biochem.* 158:203.

Waldmann, R., Nieberding, M., and Walter, U., 1987, Vasodilator-stimulated protein phosphorylation in platelets is mediated by cAMP- and cGMP-dependent protein kinases, *Eur. J. Biochem.* 167:441.

Walter, U., 1984, Cyclic cGMP-regulated enzymes and their possible physiological functions, *Adv. Cyclic Nucleotide Protein Phosphorylation Res.* 17:251.

Walter, U., 1989, Physiological role of cGMP and cGMP-dependent protein kinase in the cardiovascular system, *Rev. Physiol. Biochem. Pharmacol.* 113:42.

Walter, U., Nolte, C., Geiger, J., Schanzenbächer, P., Kochsiek, K., 1991, Inhibition of platelet function by cyclic nucleotides and cyclic nucleotide-dependent protein kinases, in "Antithrombotics: pathophysiological rationale for pharmacological interventions", Herman, A.G., ed., Kluwer Academic Publishers, Dortrecht.

Zhang, J., Fry, M.J., Waterfield, M.D., Jaken, S., Liao, L., Fox J.E.B., and Rittenhouse, S.E., 1992, Activated phosphoinositide 3-kinase associates with membrane skeleton in thrombin-exposed platelets, *J. Biol. Chem.*, 267:4686.

THE BIOLOGICAL AND PHARMACOLOGICAL ROLE OF NITRIC OXIDE

IN PLATELET FUNCTION

Marek W. Radomski and Salvador Moncada

Wellcome Research Laboratories,
Langley Court Beckenham, Kent BR3 3BS, U.K.

INTRODUCTION

Regulation of platelet haemostasis is of vital importance for homeostasis of the vascular wall. The purpose of this article is to present the physiological role of the constitutive nitric oxide (NO) synthase as a major regulatory mechanism for the control of platelet haemostasis and prevention of thrombosis. We will also describe pathological implications of expression of the inducible NO synthase and review the pharmacological aspects of the inhibitory action of NO on platelets.

ENDOTHELIUM-DEPENDENT RELAXATION AND ITS RELATION TO THE FORMATION OF NITRIC OXIDE

Furchgott and Zawadzki (1980) demonstrated that the vascular relaxation induced by acetylcholine was dependent on the presence of the endothelium and provided evidence that this effect was mediated by a labile humoral factor which they later named endothelium-derived relaxing factor (EDRF). This discovery sparked off an intensive research effort aimed at identifying the chemical structure of EDRF (Moncada et al., 1991). Furchgott (1988) suggested that EDRF may be NO because of the similarities in the pharmacological behaviour of EDRF and NO released from acidified NO_2^-. At the same time Ignarro et al. (1988) also speculated that EDRF may be NO or a closely-related species. The first clear evidence in support of this proposal came from the experiments in which the biological actions of EDRF and NO on vascular strips and on platelets measured by biological , pharmacological and chemical methods were shown to be indistinguishable. Moreover, endothelial cells released NO in quantities sufficient to account for the vasodilator and platelet inhibitory effects of EDRF. To date an overwhelming amount of evidence has accumulated in favour of EDRF being NO and reported discrepancies may simply reflect different experimental conditions which may vary from one to another laboratory (for rev. see Moncada et al., 1991).

Mechanisms of Platelet Activation and Control, Edited by
K.S. Authi *et al.*, Plenum Press, New York, 1993

CHARACTERIZATION OF NITRIC OXIDE SYNTHASE

Two forms of NO synthase, the constitutive and inducible enzyme, have been identified. Both are NADPH- and biopterin-dependent enzymes which use a guanido nitrogen atom of L-arginine (L-Arg) and in an enantiomer-specific reaction incorporate molecular oxygen into NO and L-citrulline (for ref. see Moncada et al., 1991). Studies with inhibitors of NO synthase such as N^G-monomethyl-L-arginine (L-NMMA) indicate that the endothelial constitutive NO synthase generates NO constantly and that this mediator can be considered as the endogenous nitrovasodilator (Moncada et al., 1989). The physiological stimuli for generation of NO are not yet fully understood but pulsatile flow and shear stress seem to be two of the major determinants (Pohl et al., 1986; Drexler et al., 1989). Unlike endothelial cells, platelets do not constantly generate NO and in resting platelets the synthesis of NO is not detectable. However, NO synthase becomes activated during platelet aggregation leading to the formation of NO (Radomski et al., 1990 a,b). Mechanisms which transduce the message carried by physiological stimuli to trigger the synthesis of NO have not been fully elucidated, however, since the constitutive platelet/endothelial NO synthase is strictly Ca^{2+}-dependent (at very narrow range of Ca^{2+} concentrations from 0.1 to 1.0 μM) it is likely that this divalent cation controls the activation of NO synthase *in vivo* (Radomski et al., 1990 a,b). An increase in $[Ca^{2+}_i]$ in the endothelium may result from the activation of stretch-activated cation-selective channels which are permeable for Ca^{2+} (Lansman et al.,1987; Ohno et al., 1990) whereas the same process in platelets is likely to be a consequence of platelet activation by some aggregating agents (Ware et al., 1987).

The other enzyme that produces NO is the Ca^{2+}-independent, cytokine-inducible NO synthase, whose expression requires de novo protein synthesis (for ref. see Moncada et al., 1991). This enzyme is expressed in endothelium and smooth muscle vascular cells following stimulation with endotoxin and cytokines (Radomski et al., 1990c; Rees et al.,1990). Platelets contain no DNA and only residual RNA, therefore they have a very limited capacity for synthesizing proteins. Platelets acquire their proteins either from the circulation or by transfer from the megakaryocyte which is capable of protein synthesis. We have recently found that human megakaryoblastic cells (Meg-01) possess the constitutive NO synthase and have the capacity to express the inducible NO synthase following stimulation with interleukin-1ß and tumour necrosis factor α (Lelchuk et al.,1992). Thus, it is likely that both enzymes are synthesized in megakaryocytes and then transferred into platelets.

AFFINITY TARGETS FOR NITRIC OXIDE

The Soluble Guanylate Cyclase

Under physiological conditions NO has a high binding affinity for haem iron and therefore reacts with haemoproteins including the soluble guanylate cyclase (SGC). This binding results in the conversion of magnesium guanosine 5'-triphosphate to guanosine 3',5'-monophosphate (cyclic GMP) which stimulates cyclic GMP-dependent protein kinase. The subsequent biochemical effects triggered by this nucleotide are less clear, however in platelets, they may result in inhibition of fibrinogen binding to the IIb/IIIa receptor, inhibition of phosphorylation of myosin light chains and of protein kinase C, stimulation of phosphorylation of the ß-subunit of glycoprotein I, modulation of phospholipase A_2- and C -mediated responses (for rev. see Walter, 1989 and Walter, this volume). Intracellular

Ca^{2+} is an important target for cyclic GMP-controlled platelet responses and an increase in SGC activity results in reduction of [Ca$^{2+}_i$] (Nakashima et al., 1986, Matsuoka et al., 1989). Receptor-mediated Ca^{2+} influx and mobilization are regulated by cyclic GMP in platelets, and it appears that this nucleotide is a more potent inhibitor of Ca^{2+} influx than Ca^{2+} mobilization (Morgan and Newby, 1989).

The intraplatelet concentrations of cyclic GMP are not only controlled by the synthesizing enzyme (SGC) but also by degrading enzymes (cyclic nucleotide phosphodiesterases, PDE). The separation of PDE activity from cytosolic fraction of human platelets by DEAE-cellulose chromotography yields three peaks (Hidaka and Asano, 1976). The first enzyme (FI) has a higher activity for cyclic GMP than for cyclic AMP and hydrolyses mainly cyclic GMP at low substrate levels. The second enzyme (FII) exhibits low affinity for both cyclic nucleotides and appears to belong to cyclic GMP-stimulated PDE, since low levels of cyclic GMP stimulate the rate of hydrolysis of cyclic AMP (Grant et al., 1990). The third enzyme (FIII) has a higher affinity for cyclic AMP and its activity is inhibited by low levels of cyclic GMP (Grant and Colman, 1984). Thus, NO-mediated increases in cyclic GMP can also significantly affect metabolism of cyclic AMP in platelets.

Stimulation of NO release in cultured endothelium results in the activation of SGC and elevation of cyclic GMP levels in these cells (Martin et al., 1988). Cyclic GMP causes inhibition of inositol 1,4,5-triphosphate formation, probably by a G protein-mediated mechanism (Lang and Lewis, 1991). In bovine and porcine aortic endothelial cells both cyclic GMP-stimulated PDE and cyclic AMP PDE have been identified. It is suggested that the levels of cyclic GMP in endothelium are regulated by the former enzyme (Souness et al., 1990).

Thiol-Containing Molecules

Highly reactive and unstable NO can react with thiol-containing molecules (R-SH) to form S-nitroso-thiols (R-SNO). It has been suggested that NO is stabilized by a reaction with R-SH that prolongs its half-life *in vivo* and preserves its biological activity (Stamler et al., 1992). Indeed, R-SNO are potent inhibitors of platelet aggregation *in vitro* (Mellion et al., 1983) and *in vivo* (Radomski et al., 1992). Low-molecular weight thiols such as S-nitroso-cysteine (Mellion et al., 1983) or S-nitroso-glutathione (Radomski et al, 1992) and high-molecular weight thiols such as S-nitroso-albumin (Stamler et al., 1992) are likely candidates for NO-carrying molecules, however, the generation of these compounds *in vivo* remains to be demonstrated.

Nitric oxide may also interact with thiol-containing enzymes. Recent evidence suggests that NO can S-nitrosylate thiol groups of glyceraldehyde-3-phosphate dehydrogenase leading to inactivation of this enzyme (Molina y Vedia et al., 1992).

Enzymes Containing Nonhaem Iron Coordinated to Sulphur Atoms (Fe-S)

Exposure of Fe-S groups of tumour cells and micro-organisms to NO produced by the inducible NO synthase in activated macrophages results in inhibition of the DNA replication and mitochondrial respiration in the target cells (Hibbs et al., 1990). The platelets as anuclear elements may not be affected by this action of NO and the inhibition of mitochondrial respiration has little effect on intracellular ATP levels because of a compensatory increase in anaerobic glycolysis (Mills, 1981). Moreover, high amounts of NO (10μM) protect platelets from damage via a cyclic GMP-dependent mechanism (Radomski et al., 1988). However, nuclear megakaryocytes and endothelial cells are likely to be more affected by this cytostatic effect of NO.

PHYSIOLOGICAL ROLE OF THE CONSTITUTIVE NO SYNTHASE-SOLUBLE GUANYLATE CYCLASE SYSTEM IN REGULATION OF PLATELET HAEMOSTASIS

Role of Nitric Oxide

In the endothelium, the accumulation of cyclic GMP is a result of activation of NO synthase in these cells by different stimuli. This local, autocrine effect of NO is believed to down regulate the release of NO from endothelium (Martin et al., 1988). The autocrine down-regulating role of NO synthase in platelets has recently been discovered (Radomski et al., 1990a,b). This observation reconciles seemingly paradoxical data concerning the biological significance of cyclic GMP in platelet activation. Early observations showed that platelet aggregating agents such as arachidonic acid and collagen caused an increase in the intraplatelet content of cyclic GMP (Davies et al., 1976). These, in conjunction with the known platelet inhibitory role of cyclic AMP (Marcus and Zucker, 1965), brought about the yin yang hypothesis according to which the function of cyclic GMP was to antagonize the actions of cyclic AMP (Goldberg et al., 1975). In 1981, however, it was found that NO-induced inhibition of platelet aggregation is accompanied by an increase in cyclic GMP. This led to reassessment of the yin yang hypothesis and to the suggestion that cyclic GMP causes inhibition of platelet aggregation (Mellion et al., 1981). The presence of NO synthase in platelets (Radomski et al., 1990a, Pronai et al., 1991) strongly supports this suggestion. The stimulation of platelets by aggregating agents results in activation of NO synthase most probably via Ca^{2+}- and L-Arg-dependent mechanism. The resultant NO activates SGC and increases cyclic GMP whose actions down regulate aggregation (Radomski et al., 1990a,b; Pronai et al., 1991). The efficacy of NO synthase to regulate aggregation varies such that it is most active against collagen and arachidonic acid and less active against ADP, thrombin and Ca^{2+} ionophore A23187. Interestingly, aggregation induced by low concentrations of collagen and arachidonic acid is primarily dependent on activation of the phospholipase A_2/arachidonic acid/thromboxane A_2 (TXA_2) pathway. Therefore, the physiological regulation of aggregation by the platelet NO synthase-cyclic GMP system could be limited to the inhibition of this pathway of platelet aggregation (Sane et al., 1989). The NO synthase-cyclic GMP system is not the only one capable of down regulating TXA_2 pathway of platelet aggregation. The soluble guanylate cyclase may be also activated as a result of stimulation of platelet purinoceptor with guanosine triphosphate (Laustiola et al., 1991) or by fatty acid hydroperoxides (Brune and Ullrich, 1991). In addition, Murray et al. (1990) have recently shown that activation of the TXA_2 pathway is linked to sensitization of platelet adenylate cyclase and increased formation of cyclic AMP. The biological significance of interactions between adenylate and guanylate cyclases in platelets is discussed below.

The amounts of NO available for regulation of platelet function are considerably boosted by the synthesis and release of this mediator from vascular endothelium. Cultured and fresh endothelial cells when stimulated with bradykinin release NO in quantities sufficient to inhibit platelet adhesion (Radomski et al., 1987b,c; Sneddon and Vane, 1988). Moreover, intact coronary and pulmonary vasculature release NO to inhibit platelet adhesion under constant flow conditions (Venturini et al., 1989; Pohl and Busse, 1989). Platelet aggregation *in vitro* induced by a variety of agonists is inhibited by NO released from fresh or cultured endothelial cells (for ref. see Radomski and Moncada, 1991). This NO causes also disaggregation of preformed platelet aggregates (Radomski et al., 1987d). Moreover, stimulation of NO release *in vivo* by cholinergic stimuli, or by substance P results in inhibition of platelet aggregation induced by some aggregating agents (Bhardwaj et al., 1988; Hogan et al., 1988; Humphries et al., 1990). Thus, a concerted action of endothelial and platelet NO synthases regulates platelet activation causing inhibition of adhesion and aggregation, down-regulation of aggregation and induction of disaggregation.

Synergistic Model of Regulation of Platelet Aggregation

Platelets and endothelial cells produce and/or secrete agents which directly inhibit platelet aggregation. In addition to NO, the inhibitory activity of prostacyclin, prostaglandin D_2, tissue plasminogen activator and adenosine triphosphate have been described (Moncada et al., 1976; Whittle et al., 1978; Loscalzo and Vaughan, 1987). Under basal conditions *in vivo* the synthesis and release of a single inhibitor is unlikely to account for regulation of platelet aggregation. Nitric oxide and prostacyclin synergize with each other as inhibitors of platelet aggregation and inducers of disaggregation (Radomski et al., 1987d). In addition, synergistic induction of platelet disaggregation has been recently demonstrated with the combination of glyceryl trinitrate (a NO donor), prostaglandin E_1 and tissue plasminogen activator which act via cyclic GMP, cyclic AMP and plasmin-dependent mechanisms respectively (Stamler et al., 1989). Thus, it is likely that platelet aggregation *in vivo* is regulated by synergistic interactions between inhibitors which may counterbalance the action of proaggregating agents (Ware et al., 1987). The biochemical rationale for the synergistic inhibition of platelet aggregation is unclear, however, for NO and prostacyclin it may depend on an NO-induced increase in cyclic GMP with subsequent inhibition of cyclic GMP-inhibited cyclic AMP PDE leading to an increase in cyclic AMP (Maurice and Haslam, 1990).

Interestingly, NO and prostacyclin do not synergize with each other as inhibitors of platelet adhesion to collagen fibrils. Furthermore, a decrease in prostaglandin synthesis by aspirin does not affect inhibition of platelet adhesion induced by NO (Radomski et al., 1987b,c; Venturini et al., 1989). This suggests that prostacyclin is not a part of the system which controls platelet adhesion. The interactions between NO and other factors known to inhibit adhesion such as 13-hydroxyoctadecadienoic acid and α-tocopherol (Buchanan et al., 1987; Jandak et al., 1989) as inhibitors of platelet adhesion to endothelium and other structures of the vascular wall remain to be investigated.

NITRIC OXIDE SYNTHESIS IN THROMBOTIC AND HAEMORRHAGIC DISEASE

Atherosclerosis is the most important cause of thrombotic disease. The capacity of the endothelium to synthesise NO is reduced both in experimental animal (Verbeuren et al., 1986) and coronary human atherosclerosis (Chester et al., 1990). Moreover, NO has been found to inhibit mitogen release from stimulated human platelets (Barrett et al., 1989). Mitogens such as platelet-derived growth factor released from injured or stimulated platelets may be responsible for production of atherosclerotic proliferative lesions which underlie arterial constriction, thrombosis and plaque formation (Ross and Glomset, 1976a,b). Thus, a deficient NO production could be one of the precipitating or perpetuating factors of atherosclerosis.

Decreased formation of NO can lead directly to thrombosis. Laser-induced endothelial damage *in vivo* which inhibits NO release results in platelet aggregation (Rosenblum et al., 1987). More effective release of NO may be responsible for a better patency of human arterial coronary than venous bypass grafts (Luscher et al., 1989). The inhibition of NO synthesis in the rabbit by N^G-nitro-L-arginine methyl ester (L-NAME) potentiates the pulmonary accumulation of and prolongs disaggregation of [111]indium-labelled platelets induced by sub-maximal doses of ADP, platelet-activating factor and thrombin (May et al., 1991). Moreover, in experimental coronary stenosis in rabbits a massive spreading and formation of aggregates on the surface of damaged endothelium were detected only when animals were pre-treated with L-NMMA (Herbaczynska-Cedro et al.,

1991). These data indicate that the failure of the haemostatic system to produce NO has serious thrombotic repercussions.

Interestingly, platelet thrombosis has been also implicated in the hematogenous dissemination of tumours (Gasic and Gasic, 1962). Platelets form aggregates with tumour cells in the circulation, facilitating their adhesion to vascular endothelium. There is a correlation between the ability of some tumour cells to aggregate platelets *in vitro* and their propensity for metastasis (Gasic et al., 1973). We have recently shown that the generation of NO by human colorectal adenocarcinoma cells inversely correlates with their aggregating potential, thus tumour cells with high NO synthase activity are less likely to form metastasis via a platelet aggregation-dependent mechanism (Radomski et al., 1991a).

Experimental and clinical studies have demonstrated that lipopolysaccharides of the outer membrane of Gram-negative bacteria (endotoxins) are responsible for the development of clinical manifestations of septicaemia and endotoxin shock. Severe hypotension and disturbances in haemostatic-thrombotic balance are among the major clinical symptoms. There is now evidence that these symptoms are due to increased synthesis of NO as a result of cytokine-dependent induction of NO synthase (Petros et al., 1991). Indeed, cytokine-stimulated endothelial and smooth muscle vascular cells produce amounts of NO capable of vasodilatation and inhibition of platelet adhesion and aggregation (Radomski et al., 1990c; Durante et al., 1991; Fig.1.). It is however, important to note that endotoxin stimulates the coagulation cascade (Corrigan et al., 1968) and that some endothelial cells are damaged as a consequence of the induction of NO synthase (Palmer et al., 1992). Therefore, platelet activation may also appear in the areas of injury.

The haemostatic defect occurs also in the uraemic state which is characterized by an accumulation of metabolites of the urea cycle, a prolonged bleeding time and decreased

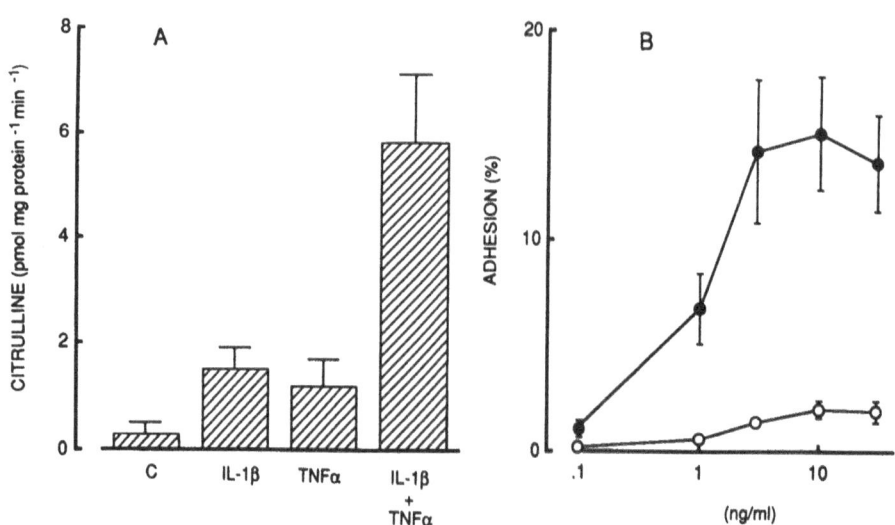

Fig.1   A. Cytokine-induced expression of the inducible NO synthase in human umbilical vein endothelial cells. Interleukin-1ß (IL-1ß, 10ng/ml) and tumour necrosis factora (TNFα, 10ng/ml) synergize (IL-1ß + TNFα) to express the inducible NO synthase determined by the production of citrulline. C -Nitric oxide synthase activity in unstimulated cells.
B. Inhibition of NO synthesis by N[G]-monomethyl-L-arginine (●) enhances IL-1ß + TNFα-induced adhesion (○) of ^{111}In-labelled platelets to human umbilical vein endothelial cells. Cytokines were incubated for 18h with endothelial cells.

platelet adhesion and aggregation. Remuzzi et al. (1990) have shown that the prolonged bleeding time of uraemic rats can be normalized by systemic administration of L-NMMA. In addition, intraplatelet cyclic GMP content is higher in regularly haemodialysed uraemic patients than in control subjects (Gordge and Neild, 1992). These observations suggest that an excessive formation of NO may disturb the haemostatic/thrombotic balance and favour bleeding diathesis.

CLINICAL PHARMACOLOGY OF THE L-ARGININE TO NITRIC OXIDE PATHWAY

Pharmacology of Nitric Oxide Formation and Action

Under physiological conditions, there is a little effect of L-Arg administered acutely on the formation of NO by intact endothelium and by resting platelets (Palmer et al., 1988; Radomski et al., 1990a). This may be due to a high plasma and intracellular content of L-Arg which down-regulates both transport and activation of NO synthase or to the presence of arginase, an enzyme which redirects L-Arg towards metabolites of the urea cycle (Villaneuva and Giret, 1980; Baydoun et al., 1990). However, transport of L-Arg into cultured endothelial cells which were previously depleted of this amino acid is up regulated and under these conditions L-Arg causes the formation of NO (Palmer et al., 1988; Baydoun et al., 1990). Nitric oxide is also formed during platelet aggregation and L-Arg effectively inhibits this process (Caren and Corbo, 1973; Houston et al., 1983; Radomski et al., 1990a). Thus, it is likely that NO synthase is activated during the course of vascular disorders and therefore an acute or chronic administration of L-Arg may have pharmacological significance.

In healthy volunteers, the short-lasting administration of L-NMMA into the brachial artery did not affect collagen-induced platelet aggregation *ex vivo* (Vallance et al., 1992). However, the pharmacological administration of inhibitors of NO synthase potentiated platelet activation induced by aggregating agents. (May et al., 1991; Herbaczynska-Cedro et al., 1991). In addition, inhibitors of NO synthase have recently been used to treat hypotension in patients with septic shock (Petros et al.,1991). Since the inhibition of NO synthesis in cytokine-stimulated human endothelium greatly increases its adhesive properties (Fig.1), platelet behaviour should be carefully monitored during administration of NO synthase inhibitors. However, a limited administration of these compounds may prove of value for stimulation of local haemostasis.

The chemical half-life of released NO is very short and it is inactivated within a few seconds . Several studies have shown that superoxide anions (O_2^-), which are often co-generated with NO largely contribute to the inactivation of this mediator (Gryglewski et al., 1986; Rubanyi et al., 1986). Superoxide anion dismutase (SOD), a scavenger of O_2^-, significantly prolongs and unmasks the antiadhesive, antiaggregating and disaggregating activity of subthreshold amounts of NO (Radomski et al., 1987a-d). Higher concentrations of SOD may also stimulate the activity of guanylate cyclase and inhibit platelet adhesion and aggregation (Salvemini et al., 1989). In addition, SOD reduces *in vitro* an oxidative stress which follows the formation of peroxynitrite from superoxide and nitric oxide radicals (Beckman et al., 1990). On the other hand, SOD initiates the formation of H_2O_2 from O_2^- and this peroxide in the presence of a transition metal may be converted to a very reactive hydroxyl radical which enhances platelet aggregation *in vitro* (Iuliano et al., 1991). All these actions may be pertinent to the effect of SOD on platelets and add to the known pharmacological profile of this agent.

Since the initial demonstration of its action on EDRF (Martin et al., 1985) up to date oxyhaemoglobin (Hb) has become one of the most widely used agents for inhibition of the biological effects of NO. The affinity of haem in Hb to NO is high and therefore this haemoprotein probably acts by reacting with NO to form methaemoglobin and inorganic nitrate which is largely devoid of the biological activity of its precursor. Low concentrations of Hb (0.1-10 μM) are needed for inactivation of platelet inhibitory amounts of NO generated by fresh, cultured endothelial cells or by isolated perfused vasculatures (Sneddon and Vane, 1988; Radomski et al., 1987a-d; Venturini et al., 1989). However, when contained within red blood cells (estimated haematocrit of 10%) the concentration of free Hb required for inactivation of platelet inhibitory effect of NO is 40 to 50 times higher (Houston et al., 1990). In addition, endogenously released NO has been shown to inhibit platelet aggregation *ex vivo* (Bhardwaj et al., 1988). Thus, it appears that under physiological circumstances the large vascular pool of Hb which is contained in red blood cells does not interact readily with NO.

Pharmacology of Nitrovasodilators

Nitrovasodilators are generally accepted as basic medication in almost all patients with angina pectoris unless there are striking contraindications (Jansen et al., 1990). Platelet activation plays an important role in the pathogenesis of ischaemic heart disorder (Fuster et al., 1987). The recognition of the potent antiplatelet properties of endogenous NO raises a possibility that the antianginal effect of nitrovasodilators, which act by releasing NO (Feelisch and Noack, 1987), may also depend on inhibition of platelet activation. Indeed, intravenous administration of glyceryl trinitrate for 24h to patients with acute myocardial infarction significantly inhibited platelet adhesion and aggregation to fibrillar collagen (Gebalska, 1990). Moreover, oral administration of isosorbide dinitrate decreased platelet reactivity in patients with coronary artery disease (Sinzinger et al., 1992). In addition, inhibition of platelet aggregation was observed following intravenous infusion of an NO donor, SIN-1, in patients with myocardial infarction (Wautier et al., 1989). Thus, clinically used nitrovasodilators are effective inhibitors of platelet function.

The antiplatelet actions of NO donors described herein cannot be separated from those on the vascular tone which may limit the use of these drugs solely as inhibitors of platelet activation. We have recently found that S-nitroso-glutathione at platelet inhibitory doses had only a small effect on the blood pressure of the conscious rat (Radomski et al., 1992). Therefore, it may be possible to design platelet-selective NO donors based on the structure of stable S-nitrosothiols such as S-nitroso-glutathione.

Finally, NO donors such as SIN-1 and sodium nitroprusside have been shown to stimulate fibrinolysis probably via inhibition of release of the plasminogen activator inhibitor from platelets (Basista et al.,1985; Korbut et al., 1991).

CONCLUDING REMARKS

The constitutive and inducible NO synthases represent a major biological system which affects platelet function (Fig.2). Nitric oxide synthesized by the constitutive NO synthase is a part of a synergistic and down-regulating pathway which controls platelet haemostasis. The constituents of this pathway are produced not only by the endothelium but also by the platelets themselves. Thus, the platelet can determine its own fate by participating in the process which controls platelet activation and ensures haemostasis. The expression of the inducible NO synthase may have both beneficial and detrimental repercussions for vascular homeostasis (Fig.2).

The biological importance of NO synthase can be further emphasized by its presence and function in haemocytes which are the multicompetent cells in the haemolymph of an arthropod, the American horseshoe crab. The formation of NO in haemocytes down-

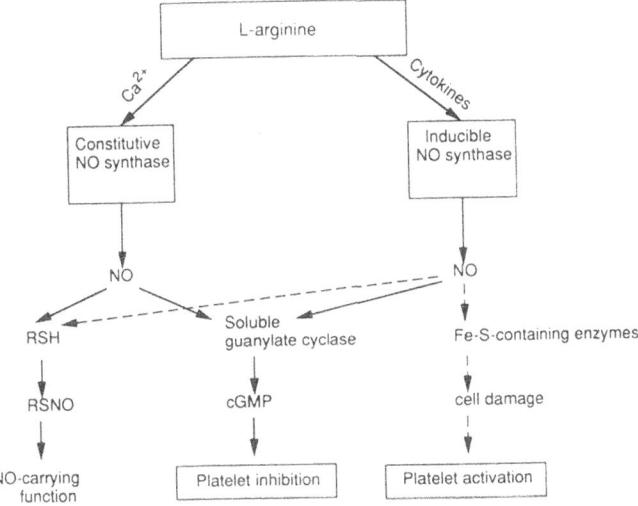

Fig.2 Biological roles of constitutive and inducible NO synthases in platelet haemostasis. Dashed lines represent actions which remain to be demonstrated.

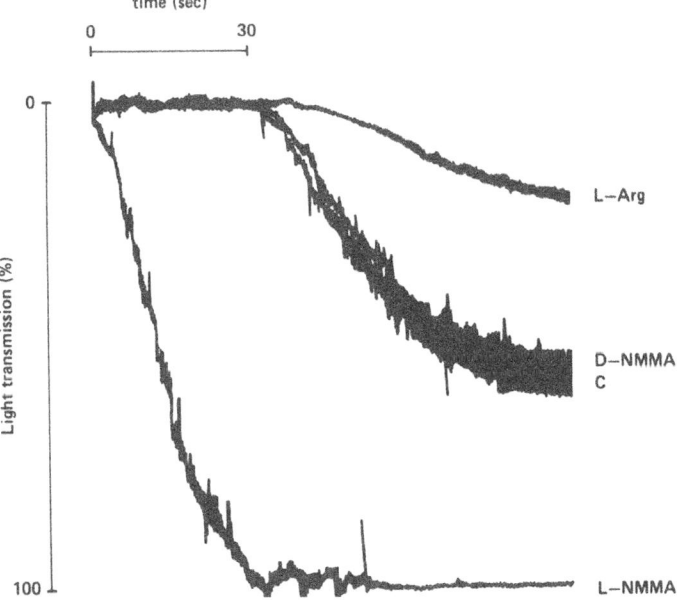

Fig.3 The effect of intracardial administration of L-arginine (L-Arg, 300mg/kg) and N^G-monomethyl-L-arginine (L-NMMA, 100mg/kg) on the *ex vivo* aggregation of haemocytes of American horseshoe crab. The spontaneous aggregation of control haemocytes (C) occured after a lag-phase of ca. 30s and was submaximal. This was not significantly different in samples from crabs treated with D-NMMA (100mg/kg), an inactive enantiomer of L-NMMA. However, aggregation of haemocytes from crabs treated with L-NMMA occured immediately and was maximal (from Radomski et al., 1991, Phil.Trans.R.Soc.Lond. B 334:129; with permission)

regulates the haemostatic function of these cells (Fig.3) in a manner similar to that in human platelets (Radomski et al., 1991b). Since this animal is a "phylogenetic relic" whose existence spans over 500 million years of evolution it is likely that NO synthase ranks among the oldest regulatory systems which have contributed to the development of animal life. Therefore, it is not surprising that the discoveries of endothelium-dependent relaxation and of NO synthase have opened a new chapter in our knowledge on the homeostasis of the vascular wall, pages of which are still being written.

REFERENCES

Barrett, M.L., Willis, A.L. and Vane, J.R., 1989, Inhibition of platelet-derived mitogen release by nitric oxide (EDRF), *Agents and Actions* 27:488.

Basista, M., Grodzinska, L. and Swies, J., 1985, The influence of molsidomine and its active metabolite SIN-1 on fibrinolysis and platelet aggregation, *Thromb. Haemost.* 54:746.

Baydoun, A.R., Emery, P.W., Pearson, J.D. and Mann, G.E., 1990, Substrate-dependent regulation of intracellular amino acid concentrations in cultured bovine aortic endothelial cells, *Biochem. Biophys. Res. Commun.* 173:940.

Beckman, J.S., Beckman, T.W., Chen, J., Marshall, P.A. and Freeman, B.A., 1990, Apparent hydroxyl radical production by peroxynitrite: Implications for endothelial injury from nitric oxide and superoxide, *Proc. Natl. Acad. Sci. USA* 87:1620.

Bhardwaj, R., Page, C.P., May, G.R. and Moore, P.K., 1988, Endothelium-derived relaxing factor inhibits platelet aggregation in human whole blood *in vitro* and in the rat *in vivo*, *Eur. J. Pharmacol.* 157:83.

Brune, B. and Ullrich, V., 1991, 12-hydroperoxyeicosatetraenoic acid inhibits main platelet functions by activation of soluble guanylate cyclase, *Mol. Pharmacol.* 39:671.

Buchanan, M.R., Richardson, M., Haas, T.A., Hirsh, J. and Madri, J.A., 1987, The basement membrane underlying the vascular endothelium is not thrombogenic: *In vivo* and *in vitro* studies with rabbit and human tissue, *Thromb. Haemost.* 58:698.

Caren, R. and Corbo, L., 1973, Response of plasma lipids and platelet aggregation to intravenous arginine, *Proc. Soc. Exp. Biol. Med.* 143:1067.

Chester, A.H., O'Neil, G.S., Moncada, S., Tadjkarimi, S. and Yacoub, M.H., 1990, Low basal and stimulated release of nitric oxide in atherosclerotic epicardial coronary arteries, *Lancet* 336:897.

Corrigan, J.J., Jr., Ray, W. and May, N., 1968, Changes in the blood coagulation system associated with septicemia, *New Engl. J. Med.* 279:851.

Davies, T., Davidson, M.M.L., Mc Clenaghan, M.D, Say, A. and Haslam,R.J., 1976, Factors affecting platelet cyclic GMP levels during aggregation induced by collagen and by arachidonic acid. *Thromb. Res.* 9:387.

Drexler, H., Zeiher, A.M., Wollschlager, H., Meinertz, T., Just, H. and Bonzel, T., 1989, Flow-dependent coronary artery dilatation in humans, *Circulation* 80:466.

Durante, W., Schini, V.B., Scott-Burden, T., Junquero, D.C., Kroll, M.H., Vanhoutte, P.M. and Schafer, A.I., 1991, Platelet inhibition by an L- arginine-derived substance released by Il-1ß-treated vascular smooth muscle cells, *Am. J. Physiol.* 261:H2024

Feelisch, M. and Noack, E.A., 1987, Correlation between nitric oxide formation during degradation of organic nitrates and activation of guanylate cyclase, *Eur. J. Pharmacol.* 139:19.

Furchgott, R.F., 1988, Studies on relaxation of rabbit aorta by sodium nitrite: the basis for the proposal that the acid-activatable inhibitory factor from retractor penis is inorganic nitrite and the endothelium-derived relaxing factor is nitric oxide, in: *"Vasodilatation: Vascular Smooth muscle, Peptides, Autonomic Nerves and Endothelium"*. P.M. Vanhoutte, ed., Raven Press, New York.

Furchgott, R.F., Zawadzki. and J.V., 1980, The obligatory role of endothelial cells in the relaxation of arterial smooth muscle by acetylcholine, *Nature* 288:373.

Fuster, V., Badimon, L., Badimon, J., Adams, P.C., Turitto, V. and Chesebro, J.H., 1987, Drugs interfering with platelet functions: mechanisms and clinical relevance, in: *Thrombosis and Haemostasis 1987* eds. Verstraete, M., Vermylen, J., Lijnen, R., Arnout, J., University Press, Leuven.

Gasic, G.J. and Gasic, T.B., 1962, Removal of sialic acid from the cell coat in tumor cells and vascular endothelium, and its effect on metastasis, *Proc. Natl. Acad. Sci. USA* 48:1172.

Gasic, G.J., Gasic, T.B., Galanti, N., Johnson, T. and Murphy, S., 1973, Platelet-tumor cell interactions in mice. The role of platelets in the spread of malignant disease, *Int. J. Cancer* 11:704.

Gebalska, J., 1990, Platelet adhesion and aggregation in relation to clinical course of acute myocardial infarction, M.D. thesis, Warsaw (in Polish).

Goldberg, N.D., Haddox, M.K., Nicol, S.E., Glass, D.B, Sanford, C.H., Kuehl, Jr., F.A. and Estensen, R., 1975, Biological regulation through opposing influences of cyclic GMP and cyclic AMP: The yin yang hypothesis, in: *Advances in Cyclic Nucleotide Research* vol.5., Drummond, G.I., Greengard, P., Robison, G.A., eds. Raven Press, New York .

Gordge, M.P. and Neild, G.H., 1992, Impaired activity of soluble guanylate cyclase in platelets from haemodialysis patients, *Br.J.Haematol.* 80S1:7

Grant, P.G. and Colman, R.W., 1984, Purification and characterization of a human platelet cyclic nucleotide phosphodiesterase, *Biochemistry* 23:1801.

Grant, P.G., Mannarino, A.F and Colman, R.W., 1990, Purification and characterization of a cyclic GMP-stimulated cyclic nucleotide phosphodiesterase from the cytosol of human platelets. *Thromb. Res.* 59:105.

Gryglewski, R.J., Palmer, R.M.J. and Moncada, S., 1986, Superoxide anion is involved in the breakdown of endothelium-derived vascular relaxing factor, *Nature* 320:454.

Herbaczynska-Cedro, K., Lembowich, K. and Pytel, B., 1991, N^G-monomethyl-L-arginine increases platelet deposition on damaged endothelium *in vivo*. A scanning electron microscopy study, *Thromb. Res.* 64:1.

Hibbs, J.B., Jr., Taintor, R.R., Vavrin, Z., Granger, D.L., Drapier, J.C., Amber, I,J. and Lancaster, J.R., Jr.,1990, Synthesis of nitric oxide from a terminal guanidino atom of L- arginine: a molecular mechanism regulating cellular proliferation that targets intracellular iron, in: *Nitric oxide from L-Arginine: A Bioregulatory System,* ed. Moncada, S., Higgs, E.A., Elsevier, Amsterdam.

Hidaka, H. and Asano, T., 1976, Human blood platelet 3':5'-cyclic nucleotide phosphodiesterase. *Biochim. Biophys. Acta* 429:485.

Hogan, J.C., Lewis, M.J. and Henderson, A.H., 1988, *In vivo* EDRF activity influences platelet function, *Br. J. Pharmacol.* 94:1020.

Houston, D.S., Gerrard, J.M., NcCrea, J., Glover, S. and Butler, A.M., 1983, The influence of amines on various platelet responses, *Biochim. Biophys. Acta* 734:267.

Houston, D.S., Robinson, P. and Gerrard, J.M., 1990, Inhibition of intravascular platelet aggregation by endothelium-derived relaxing factor: reversal by red blood cells, *Blood* 76:953.

Humphries, R.G., Tomlinson, W., O'Connor, S.E. and Leff, P., 1990, Inhibition of collagen- and ADP-induced platelet aggregation by substance P *in vivo*: Involvement of endothelium-derived relaxing factor, *J. Cardiovasc. Pharmacol.* 16:292.

Ignarro, L.J., Byrns, R.E. and Wood, K.S., 1988, Biochemical and pharmacological properties of endothelium-derived relaxing factor and its similarity to nitric oxide radical, in: *"Vasodilatation: Vascular Smooth muscle, Peptides, Autonomic Nerves and Endothelium"*. P.M. Vanhoutte, ed., Raven Press, New York.

Iuliano, L., Pratico, D., Ghiselli, A., Bonavita, M.S. and Violi, F., 1991, Superoxide dismutase triggers activation of "primed" platelets, *Arch. Biochem. Biophys.* 289:180.

Jandak, J., Steiner, M. and Richardson, P.D., 1989, Alpha-tocopherol, an effective inhibitor of platelet adhesion, *Blood* 73:141.

Jansen, W., Prenze, R., Kumper, H. and Tauchert, M., 1990, Interval treatment of coronary artery disease with sustained-release isosorbide-5-mononitrate, *Am. J. Cardiol.* 65:16J.

Korbut, R., Lidbury, P.S. and Vane, J.R., 1991, Prolongation of fibrinolytic activity of tissue plasminogen activator by nitrovasodilators, *Lancet* 335:669.

Lang, D. and Lewis, M.J., 1991, Inhibition of inositol 1,4,5-triphosphate formation by cyclic GMP in cultured aortic endothelial cells of the pig, *Br. J. Pharmacol.* 102:277.

Lansman, J.B., Hallam, T.J. and Rink, T.J., 1987, Single stretch-activated ion channels in vascular endothelial cells as mechanotransducers? *Nature* 325:811.

Laustiola, K.E., Vuorinen, P., Porsti, I., Metsa-Ketela, T. and Vapaatalo, H., 1991, Exogenous GTP enhances the effects of sodium nitrite on cyclic GMP accumulation, vascular smooth muscle relaxation and platelet aggregation, *Pharmacol. Toxicol.* 68:60.

Lelchuk, R., Radomski, M.W., Martin, J.F. and Moncada, S., 1992, Constitutive and inducible nitric oxide synthases in human megakaryoblastic cells., *J. Pharmacol. Exp. Ther.*,262;1220.

Loscalzo, J. and Vaughan, D.E., 1987, Tissue Plasminogen activator promotes platelet disaggregation in plasma, *J. Clin. Invest.* 79:1749.

Luscher, T.F., Yang, Z., Diederich, D. and Buhler, F.R., 1989, Endothelium-derived vasoactive substances: Potential role in hypertension, atherosclerosis, and vascular occlusion, *J. Cardiovasc. Pharmacol.* 14:S63.

Marcus, A.J. and Zucker, M.B., 1965, The physiology of blood platelets, Grune & Stratton Inc. New York, London.

Martin, W., Villani, G.M., Jothianandan, D. and Furchgott, R. F., 1985, Selective blockade of endothelium-dependent and glyceryl trinitrate-induced relaxation by hemoglobin and by methylene blue in the rabbit aorta, *J. Pharmacol. Exp. Ther.* 232:708.

Martin, W., White, D.G. and Henderson, A.H., 1988, Endothelium-derived relaxing factor and atriopeptin II elevate cyclic GMP levels in pig aortic endothelial cells. *Br. J. Pharmacol.* 93:229.

Matsuoka, I., Nakahata, N. and Nakanishi, H., 1989, Inhibitory effect of 8-bromo cyclic GMP on an extracellular Ca^{2+}-dependent arachidonic acid liberation in collagen -stimulated rabbit platelets, *Biochem. Pharmacol.* 38:1841.

Maurice, D.H. and Haslam, R.J., 1990, Molecular basis of the synergistic inhibition of platelet function by nitrovasodilators and activators of adenylate cyclase: Inhibition of cyclic AMP breakdown by cyclic GMP, *Mol. Pharmacol.* 37:671.

May, G.R., Crook, P., Moore P.K. and Page, C.P., 1991, The role of nitric oxide as an endogenous regulator of platelet and neutrophil activation within the pulmonary circulation of the rabbit, *Br. J. Pharmacol.* 102:759.

Mellion, B.T., Ignarro, L.J., Ohlstein, E.H., Pontecorvo, E. G., Hyman, A.L. and Kadowitz, P.J., 1981, Evidence for the inhibitory role of guanosine 3',5'-monophosphate in ADP - induced human platelet aggregation in the presence of nitric oxide and related vasodilators, *Blood* 57:946.

Mellion, B.T., Ignarro, L.J., Myers, C.B., Ohlstein, E.H., Ballot, B.A., Hymana, A.L. and Kadowitz, P.J., 1983, Inhibition of human platelet aggregation by S-nitrosothiols. Heme-dependent activation of soluble guanylate cyclase and stimulation of cyclic GMP accumulation, *Mol. Pharmacol.* 23:653.

Mills, D.C.B., 1981, The basic biochemistry of the platelet, in: Haemostasis and Thrombosis, ed. Bloom, A.L., Thomas, D.P., Churchill Livingstone, Edinburgh, London, Melbourne , New York.

Molina y Vedia, L., McDonald, B., Brune, B. and Lapetina, E.G., 1992, Nitric oxide-induced S-nitrosylation of glyceraldehyde-3-phosphate dehydrogenase inhibits enzymatic activity and increases endogenous ADP-ribosylation, *J. Biol. Chem.*, 267; 24929.

Moncada, S., Gryglewski, R.J., Bunting, S. and Vane, J.R., 1976, An enzyme isolated from arteries transforms prostaglandin endoperoxides to an unstable substance that inhibits platelet aggregation, *Nature* 263:663.

Moncada, S., Palmer, R.M.J. and Higgs, E.A., 1989, Biosynthesis of nitric oxide from L-arginine. A pathway for the regulation of cell function and communication, *Biochem. Pharmacol.* 38:1709.

Moncada, S., Palmer, R.M.J. and Higgs, E.A., 1991, Nitric oxide: physiology, pathophysiology and pharmacology. *Pharmacol. Rev.* 43:109

Morgan, R.O. and Newby, A.C., 1989, Nitroprusside differentially inhibits ADP-stimulated calcium influx and mobilization in human platelets, *Biochem. J.* 258:447.

Murray, R., Shipp, E. and FitzGerald, G.A., 1990, Prostaglandin endoperoxide/thromboxane A_2 receptor desensitization. Cross-talk with adenylate cyclase in human platelets. *J. Biol. Chem.* 265:21670.

Nakashima, S., Tohmatsu, T., Hattori, H., Okano, Y. and Nozawa, Y., 1986, Inhibitory action of cyclic GMP on secretion, polyphosphoinositide hydrolysis and calcium mobilization in thrombin-stimulated human platelet, *Biochem. Biophys. Res. Commun.* 135:109.

Ohno, M., Ochiai, M., Taguchi, J., Hara, K., Akatsuka, N. and Kurokawa, K., 1990, Stretch may enhance the release of endothelium-derived relaxing factor in rabbit aorta, *Biochem. Biophys. Res. Commun.* 173:1038.

Palmer, R.M.J., Ashton, D.S. and Moncada, S., 1988, Vascular endothelial cells synthesize nitric oxide from L-arginine, *Nature* 333:664.

Palmer, R.M.J., Bridge, L., Foxwell, N.A. and Moncada, S., 1992, The role of nitric oxide in endothelial cell damage and its inhibition by glucocorticoids, *Br. J. Pharmacol.* 105:11.

Petros, A., Bennett, D. and Vallance, P., 1991, Effect of nitric oxide synthase inhibitors on hypotension in patients with septic shock, *Lancet* 338:1557.

Pohl, U. and Busse, R., 1989, EDRF increases cyclic GMP in platelets during passage through the coronary vascular bed, *Circ. Res.* 65:1798.

Pohl, U., Busse, R., Kuon, E. and Bassenge, E., 1986, Pulsatile perfusion stimulates the release of endothelial autacoids, *J. Appl. Cardiol.* 1:215.

Pronai, L., Ichimori, K., Nozaki, H., Nakazawa, H., Okino, H., Carmichael, A.J. and Arroyo, C.M., 1991, Investigation of the existence and biological role of L-arginine/nitric oxide pathway in human platelets by spin-trapping/EPR studies, *Eur. J. Biochem.* 202:923.

Radomski, M.W. and Moncada,S., 1991, Role of nitric oxide in endothelial cell-platelet interactions, in: Antithrombotics. Pathophysiological Rationale for Pharmacological Interventions, ed. Herman, A.G., Kluwer Academic Publishers, Dordrecht.

Radomski, M.W., Palmer, R.M.J. and Moncada, S., 1987a, Comparative pharmacology of endothelium-derived relaxing factor, nitric oxide and prostacyclin in platelets, *Br. J. Pharmacol.* 92:181.

Radomski, M.W., Palmer, R.M.J. and Moncada, S., 1987b, Endogenous nitric oxide inhibits human platelet adhesion to vascular endothelium, *Lancet* ii:1057.

Radomski, M.W., Palmer, R.M.J. and Moncada, S., 1987c, The role of nitric oxide and cGMP in platelet adhesion to vascular endothelium, *Biochem. Biophys. Res. Commun.* 148:1482.

Radomski, M.W., Palmer, R.M.J. and Moncada, S., 1987d, The anti-aggregating properties of vascular endothelium: interactions between prostacyclin and nitric oxide, *Br. J. Pharmacol.* 92:639.

Radomski, M.W., Palmer, R.M.J. and Moncada,S., 1988, Isolation and washing of human platelets with nitric oxide, *Thromb. Res.* 50:537.

Radomski, M.W., Palmer, R.M.J. and Moncada, S., 1990a, An L-Arginine/nitric oxide pathway present in human platelets regulates aggregation, *Proc. Natl. Acad. Sci. USA* 87:5193.

Radomski, M.W., Palmer, R.M.J. and Moncada, S., 1990b, Characterization of the L-arginine: nitric oxide pathway in human platelets, *Br. J. Pharmacol.* 101:325.

Radomski, M.W., Palmer, R.M.J. and Moncada, S., 1990c, Glucocorticoids inhibit the expression of an inducible, but not the constitutive, nitric oxide synthase in vascular endothelial cells, *Proc. Natl. Acad. Sci. USA* 87:10043.

Radomski, M.W., Jenkins, D.C., Holmes, L.and Moncada,S., 1991a, Human colorectal adenocarcinoma cells: differential nitric oxide synthesis determines their ability to aggregate platelets, *Cancer Res.* 51:6073

Radomski, M.W., Martin, J.F. and Moncada, S., 1991b, Synthesis of nitric oxide by the haemocytes of the American horseshoe crab (Limulus polyphemus), *Phil. Trans. R. Soc. Lond.* B 334:129

Radomski, M.W., Rees, D.D., Dutra, A. and Moncada,S., 1992, S-nitroso-glutathione inhibits platelet activation *in vitro* and *in vivo*, *Br. J. Pharmacol.* 107; 745.

Rees, D.D., Cellek, S., Palmer, R.M.J. and Moncada, S., 1990, Dexamethasone prevents the induction by endotoxin of a nitric oxide synthase and the associated effects on vascular tone: an insight into endotoxin shock, *Biochem. Biophys. Res. Commun.* 173:541.

Remuzzi, G., Perico, N., Zoja, C., Corna, D., Macconi, D. and Vigano, G., 1990, Role of endothelium-derived nitric oxide in the bleeding tendency of uremia, *J. Clin. Invest.* 86:1768.

Rosenblum, W.I., Nelson, G.H. and Povlishock, J.T., 1987, Laser-induced endothelial damage inhibits endothelium-dependent relaxation in the cerebral microcirculation of the mouse, *Circ. Res.* 60:169.

Ross, R. and Glomset J A., 1976a, The pathogenesis of atherosclerosis (First of two parts), *New Engl. J. Med.* 295:369.

Ross, R. and Glomset, J.A., 1976b, The pathogenesis of atherosclerosis (Second of two parts), *New Engl. J. Med.* 295:420.

Rubanyi, G.M. and Vanhoutte, P.M., 1986, Superoxide anions and hyperoxia inactivate endothelium-derived relaxing factor, *Am. J. Physiol.* 250:H822.

Salvemini, D., de Nucci, G., Sneddon, J.M. and Vane, J.R., 1989, Superoxide anions enhance platelet adhesion and aggregation, *Br. J. Pharmacol.* 97:1145.

Sane, D.C., Bielawska, A., Greenberg, C.S. and Hannun, Y.A., 1989, Cyclic GMP analogs inhibit g-thrombin-induced arachidonic acid release in human platelets, *Biochem. Biophys. Res. Commun.* 165:708.

Sinzinger, H., Virgolini, I., O'Grady, J., Rauscha, F. and Fitscha, P., 1992, Modification of platelet function by isosorbide dinitrate in patients with coronary artery disease, *Thromb. Res.* 65:323.

Sneddon, J.M. and Vane, J.R., 1988, Endothelium-derived relaxing factor reduces platelet adhesion to bovine endothelial cells, *Proc. Natl. Acad. Sci. USA* 85:2800.

Souness, J.E, Diocee, B.K., Martin, W. and Moodie, S.A., 1990, Pig aortic endothelial-cell cyclic nucleotide phosphodiesterases. Use of phosphodiesterase inhibitors to evaluate their roles in regulating cyclic nucleotide levels in intact cells, *Biochem. J.* 266:127.

Stamler, J.S., Vaughan, D.E. and Loscalzo, J., 1989, Synergistic disaggregation of platelets by tissue-type plasminogen activator, prostaglandin E_1, and glyceryl trinitrate, *Circ. Res.* 65:796.

Stamler, J.S., Simon, D.I., Osborne, J.A., Mullins, M.E., Jaraki, O., Michel, T., Singel, D.J. and Loscalzo, J., 1992, S-nitrosylation of proteins with nitric oxide: synthesis and characterization of biologically active compounds, *Proc. Natl. Acad. Sci. USA* 89:444.

Vallance, P., Benjamin, N. and Collier, J., 1992, The effect of endothelium-derived nitric oxide on *ex vivo* whole blood platelet aggregation in man, *Eur. J. Clin. Pharmacol.* 42:37.

Venturini, C.M., Del Vecchio, P.J. and Kaplan, J.E., 1989, Thrombin induced platelet adhesion to endothelium is modified by endothelial derived relaxing factor (EDRF). *Biochem. Biophys. Res. Commun.* 159:349.

Verbeuren, T.J., Jordaens, F.H., Zonnekeyn, L.L., Va Hove, C.E, Coene, M.C. and Herman, A.G., 1986, Effect of hypercholesterolemia on vascular reactivity in the rabbit. I. Endothelium-dependent and endothelium-independent contractions and relaxations in isolated arteries of control and hypercholesterolemic rabbits, *Circ. Res.* 58:552.

Villaneuva, V.R. and Giret, M., 1980, Human platelet arginase, *Mol. Cell. Biochem.* 33:97.

Walter, U, 1989, Physiological role of cGMP and cGMP-dependent protein kinase in the cardiovascular system, *Rev. Physiol. Biochem. Pharmacol.* 113:41.

Ware, J.A., Smith, M. and Salzman, E.W., 1987, Synergism of platelet-aggregating agents. Role of elevation of cytoplasmic calcium, *J. Clin. Invest.* 80:267.

Wautier, J.L., Weill, D., Kadeva, H., Maclouf, J. and Soria, C., 1989, Modulation of platelet function by SIN-1A, a metabolite of molsidomine, *J. Cardiovasc. Pharmacol.* 14:S111.

Whittle, B.J.R., Moncada, S. and Vane, J.R., 1978, Comparison of prostacyclin (PGI$_2$), prostaglandin E$_1$ and D$_2$ on platelet aggregation in different species, *Prostaglandins* 16:373.

CONTRIBUTORS

Dr. Kaiwant S. Authi
Platelet Section
Thrombosis Research Institute
Manresa Road
Chelsea, London SW3 6LR
UNITED KINGDOM

Dr. Lawrence F. Bass
Hematology-Oncology Section
Silverstein 7
The University of Pennsylvania
3400 Spruce Street
Philadelphia, PA 19104-4283 USA

Dr. Edourd M. Bevers
Department of Biochemistry
University of Limburg
P.O. Box 616
6200 MD Maastrict
THE NETHERLANDS

Dr. Kenneth J. Clemetson
Theodor Kocher Institut
University of Bern
Freiestrasse 1, P.O. Box 99
CH-3012 Bern 9 SWITZERLAND

Prof. Maurice B. Feinstein
Department of Pharmacology
University of Connecticut Health Ctr.
Farmington, Connecticut 06032 USA

Dr. Joan E.B. Fox
The Gladstone Foundation Laboratories
   for Cardiovascular Diseases
Department of Pathology
P.O. Box 40608
San Francisco, CA 94140-1608 USA

Prof. Adrian R.L. Gear
Department of Biochemistry
University of Virginia
School of Medicine
Box 440, Charlottesville
Virginia 22908 USA

Dr. Jon M. Gerrard
Manitoba Institute of Cell Biology
Dept. of Pediatrics
University of Manitoba
100 Olivia Street, Winnipeg
Manitoba, RE 0V9 CANADA

Dr. Richard J. Haslam
Department of Pathology
McMaster University, Hamilton
Ontario, L8N 3Z5 CANADA

Dr. Eduardo G. Lapetina
Division of Cell Biology
Burroughs Wellcome Co.
3030 Cornwallis Road
Research Triangel Park
North Carolina 27709 USA

Dr. Gerard Mauco
INSERM Unite 326
Phospholipides Membranaires
Signalisation Cellulaire et Lipoproteines
Hopital Pupan
31059 Toulouse
Cedex FRANCE

Prof. Yoshinori Nozawa
Department of Biochemistry
Gifu University School of Medicine
Tsukasamachi-40
Gifu 500 JAPAN

Dr. Marek Radomski
The Wellcome Research Laboratories
Langley Court, South Eden Park Road
Beckenham, Kent BR3 3BS
UNITED KINGDOM

Dr. Stewart O. Sage
The Physiological Laboratory
University of Cambridge
Downing Street
Cambridge CB2 9EG
UNITED KINGDOM

Prof. Michael Scrutton
Division of Bimolecular Sciences
Kings College London
Campden Hill Road
London W8 7AH UNITED KINGDOM

Dr. Wolfgang Siess
Institut fur Phophylaxe und
  Epidemiologie der Kreislaufkranheiten b.d.
Universitat Munchen
Pettenkoferstrasse 9
800 Munchen 2 GERMANY

Prof. Ulrich Walter
Department of Internal Medicine
University of Wurzburg
Josef - Schneider Str. 2
8700 Wurzburg GERMANY (W)

Dr. Steve P. Watson
Department of Pharmacology
Mansfield Road
Oxford OX1 3QT UNITED KINGDOM

Dr. Gilbert C. White II
Center for Thrombosis and Haemostasis
Department of Medicine
Dental Research Center
The University of North Carolina
Chapel Hill, NC 27599 USA

Dr. J. Michael Wilkinson
Department of Biochemistry
  and Cell Biology, Hunterian Institute
Royal College of Surgeons
35-43 Lincoln's Inn Fields
London WC2A 3PN
UNITED KINGDOM

INDEX

Acetylcholine, 251
N-Acetylglucosamine, 96
Acid hydrolase, 149
Actin, 18, 43, 44, 86, 142, 169, 171, 175–184
α-Actinin, 175
Activation, *see* Platelet
Adenocarcinoma,colorectal, 256
Adenosine diphosphate (ADP), 2–6, 20, 22, 57–65,
    69–73, 76–80, 83, 84, 95, 97, 98, 105,
    115, 132–135, 141, 149, 191, 196, 211–
    214, 225, 237, 238, 243
Adenosine monophosphate,cyclic (cAMP), 7, 10–
    11, 17–19, 28, 29, 51, 89, 90, 92, 94,
    134, 190, 193, 229–239, 242, 253, 254
Adenosine triphosphatase (ATPase), 84–89, 93, 96
Adenosine triphosphate (ATP), 9–10, 93, 96, 136,
    140–143, 149, 152, 153, 159, 192, 193,
    201, 202, 255
Adenylate cyclase, 229, 234, 254
Adenylylcyclase, 17, 21, 26, 29, 238
ADP, *see* Adenosine diphosphate
Adrenaline, 2–5, 105
β-Adrenergic receptor kinase (beta-ARK), 26–27
Annexin, 169
Antibody, 87, 88, 91, 93, 130, 132, 143
    monoclonal and platelet activation, 221–228
        list of twenty-two, 222
Antihistamine and bleeding disorder, 214
Aggregation, 2, 5–7, 18, 119, 126, 132, 140–145,
    221, 255
Aggregometry by quenched flow, 59–60
Agonist, *see* Platelet
Aequorin, 3, 6
Aminoguanidine, 213
Aminophospholipid, 201, 203
Aminophospholipid translocase, 198, 201–203
cAMP, *see* Adenosine monophosphate,cyclic
Antibody, *see* Platelet
Apyrase, 4, 132
Arachidonate, Arachidonic acid
Arachidonic acid, 8, 10, 18, 19, 86, 94, 133–136,
    139, 140, 157, 160, 161, 167, 213
L-Arginine, 257–259
Artery disease, coronary, 258
    and isosorbide dinitrate, 258
Aspirin, 132, 135, 213, 224, 225, 255
Astemizole, 212

Atherosclerosis, 255
    and nitric oxide, 255
ATPase, *see* Adenosine triphosphatase
ATP, *see* Adenosine triphosphate
Autophosphorylation, 123, 136

BAPTA, 157–158, 162
Barium, 74, 76
Bernard–Soulier syndrome, 176–177
Bleeding disorders and antihistamine, 214
Bradykinin, 254

Caffeine, 94, 95
Calcium, 1–15, 18, 57, 58, 60–62, 65, 69–104, 106,
    109, 119, 122, 124, 127, 133, 140, 149–
    158, 160, 166, 176, 191, 197–202, 212,
    223, 225, 237, 242–244, 252, 253
    homeostasis, 83–104 *see also* Platelet
    ionophore A*23187*, 51, 83, 106, 109, 151, 177,
        191, 196, 199–2023, 210–214, 230, 231,
        234
Calmodulin, 85, 96, 243
Calpain, 177–182, 200–202
Calreticulin, 91
Calyculin, 25
γ-Carboxyglutamic acid, 196
Catecholamine, 238
Cell
    calcium-driven, *see* Platelet
    fusion of man and mouse, 209
    lines, 21, 96, 97, 132, 133, 229
Cholera toxin, 28
Choline, 7–9, 154
Choline phosphate, 7–9
Chlorcyclizine, 212
Chymotrypsin, 6
Cimetidine, 212
Coagulation cascade, 195, 256
Collagen, 1, 2, 4, 6–9, 18, 20, 83, 94, 105, 116, 133,
    142, 191, 196, 197, 202, 210, 211, 225,
    237, 255, 257,258
Concanavalin A, 144
Cycloheximide, 25
Cyproheptidine, 212
Cytochalasin, 179, 182, 183
Cytochrome P-*450*, 75, 76, 96

Cytoskeleton, 43–44, 126, 142–145, 165–194

Daidzin, 137
Dami cell thrombin receptor, 21, 26
Dense tubular system (DTS), 87
Dephosphorylation, 27, 91, 132, 140, 233, 234
1,2-Diacylglycerol (DAG), 4, 7–9, 18, 19, 52, 80,
        83, 105, 107, 130, 139, 149, 152, 156,
        158, 160, 162, 165–168, 171, 243
Diamide, 197, 199
Diamine oxidase, 213
Dibucaine, 177, 196, 199
2,5-Di-(t-butyl)-*1,4*-benzohydroquinone, 73, 74, 88,
        93–96
Dihydropyridine, 75
1,2-Dioctanoin, 6, 8
Dioctanoylglycerol, 99, 107
Dioctanoylphosphatidic acid, 107
Diolein, 159
Dioleoylthiophosphatidic acid, 159
Disease *see* separate syndromes
    atherosclerosis, 255
    hemorrhagic, 255–257
    thrombotic, 255–257
DPPE, 211–213

Econazole, 75, 76, 96
EDTA, 141
EGTA, 3, 5, 157, 179, 225
Electropermeabilization, 149–164
Electrophoresis, 121–125
Electrophysiology, 73
Electroporation, 59
Endothelium-derived relaxing factor (EDRF), 237–
    239, 251
Endothelium, vascular, 254
Endotoxin, bacterial, 252, 256
Enolase, 143
Enterotoxin of *Escherichia coli*, 238
Epinephrine, 17–20, 22, 133, 137, 141, 191, 196,
    200, 234, 237
Erbstatin, 136, 137
Erythroleukemia, human, 51–53
Erythropoietin, 53
Ethanol, 155–157, 160–161
Exocytosis, 154, 157
Ezrin, 132

Factor
    VIIIR, 6
    IX, 196
    X, 1, 196
    activating, *see* Platelet
Fibrin, 143, 195
Fibrinogen, 4, 5, 8, 18, 19, 57, 79, 106, 115, 120,
    135, 140–144, 149, 171, 178, 181, 183,
    195, 213, 223, 226, 252
    receptor, *see* Glycoprotein IIb-IIIa
Fibroblast, 178, 238
Fibronectin, 141, 178
Filamin, 243
Filopodia, 175

Flow cytometry, fluorescent, 200
Flow technique, quenched, 57–67
Fluorimetry, 122
    by stopped flow, 70–72, 76
Fodrin, 131, 176
Forskolin, 134, 229
Function of platelet, *see* Platelet

GAP, 50, 51
Gelsolin, 43, 44, 169, 176
Genes, 89, 91
Genistein, 53, 136–138
Glanzmann's thrombasthenia, 124, 141, 190
β-Glucuronidase, 157
Glutathione, 135
Glycerol, 160
Glyceryl trinitrate, 255, 258
Glycoprotein, 86, 165–174
    Ib, 6, 18, 20, 27, 28
    Ib-IX, 176, 177
    IIb-IIIa, 5, 18, 57, 106, 115, 119, 120, 123–126,
        131, 135, 140–144, 165–184, 189, 190,
        221, 225, 231, 252
G protein, 7, 9, 10, 12, 17–36, 83, 111, 132, 133,
    187–194
Granule, 1, 18, 19, 149, 150, 153
Guanine nucleotide, 17, 150–152, 160
Guanosine monophosphate,cyclic (cGMP), 237–
    239, 242, 253, 254
Guanosine triphosphatase, 50–53, 187
Guanosine triphosphate (GTP), 94, 151–158, 165
Guanosine triphosphate binding proteins, 39–45
Guanylate cyclase, 234, 238, 252–257
Guanylin, 238
Guanylyl cyclase, 238

Heart disorder, ischemic, 258
Hemocyte, 259
Hemostasis, 251, 254–257, 259
Heparin, 94, 95
Hirudin, 21, 28, 127
Histamine, 209–219
Histamine methyltransferase, 213
Histidine decarboxylase, 211, 212, 216
Homeostasis of calcium, 83–104
α-Hydrazohistidine, 214
Hydrogen peroxide, 111–116, 135
5-Hydroxytryptamine, 2, 106, 109
17-Hydroxywortmannin, 119–128

Iloprost, 52, 234
Immunoprecipitation, 133
Indomethacin, 4, 93, 94, 112, 224, 225
Injury, vascular, 18
Inositol, 59, 83, 168
    lipid metabolism, 165–174
Inositol phosphate, 59, 60, 113, 152–154
Inositol phospholipid, 57
Inositol *1,3,4,5*-tetrakisphosphate, 75, 85
Inositol *1,4,5*-trisphosphate, 18, 19, 57, 61–65, 83, 84,
    90–95, 97, 105, 130, 149, 166, 171, 243
Insulin receptor, 129, 187

Integrin, 141, 144, 165, 170, 175, 178, 183, *see also*
    Glycoprotein
Interconversion, 58
Interleukin, 252, 256
Ionomycin, 86, 225
Ionophore, *see* Calcium
Islet cell, pancreatic, 158
Isoflavone, 136
Isosorbide dinitrate, 258

Ligand, tethered, 132, 143
Lysosome, 149–151

Manganese, 70, 71, 76, 96, 97
MAP.2 protein, 131
Mast cell, 157, 158
Mastoparan, 141, 233
Megakaryocyte, 124, 131, 252
    fragment, anucleate, *see* Platelet
Membrane, 121, 133, 175–185, 198
Messenger
    intracellular, 209–219
    second, 95–97
        and channels, 76–77
Metastasis, 256
Methemoglobin, 258
α-Methylhistidine, 215
Miconazole, 75
Microparticles, 198, 199
Microvesicle of platelet, 198–201
Molybdate, 134, 137
Monoacylglycerol, 107
Monocyte, cardiac, 238
Myosin, 86, 142
    light chain, 18, 107, 108, 124, 158–160, 230, 231
        kinase, 119, 120, 237
        phosphorylation, 243
Myristoylation, 129

Neomycin, 94, 95
Neuraminidase, 86
Neutrophil, 150, 157, 158, 168, 209, 214
Nexin, 21
Nitric oxide, 251–264
    action, 257–258
    affinity targets, 252–253
    formation, 257–258
    half life, 257
    pharmacology, 257–258
    in platelet, 251–264
Nitric oxide synthetase, 251, 252, 259
Nitroprusside, sodium, 134, 237, 240, 242, 258
S-Nitrosoglutathione, 258
S-Nitrosothiol, 258
Nitrovasodilator, 238, 239, 243, 252, 258
Nordihydroguaiaretic acid, 76
Nucleotide, cyclic, 237–239
Nucleotide phosphodiesterase,cyclic, 253
Nystatin, 72

Okadaic acid, 25, 111, 114, 233
1-Oleoyl 2-acetylglycerol, 126, 150

Oncogene, 129
    product pp60, 168–170, 181, 182, *see also* Tyro-
        sine kinase
Orthophosphate, 110
Orthovanadate, 134–137
Oscillation and calcium, 80–81
Oxyhemoglobin, 257

PAF, 1, 4, 75, 76, 83, 97, 131, 133, 138, 196
Peroxynitrite, 257
Pertussis toxin, 10, 17, 19, 28–30
Pervanadate, 131, 134–140
Phenyltoloxamine, 212
Phorbol ester, 8, 44, 105, 125, 140, 230, 231
    dibutyrate, 51, 106, 109–111
    12-myristate *13*-acetate (PMA), 138–141, 150–
        158, 161, 191, 210–213
Phosphatase, 105–118, 124
Phosphatidic acid, 7, 58, 107, 154, 165, 183–233
Phosphatidyl alcohol, 7
Phosphatidyl choline, 7, 8, 29, 154, 160, 161, 198
Phosphatidyl ethanol, 154
Phosphatidylinositol, 161, 167, 196, 233
    metabolism, 58
Phosphatidylinositol-*4,5*- biphosphate, 7, 18, 52, 57,
        83, 105
Phosphatidylinositol-*4*-phosphate, 7
Phosphatidylserine, 1, 2, 4, 158, 160, 196, 198–203
Phosphocholinidase C, 7, 8
Phosphodiesterase, 29, 238, 239
Phosphoinositidase C, 7–12
Phosphoinositide, 20, 37–47, 57–67, 160
    hydrolysis, 20, 29, 30, 37, 44
Phosphoinositide-*3*-kinase, 181, 182
Phospholamban, 90
Phospholipase
    A2, 5, 10, 18, 19, 29–31, 93, 197, 213, 242
    C, 17–21, 26, 29, 30, 37–47, 52, 53, 58, 83, 113–
        115, 120, 127, 129, 149–171, 187, 213,
        229, 237
        identification, 38–39
        isoforms, 38–39
        phosphoinositide-specific, 37–47
        purification, 38–39
        and thrombin, 38
    D, 7, 8, 150, 149–164
Phospholipid, 18, 165–174, 213
    asymmetry, 198
    effect, classical, 165–168, *see also* Phospholipase C
Phosphoprotein phosphatase, 160
Phosphorylation, 26, 27, 44–45, 52, 53, 86–94, 120,
        122, 126, 166, 192, 243
    cascade, 119, 120
    of protein, 105–118
    of protein tyrosine, 28, 44, 111–116
    and thrombin desensitization, 26
    of tyrosine, 129–148
Phosphothreonine, 124, 125
Phosphotyrosine, 53
Plaque formation, 255
Plasma membrane, 85–87, 90, 129, 175–185
    preparation, 86

Platelet
activating factor, 17, 20, 105, 161, 210, 211, 237
activation, 18, 19, 61, 62, 83, 105–118, 142–145,
221–228
by agonist receptor, 17–36
by antibody, monoclonal, 221–228
by G protein, 17–36
mediators of, 17–36
of thrombin receptor, 23–25
aggregation, 2, 119, 142–145, 221, 255
kinetics, 59–60
agonist, 1, 4–6
adrenaline as, 4–5
list of, 229
receptor, 17–36
response, 1–2
antibody, monoclonal, 105, 221–228
and activation, 221–228
binding of, 221–228
arachidonate mobilization, *see* Arachidonic acid
adhesion, 1–2
adrenaline agonist, 4–5
calcium, 1–15
homeostasis, 83–104
influx, 69–82
signal, organization of, 78–81
signalling, 57–67
as cell, calcium-driven, 1–15
collagen agonist, 6
cytoskeleton, 43–44, 142–145, 165–194
1,2-diacylglycerol formation, 7–9
"dust", *see* microparticle
electrophysiology, 73
electroporation, 59
factor, activating, *see* activating factor
fatty acids, 167
flow technique, quenched, 57–67
function, 49–55, 58–60, 175–194, 229–235, 251–
264
glycoprotein
IIb, 165–174
IIIa, 165–174
G protein and activation of, 17–36
granules, 1, 18, 19, 149
guanosine triphosphate-binding protein, 39, 40
hemostasis, 251, 254–257, 259
histamine, 209–219
homeostasis, 251
17-hydroxywortmannin, 119–128
inhibition, 11–12, 229–235
inhibitors, 105–118
inositol lipid metabolism, 165–174
inositol-loading, 59
isolation, 59, 121–122
lysosome activation of, 149
mediator of activation, 17–36
membrane
preparation, 121
protein, 133
skeleton, 175–185
messenger, intracellular, 209–219
microparticle, 198

Platelet (*cont'd*)
microsome, 90
microvesicle, 198–201
nitric oxide, 251–264
organelle, secretory, 149
granules, 149
permeabilized, 149–164, *see also* Saponin
phosphatase, 105–118
phosphoinositide metabolism, 57–67
and phospholipase C, 37–47
phospholipase C, 37–47, 149–164
phospholipid, 165–174
phosphorylase D, 149–164
plasma membrane, 175–185
pools, intracellular, 83–104
procoagulant, 195–207
procoagulation, 195–207
protein, 130
labelled, 121–122
of membrane, 133
phosphorylation, 105–118, 229–235
rap*1*b, 49–55, 187–194
protein kinase, 237–249
inhibitor, 105–118
VASP substrate of, 237–249
protein phosphatase inhibitor, 105–118
protein rap*1*b, 49–55, 187–194
reactivity, 2
regulation, 237–249
of function, 175–185
response pattern, 1–4, 11–12
secretion, 149–164
and phospholipase C, 149–164
serine kinase, 119–128
signal
organization, 69–82
transduction, 17–19, 119–128
stimulation by thrombin, 43
storage granule, 1
threonine kinase, 119–128
thrombin, 119–128, *see also* Thrombin
receptor, 23–25
for stimulation, 43
thrombosis, 256
tyrosine kinase, 130–132
tyrosine phosphorylation, 129–148
unstimulated, 142–145
VASP substrate of protein kinase, 237–249
Pleckstrin, 9, 94, 107, 120, 124, 125, 150–153, 158,
159
phosphorylation, 243
Polyacrylamide gel electrophoresis, 121
Polyamine, 215, 216
Procoagulant in platelet, 195–207
Profilactin, 169
Profilin, 44, 169, 171
Prolactin, 215
Promethazine, 212
Prostacyclin, 51, 134, 140, 193, 229, 242, 255
Prostaglandin, 18, 41, 229, 238, 255
Protein, 121–122, 130
cytoskeletal, 131

Protein (*cont'd*)
  MAP-2, 131
  phosphorylation, 44–45, 158–160
  rap*1*b, 18, 41, 49–55, 90, 187–194, 231–234
  ras, 49–55
  src, 129–131, 144
  tau, 131
Protein kinase, 41, 85, 89, 90, 94, 105–120, 123,
      237–249
  A, 18, 52, 229, 231, 232
  adenosine monophosphate, cyclic, dependent, 238
  C, 4, 7–9, 17, 19, 24, 26, 51, 52, 105–118, 149–
      152, 158–161, 166, 181, 182, 211–213,
      229, 230
    inhibitor, 107–111
  inhibitor, 105–118
  in platelet, human, 240–243
  regulation, 241
Protein phosphatase, 105–118
  inhibitor, 105–118
Protein phosphorylation, 105–118, 213, 229–235,
      238, 239
Protein tyrosine kinase, 129, 182, 237
Prothrombin, 1, 196
Prothrombinase, 83, 196, 198, 200, 201
Protooncogene
  ras, 187
  c-*src*, 129
Purkinje cell, cerebellar, 238
Putrescine, 215
Pyridyldithioethanolamine, 197, 199
Pyrilamine, 212

R*59022*, 107
  as enzyme inhibitor, 107
R*59949*, 107
Ranitidine, 212
Rap*1*b, *see* Protein
Reticulum
  endoplasmic, 94
  sarcoplasmic, 87
Retrovirus, transforming, 129
Rhesus factor, 201
Rhodopsin kinase, 26
Ristocetin, 6, 144
Ro *31-8220*, 106–111, 114–116
  structure, 108
Rous sarcoma virus, 129
  protein tyrosine kinase discovered, 129
Ryanodine, 75, 95
  receptor, 84, 91, 94

Saponin, 87, 88, 92–95, 110–112, 137, 212
Scott syndrome (bleeding disorder), 200–201
Sea urchin egg homogenate, 95
Sendai virus, 157
Serine kinase, 119–128
Serotonin, 58, 80, 149–153, 156, 157, 161, 191, 212,
      224
Serratia marcescens protease, 20
Shock, septic, 257
Sialic acid, 96

Signal transduction, 4–5, 17–19, 119–130, 149
Sodium, 72
Sodium nitroprusside, *see* Nitroprusside
Spectrin, 182
Spermidine, 215
Spermine, 215
Sphingomyelin, 86, 198
Sphingomyelinase, 197
Src protein, 129, 130
Staurosporine, 107–110, 116, 119–124, 136, 139,
      140, 151–153
  analog Ro *31-8220*, 107–110
  structure, 108
Stearic acid, 107, 167
1-Stearoyl-2-arachidonylglycerol, 7
Stenosis, coronary, in rabbit, 255
Streptomycin, 94
Stress fiber, 179, 182
Substance P, 254
Superoxide anion dismutase, 257
System, dense tubular (DTS), 87

Talin, 131, 132, 142, 182, 183
Tenase, 196, 198, 201
Tetracaine, 196
Tetrakisphosphate, 65
Thapsigargin, 73, 88, 96, 214
Theophylline, 51
Thiol-containing molecules, 253
Threonine kinase, 119–128
Thrombin, 1, 4, 7–9, 11, 17, 18, 20–26, 29, 43, 44,
      51–53, 57–65, 75–78, 83, 93, 96–98,
      105–111, 113, 115, 119–128, 132–135,
      138–140, 149–152, 160, 161, 166–171,
      180, 188–197, 202, 203, 210–212, 237,
      238, 241
  receptor, 20–28, 132
β-Thromboglobulin, 149–153
Thromboplastin, 196
Thrombosis, 255, 256
  and nitric oxide, 255
Thrombospondin, 149, 180
Thromboxane A-2, 1–5, 17–20, 38, 83, 93, 94, 97,
      105, 132, 133, 139, 149, 210–213, 225,
      237, 241
  B-2, 223, 224
α-Tocopherol, 255
Transducin, 29
Translocation, enzymic, 131
Transmembrane signaling, 180–182
  ligand-induced, 180–182
  receptor, 129
Transphosphatidylation, 7
Tropomyosin, 175
Trypsin, 22, 28
Tubular system, dense (DTS), 87
Tubulin, 131
Tumor
  cell growth, 215
  dissemination, 256
Tumor necrosis factor, 252, 256
Tyrosine kinase, 130–132

Tyrosine phosphorylation, 28, 44, 53, 75, 96, 108, 111–116, 129–148, 170, 171, 213
    method, 132
    in platelet, 130–132
Tyrosine kinase, 19, 53, 75, 106, 111, 119, 168, 169, 178, 182, 213
    inhibitor, 136–140
Tyrosine phosphatase, 75, 111
Tyrphostin, 136–142

Vanadate, 111–116
Vanadyl hydroperoxide, 135
Vasodilatation, 256
Vasodilator, 233–234

Vasodilator-stimulated phosphoprotein (VASP), 237–249
Vasopressin, 1, 75, 83, 105, 133, 138, 237
VASP, *see* Vasodilator-stimulated phosphoprotein
Vesicle, adrenal, bovine, 97
Vinculin, 131, 142, 183
von Willebrand, factor, 178

Western blot analysis, 121, 123, 125, 131, 143, 179, 180, 189–192
Wheat germ agglutinin, 131
Wortmannin, 120

Xenopus laevis oocyte, 21

The manufacturer's authorised representative in the EU is Springer
Nature Customer Service Centre GmbH, Europaplatz 3, 69115 Heidelberg,
Germany. If you have any concerns regarding our products, please
contact ProductSafety@springernature.com

Printed and bound by CPI Group (UK) Ltd, Croydon, CR0 4YY
23/04/2026
02095632-0004